Energy Balance and Cancer

Series Editor:
Nathan A. Berger
Case Western Reserve University
Cleveland, OH, USA

For further volumes:
http://www.springer.com/series/8282

Nathan A. Berger
Editor

Murine Models, Energy Balance, and Cancer

Springer

Editor
Nathan A. Berger
Center for Science, Health & Society
Case Western Reserve University School
of Medicine
Cleveland, Ohio
USA

ISSN 2199-2622 ISSN 2199-2630 (electronic)
Energy Balance and Cancer
ISBN 978-3-319-38355-2 ISBN 978-3-319-16733-6 (eBook)
DOI 10.1007/978-3-319-16733-6

Springer Cham Heidelberg New York Dordrecht London

Softcover reprint of the hardcover 1st edition 2015

Printed on acid-free paper

Springer International Publishing AG Switzerland is part of Springer Science+Business Media (www.springer.com)

Preface

Overweight and obesity have increased on a worldwide basis, reaching 60–70 % of the adult population in developed countries and the incidence continues to increase in developing countries. In addition to having devastating psychosocial impact and causing severe debilitation and death at the individual level, this obesity pandemic along with an associated increase in multiple malignancies poses a major series of public health problems that challenge both health care systems and health care costs. Cohort and case control studies in humans have contributed significantly to understanding the epidemiology and clinical correlates of these problems. Likewise, great insights have been gained from biochemical, molecular biologic, genetic, physiologic, and pathologic analysis of human tissues and fluids, and in some cases, it has been possible to conduct randomized controlled trials in humans to study cause and control. The latter, however, have usually been short-term relative to the long-term problems of obesity development and control and its relation to cancer. Moreover, obesity and its comorbidities have been difficult to study in humans due to difficulties in sustaining controlled diets, environments, and behavior modifications as well as ethical and technical considerations that preclude performance of human experiments where harmful effects such as the development of obesity and/or cancer may be an important component of the endpoint.

In contrast to the research limitations imposed by human studies, animal studies offer many advantages, and murine systems have been particularly useful to study the relation of cancer with diet, obesity, and other factors affecting energy balance. These models are useful, first because rodents are similar to humans in that they develop a series of diseases characteristic of humans including diabetes, hypertension, obesity, cardiovascular disease, autoimmune disorders, and a variety of malignancies including carcinomas, sarcomas, lymphomas, and leukemias. In addition, they are small, easy to handle, easy to control, and manipulate their diet and environment. Another important attribute of murine models is that a series of immunodefficient strains, athymic "nude" mice, and severe combined immunodefficient (SCID) mice, lacking major components of the immune system are available, which make it easy to transplant tumors as pieces, minces, or cell suspensions, between animals or from tumor cell lines and across species, including from humans to mice, to study the consequences of variations in energy balance such as diet and exercise.

Transplanted tumors are useful also for the studies of new therapeutic agents alone and in combination.

Like that of humans, the entire mouse genome has been sequenced and areas of similarity and differences can be compared. The ability to develop inbred strains of mice provides the basis for statistically significant genetic and phenotypic expression of normal and abnormal traits and their association with specific genes and their products. Moreover, the ability to genetically modify mice by making transgenic, knockout, knockin, and consomic chromosome animals, provides the basis for developing a variety of sophisticated and valuable tools to elucidate the independent and combined contributions of specific factors to the obesity–cancer relation. In addition, their short reproduction time makes it easy to study both genetic and epigenetic influences. The value of these models is further enhanced by technology to identify DNA sequence changes that result in cancer driver genes in humans and then reproduce them in mice to study the interaction of obesity with a variety of other genetic, epigenetic, and environmental modifiers such as diet, exercise, diurnal rhythm, and others on cancer. Likewise, a series of devices including tread mills, running wheels, swimming apparatus, and motion detectors have been developed for mice to encourage and provide opportunities for exercise, exercise training, and measurement of physical activity, endurance, and energy consumption.

This volume in the *Energy Balance and Cancer* series discusses many of the leading murine models used to study the mechanisms and markers linking obesity to cancer, their modification by environment, and how they may continue to be used to further elucidate these relations and explore preclinical aspects of prevention and/or therapeutic interventions.

In Chap. 1, Luciano DiTacchio and Kacee A. DiTacchio (University of Kansas) and Satchidananda Panda (Salk Institute for Biological Sciences) discuss the use of murine models to study the relevance of normal and disrupted circadian rhythm in cancer. In Chap. 2, Lei Cao (Ohio State University) examines the dramatic effect of environmental manipulation and consequent alterations in neuropeptides as mediators of the impact of energy balance and cancer. Ellen Heber-Katz (Wistar Institute) and Robert Naviaux (University of California San Diego), in Chap. 3, discuss the MRL mouse model of regeneration and cancer, while in Chap. 4, Darlene Berryman, Vivian Lesende, Lara Householder, Edward List, John Kopchick (Ohio University) discuss mouse lines with altered growth hormone secretion affecting fat metabolism and cancer. In Chap. 5, Jonathan Tucci and Steven Mittelman (University of Southern California) discuss mouse models to study hematologic malignancies, while in Chap. 6, Stephen Hursting, Emily Rossi, Laura Bowers (University of North Carolina at Chapel Hill), and Laura Lashinger (University of Texas at Austin) summarize lessons learned from genetically engineered mice to study the effect of insulin-like growth factor (IGF)-1 on energy balance and cancer. Praveena Thiagarajan and Ofer Reizes (Cleveland Clinic) report, in Chap. 6, the exciting use of murine models to study leptin effects in breast cancer stem cells. In Chap. 8, Zara Zelenko, Derek LeRoith, and Emily Gallagher (Ican School of Medicine at Mount Sinai) discuss the use of murine models to study the interrelated effects of diabetes, insulin, and IGFs on cancer. In Chap. 9, Henry Thompson

(Colorado State University) provides a conceptually challenging discussion of the impact of energy balance on chemically induced mammary carcinogenesis in the rat. Hui-Hua Chang, Guido Eibl, and Enrique Rozengurt (David Geffen School of Medicine at University of California Los Angeles) discuss the recent advances in mouse models to study the effects of energy balance on pancreatic cancer in Chap. 10. Chapter 11, by Emily Benesh and Kelle Moley (Washington University) provides an interesting epigenetic examination of how maternal energy balance affects prostate cancer in offspring. Dipali Sharma (Johns Hopkins University) and Neeraj Saxena (University of Maryland), in Chap. 12, provide a comprehensive review of mouse models to study the nonalcoholic fatty liver disease, nonalcoholic steatohepatitis and hepatocellular carcinoma and the effect of natural product. In Chap. 13, Abraham Schneider (University of Maryland) provides an important examination of the use of mouse models to study metformin in carcinogenesis.

Overall, this volume provides a series of diverse murine models for studying many of the malignancies affected by energy balance in humans and a variety of techniques by which these studies may be approached. The volume should be of high interest to all the investigators concerned with the relations between energy balance and cancer, and serve as a useful guide to those interested in performing controlled research in model systems relative to these processes.

Cleveland, Ohio Nathan A. Berger

Contents

About the Editor

Nathan A. Berger is a Distinguished University Professor. He is the Hanna-Payne Professor of experimental medicine, and the Director of the Center for Science, Health, and Society. He is Professor of medicine, biochemistry, genetics, and oncology at Case Western Reserve University, School of Medicine. He is a member of many professional societies, including the American Society of Hematology, the American Society for Biochemistry and Molecular Biology, the American Society of Clinical Oncology, the American Association for Cancer Research, the American Society of Clinical Investigation, and the Association of American Physicians. Dr. Berger, who serves on many national peer review panels and committees for the National Cancer Institute, has published over 200 scientific papers, which have received 5,443 citations.

Contributors

Emily C. Benesh Department of Obstetrics and Gynecology, Washington University School of Medicine, St. Louis, MO, USA

Darlene E. Berryman Edison Biotechnology Institute, Ohio University, Athens, OH, USA; School of Applied Health Sciences and Wellness, College of Health Sciences and Professions, Ohio University, Athens, OH, USA; Department of Biomedical Sciences, Heritage College of Osteopathic Medicine, Ohio University, Athens, OH, USA; Konneker Research Laboratories, Ohio University, Athens, OH, USA

Laura W. Bowers Department of Nutrition, University of North Carolina at Chapel Hill, Chapel Hill, NC, USA

Lei Cao College of Medicine, The Ohio State University, Columbus, OH, USA

Hui-Hua Chang Department of Surgery, David Geffen School of Medicine at UCLA, Los Angeles, CA, USA

Kacee A. DiTacchio Department of Pharmacology, Toxicology & Therapeutics, University of Kansas Medical Center, Kansas City, KS, USA

Luciano DiTacchio Department of Pharmacology, Toxicology & Therapeutics, University of Kansas Medical Center, Kansas City, KS, USA

Guido Eibl Department of Surgery, David Geffen School of Medicine at UCLA, Los Angeles, CA, USA; CURE: Digestive Diseases Research Center, David Geffen School of Medicine at UCLA, Los Angeles, CA, USA

Emily J. Gallagher Division of Endocrinology, Diabetes and Bone Diseases, Icahn School of Medicine at Mount Sinai, New York, One Gustave L. Levy PlaceNY, USA

Ellen Heber-Katz The Wistar Institute, Philadelphia, PA, USA

Lara Householder Edison Biotechnology Institute, Ohio University, Athens, OH, USA; School of Applied Health Sciences and Wellness, College of Health Sciences and Professions, Ohio University, Athens, OH, USA; Konneker Research Laboratories, Ohio University, Athens, OH, USA

Stephen D. Hursting Department of Nutrition, University of North Carolina at Chapel Hill, Chapel Hill, NC, USA

John J. Kopchick Edison Biotechnology Institute, Ohio University, Athens, OH, USA; Department of Biomedical Sciences, Heritage College of Osteopathic Medicine, Ohio University, Athens, OH, USA; Konneker Research Laboratories, Ohio University, Athens, OH, USA

Laura M. Lashinger Department of Nutritional Sciences, The University of Texas at Austin, Austin, TX, USA

Derek LeRoith Division of Endocrinology, Diabetes and Bone Diseases, Icahn School of Medicine at Mount Sinai, New York, One Gustave L. Levy PlaceNY, USA

Vivian Lesende Edison Biotechnology Institute, Ohio University, Athens, OH, USA; Konneker Research Laboratories, Ohio University, Athens, OH, USA

Edward O. List Edison Biotechnology Institute, Ohio University, Athens, OH, USA; Konneker Research Laboratories, Ohio University, Athens, OH, USA

Steven D. Mittelman Departments of Pediatrics and Physiology & Biophysics, Keck School of Medicine, University of Southern California, Los Angeles, CA, USA

Kelle H. Moley James P. Crane Professor, Department of Obstetrics and Gynecology, Washington University in Saint Louis, School of Medicine, Saint Louis, MO, USA

Robert K. Naviaux UCSD School of Medicine, San Diego, CA, USA

Satchidananda Panda Regulatory Biology Laboratory, Salk Institute for Biological Studies, La Jolla, CA, USA

Ofer Reizes Department of Cellular & Molecular Medicine, Lerner Research Institute, Cleveland Clinic Foundation, Cleveland, OH, USA

Emily L. Rossi Department of Nutrition, University of North Carolina at Chapel Hill, Chapel Hill, NC, USA

Enrique Rozengurt Division of Digestive Diseases, Department of Medicine, David Geffen School of Medicine at UCLA, Los Angeles, CA, USA; CURE: Digestive Diseases Research Center, David Geffen School of Medicine at UCLA, Los Angeles, CA, USA

Neeraj K. Saxena Department of Medicine, University of Maryland School of Medicine, Baltimore, MD, USA

Abraham Schneider School of Dentistry, Department of Oncology and Diagnostic Sciences, University of Maryland, Baltimore, MD, USA

Dipali Sharma Department of Oncology, The Sidney Kimmel Comprehensive Cancer Center, Johns Hopkins University School of Medicine, Baltimore, MD, USA

Praveena S. Thiagarajan Department of Cellular & Molecular Medicine, Lerner Research Institute, Cleveland Clinic Foundation, Cleveland, OH, USA

Henry J. Thompson Cancer Prevention Laboratory, Colorado State University, Fort Collins, CO, USA

Jonathan Tucci Department of Pediatrics, Keck School of Medicine, University of Southern California, Los Angeles, CA, USA

Zara Zelenko Division of Endocrinology, Diabetes and Bone Diseases, Icahn School of Medicine at Mount Sinai, New York, One Gustave L. Levy PlaceNY, USA

Chapter 1
Relevance of Circadian Rhythm in Cancer

Luciano DiTacchio, Kacee A. DiTacchio and Satchidananda Panda

Abstract Circadian rhythms are patterns of behavior, physiology, and metabolism that occur within a period of approximately 24 h. These rhythms are generated endogenously, but synchronize to external cues, thus enabling organisms to beneficially align physiological processes to the inherently dynamic, yet predictable, seasonal changes in the day–night cycle. The cell autonomous circadian oscillator temporally coordinates cellular processes, including metabolism, proliferation, cell signaling, organelle function, proteostasis, and DNA damage repair to sustain cellular homeostasis. It is hypothesized that the circadian oscillators evolved as a "flight from light" mechanism to minimize UV damage to single stranded DNA by restricting DNA replication to the nighttime. Support for this hypothesis is accumulating with the recent observation that the circadian rhythm and cell cycle are intimately coupled to each other, so that specific phases of cell cycle occur at a defined phase of the circadian oscillator at single-cell level [1]. Furthermore, chronic circadian disruption perturbs cellular homeostasis and predisposes to cancer. Conversely, numerous cancer cell lines display severe circadian alterations, which likely contribute to aggressive proliferation of the tumor. Hence, it is becoming increasingly important to understand the relevance of circadian rhythm for optimizing fitness

---

S. Panda (✉) · L. DiTacchio · K. A. DiTacchio
Department of Pharmacology, Toxicology & Therapeutics,
University of Kansas Medical Center, 4095 Hemenway
Bldg, 3901 Rainbow Boulevard, Kansas City, KS 66160, USA
e-mail: lditacchio@kumc.edu

K. A. DiTacchio
e-mail: kditacchio@kumc.edu

S. Panda
Regulatory Biology Laboratory, Salk Institute for Biological Studies,
La Jolla, CA 92037, USA
e-mail: panda@salk.edu

N. A. Berger (ed.), *Murine Models, Energy Balance, and Cancer,*
Energy Balance and Cancer, DOI 10.1007/978-3-319-16733-6_1

under natural conditions and its utility and adaptability in the modern world so that the knowledge can be better leveraged for the prevention and treatment of cancer.

Keywords Circadian rhythms · Circadian clock · Circadian clock cell cycle regulation · Circadian clock xenobiotic detoxification · Circadian clock DNA damage response · Circadian clock energy metabolism · CLOCK-BMAL1 · Chronodisruption · Melatonin

Circadian rhythms evolved in a predictable environment of light:dark and the associated daily rhythm in access to food. Accordingly, at the organism level, there are mechanisms to entrain the circadian system to changes in light and food availability in different seasons. However, in the modern anthropogenic world, the use of electrical lighting and abundance of food availability throughout 24 h during the day:night cycle can cause repetitive perturbation of the circadian clock and increase susceptibility to cancer. In this chapter we will first introduce the circadian clock and its molecular mechanisms in mammals, followed by an overview of the epidemiological and experimental evidence linking circadian dysfunction to cancer, especially considering evidence obtained from animal models. Then, we will cover four broadly defined aspects of physiology that are regulated by the clock, and within which we find the mechanisms upon which the hallmarks of cancer arise: (1) cell cycle regulation and DNA damage response, (2) xenobiotic detoxification, (3) endocrine function, and (4) energy metabolism.

These discussions will lay the framework to understand how circadian disruption can increase cancer risk and shed light on novel lifestyle or pharmacological cancer treatments that can build on a better understanding of the circadian cancer link.

Mechanisms of the Clock

In mammals, the circadian system is based on a cell-autonomous and self-sustaining molecular oscillator. At the molecular level the circadian oscillator is a genetic circuit composed of two interlocking transcription–translation feedback loops that revolve around the transcription factors CLOCK and BMAL1 (and their respective homologs NPAS2 and BMAL2). CLOCK and BMAL1 function as heterodimers that bind to E-box elements at promoter regions of their target genes. CLOCK-BMAL1 drive the expression of the *Cryptochrome* (*Cry1* and *Cry2*), and *Period* (*Per1*, *Per2*) genes. In turn, CRYs and PERs form complexes that repress CLOCK-BMAL1 activity, ultimately suppressing their own expression. Eventually, CRYs and PERs are degraded, relieving CLOCK-BMAL1 repression and beginning the cycle anew. Essentially, the interplay between CLOCK/NPAS2, BMAL1/2, CRYs and PERs gives rise to the self-sustained, near-24 h, alternating activation–repression cycles that are the proverbial ticking of the clock.

In a second loop, CLOCK and BMAL1 activate the *Rev-Erb-* and *Ror*-class nuclear hormone receptors. REV-ERBs and RORs are, respectively, transcriptional repressors and activators that bind to ROR elements (ROREs) present in the *Bmal1* gene promoter, and whose interplay enforces rhythmic expression of *Bmal1*. Additionally, REV-ERBs and RORs also act on fine-tuning the rhythmic expression of additional clock components. These rhythms are then propagated to off-clock genes generating large-scale transcription rhythms (up to 15 % of expressed genes in any given tissue), which eventually manifest as overt physiological rhythms.

These transcription–translation loops form the core of the molecular clock, but the mechanisms that underlie the complete circadian oscillator are far more complex. For instance, the duration of the cycle needs to have a period of completion close to 24 h. One feature of this process is that the translation of circadian mRNAs does not follow immediately after transcription; instead, there is an approximate 2–4 h delay between generation of the mRNA and protein synthesis. This delay is in part regulated via miRNA-dependent mechanisms, consistent with findings that DICER deficient cells exhibited a shortened circadian period length and had an associated acceleration on PER1 and PER2 translation [2]. In addition, the timing of individual steps and of the overall circadian cycle involves the extensive post-translational regulation of oscillator components. All known oscillator components are subject to posttranslational modifications of which most extensively studied and thus the best understood is protein phosphorylation. For example, casein kinase Iε (CKIε) phosphorylates the period proteins to regulate their degradation by the β-transducin repeat containing protein 1 (β-TrCP1)-SCF complex. In contrast, PER2 stability is enhanced by phosphorylation by casein kinase 2 (CK2). CRYs stability is similarly regulated by phosphorylation by the AMP-activated protein kinase (AMPK), an event that promotes CRY association with the F-box component of the SCF-FBXL3 ubiquitin ligase complex. In addition, phosphorylation also regulates subcellular localization of clock proteins, as well as their activity. For instance, CKIε phosphorylates PER1 and BMAL1, regulating the nuclear entry of the former and the transcriptional activity of the latter. Glycogen synthase kinase 3b (GSK3β) phosphorylates REV-ERBα, BMAL1 and CLOCK, PER2, and CRY2 with differing impacts on stability, activity, and subcellular localization [3–7]. Besides phosphorylation, clock components also undergo acetylation, O-GlcNacylation, ubiquitylation, and sumoylation, which have a range of effects as those mentioned [8, 9].

A fundamental feature of genes that are circadianly expressed is that they exhibit robust rhythms in chromatin regulation, including epigenetic modifications. As such, chromatin regulators such as histone lysine deacetylases, methyltransferases, and demethylases are components of the transcriptional machinery. Altogether, the combined action of signaling pathways, chromatin regulators, and non-clock transcription factors enable the fine-tuning of circadian rhythms and their entrainment by environmental cues.

In spite of their cell-autonomous nature, cellular-level oscillators are orchestrated in a tissue-specific manner in order to give rise to tissue- and organ-level physiological rhythms. Interestingly, sustained tissue-level circadian rhythms are dependent on the phase alignment of their individual cellular-level oscillators, yet their synchronization (or entrainment) is not a tissue-autonomous property. Thus, the circadian system is a hierarchical network, composed of peripheral oscillators that are entrained by a master circadian pacemaker: the suprachiasmatic nucleus (SCN) of the hypothalamus [10–12]. The SCN is a neural structure composed of 20,000 cells bilaterally situated above the optic chiasm that has the unique property of being able to impose coordinated self-sustaining circadian oscillations at the tissue level. In addition, the SCN is innervated by the retinohypothalamic tract through which it receives photic cues that enable it to set its circadian phase to match the local geographic time. The SCN then relays this information to peripheral oscillators via neural, humoral, and metabolic signals in order to maintain the appropriate phase relationships between the different organs and the time of day.

Cancer and the Clock

In 2007, the World Health Organization (WHO) and the International Agency for Research on Cancer (IARC) classified shift work as a possible carcinogen, a decision that was largely based on epidemiological data along with results from animal studies.

Chronodisruption, specifically night work and rotating shift work, is associated with increased incidence of certain cancers. Numerous studies have identified a link between shift work and breast cancer. A meta-analysis of 13 independent epidemiological studies conducted between 1995 and 2005 estimated that the risk of breast cancer in shift workers (nurses and flight attendants) was up to 48 % higher than in control populations [13]. Interestingly, a nested case-controlled study of women serving in the Danish military found that chronotype had a significant impact on cancer incidence, with rates being higher in evening-type populations after controlling for shift work [14]. Furthermore, the same study found that morning-type workers subjected to shift work were nearly twice more likely to develop breast cancer than the control group. Similarly, chronodisruption increases the incidences of endometrial (42 % in nonobese and 100 % in obese night-shift workers), colorectal (35 %), and possibly prostate cancers [15–21]. Interestingly, blind subjects with total absence of photoreception, and thus less sensitive to circadian stress, have reduced cancer rates in comparison to those that retain at least some light sensitivity [21, 22]. Consistently, the survival of patients with cancer of the breast, colon, or lung, is associated with marked circadian rhythmicity [23].

The importance of the circadian oscillator to cancer is further supported by epidemiological analysis of circadian genetic variability in human populations, analysis of clock gene expression in cancers in vivo and in cell models, and experimental data. In humans, several SNPs (SNP) that occur in circadian clock genes are associated with cancer incidence, progression, and survival. For example, in a study of Chinese men, individuals that harbored a SNP in *Npas2* had a decreased risk of developing prostate cancer, whereas those with an SNP in *Cry2* had nearly double the risk of those that did not [24]. Similarly, other circadian polymorphisms have been linked to increased postmenopausal breast cancer risk (*Cry2*), reduced risk of developing non-Hodgkins lymphoma (*Npas2*), increased occurrence and development of non-small cell lung cancer, and increased survival in colorectal and hepatocellular carcinoma (*Clock* and *Per3*, respectively) [25–30].

In addition to these SNP associations, the expression of clock genes is dysregulated in tumors in vivo and in cancer cell culture models. *Per2* and *Cry1* gene expression is upregulated in gastric cancer, with *Cry1* heightened expression correlating with disease progression [31]. In ovarian cancer cell lines, the *Bmal1* gene is silenced via DNA methylation [32], whereas it is down-regulated in chronic lymphocytic leukemia patients along with *Per1* and *Per2* [33]. Similarly, in colorectal carcinoma *Cry1* and *Clock* are overexpressed with their levels being correlated with progression of the disease, and decreased survival, whereas lower expression of *Bmal1*, *Cry2*, *Per1*, and *Per3* was reduced in tumor samples [34–36].

The data derived from genetic animal models is also consistent with human observations. Clock mutant mice, which harbor a mutant gene allele that gives rise to a defective CLOCK protein, have decreased proliferation and increased apoptosis rates in lymphocytes, although without increased susceptibility to low-dose γ-irradiation induced cancer [37]. Double genetic ablation of *Cry1* and *Cry2* delays tumorigenesis onset and increases the lifespan of *p53* mutant mice [38]. Mice that are homozygote for the *Per2* mutant allele (*Per2*$^{m/m}$) have a 100 % increase in [39] the number of intestinal and colonic polyps when occurring in a model of colorectal cancer (Apc (Min/+) mice) background [40]. Consistently, siRNA-mediated decreases in *Per2* levels in colon cancer cell lines resulted in heightened β-catenin expression and increased proliferation, with similar effects observed with *Per1* [41]. Interestingly, mice that carry a transgenic *Per2* allele that occurs in familial advanced sleep phase syndrome (FASPS) show increased cancer risk and incidence [42]. Furthermore, *Bmal1, Per1, Per2, Cry1, or Cry2* genetic-null mice have predisposition to both spontaneous and irradiated-induced cancer [43].

Finally, nongenetic chronodisruptive paradigms have found links between tumor formation and progression in mice models. Tumor growth was greater in mice with bilateral SCN lesions than in sham control cohorts [44]. Under a jet lag paradigm, in which mice with tumor xenografts were subject to repeated 8 h light shifts, the rate of tumor growth also increased, with a mitigation of the effect occurring when meal time was matched to the light shifts; food-derived cues are dominant peripheral entraining signals, which reinforce the phase relationship between the SCN and the periphery [45]. In addition, chronic jet lag promotes metastases [46]. Similarly, when circadian mutants or wild-type mice are subjected to jet lag, their cancer risk is greatly increased [43]. Finally, rats chronically exposed to light-at-night, a fundamental characteristic of shift work and night work, increased the incidence of spontaneous tumorigenesis [47].

So, what mechanistic link exists between circadian disruption and cancer? For cancer to arise, normal cells must acquire certain characteristics, or "hallmarks," in order to become tumorigenic. Specifically, these hallmarks are (1) self-sustained proliferation, (2) evasion of growth suppressive mechanisms, (3) apoptosis resistance, (4) induction of angiogenesis, (5) replicative immortality, (6) activation of tissue invasion and metastasis, (7) immune system evasion, and (8) reprogramming of cellular energy metabolism. The mechanisms that underlie all these hallmarks are under circadian control: cell cycle regulation and DNA damage repair, endocrine function, inflammation and apoptosis, xenobiotic detoxification, modulation of immune function, and energy metabolism are under extensive circadian control [48–51]. Thus, by impairing circadian oscillator function, chronodisruptive agents lead to a wide-spread dysregulation of physiological processes that are conducive for cancer development and progression.

The Circadian Clock, Cell Cycle Regulation and DNA Damage Response

The circadian clock and the cell cycle are now known to be extensively interconnected at the molecular level. The first and most obvious link is that many key cell cycle regulators—including *wee1, p21-Waf1, p20, cdc2, cyclin B1, cyclin D1,* and *c-Myc*—are, in fact, clock-controlled genes (CCGs). Consistently, the circadian clocks from cyanobacteria to mammals are well-known to gate the different stages of the cell cycle, and thus impose temporal regulation to its progression. In mammals, for instance, S-phase and mitotic index in the oral mucosa, corneal epithelium, digestive tract and even bone marrow exhibit circadian fluctuations [52–56]. In a striking example of cell cycle gating by the clock, the timing of M-phase entry following partial hepatectomy is dictated by the time of day of the procedure [57].

Similarly, the circadian clock impinges on the DNA damage response. PER1 physically associates with the cell cycle checkpoint proteins, ataxia telangiectasia mutated (ATM) and CHK2, sensitizes cells to apoptosis when overexpressed and, conversely, confers protection to irradiation-induced cell death [58]. In humans, the timeless protein (TIM), a putative oscillator component, interacts with CRY2 and CHK1, possibly helping further the crosstalk between circadian and cell cycle regulatory mechanisms [59]. Further, BMAL1 is necessary for p53-mediated activation of *p21CIP*; BMAL1 shRNA knockdown decreased the induction of both *p53* and *p21CIP* in response to γ-irradiation, and rendered cells insensitive to p19ARF-induced cell cycle arrest [60]. Intriguingly, CLOCK protein is localized to DNA damage sites following UV irradiation [61].

The Circadian Clock and Detoxification

Xenobiotics are environmentally derived molecules and compounds that are taken up by organisms as a result of the processes required for life. Many xenobiotics, as well as endobiotic metabolites, are toxicants that have deleterious effects on the genome and, thus, organisms have evolved mechanisms that enable their removal and/or neutralization. As with other physiological processes, xenobiotic detoxification metabolism is profoundly influenced by the circadian system. For example, the constitutive androstane receptor (CAR), retinoic X receptor (RXR), and small heterodimer partner (SHP) are nuclear hormone receptors that are involved in the activation of detoxification programs and whose expression pattern is robustly circadian. DBP, TEF, and HLF PAR-bZIP are transcription factors that are direct outputs of the circadian oscillator involved in regulating drug metabolism and detoxification. In triple DBP;TEF;HLF genetic-null mice, CAR-targets and CAR itself are expressed at very low levels [62]. Two CAR targets that are downregulated in these mice are 5-aminolevulinic acid synthase 1 (*Alas1*) gene, which is also regulated by NPAS2, and P450 oxidoreductase (POR). Both ALAS1 and POR are required for CYP activity; ALAS1 is the rate-limiting enzyme in the heme biosynthetic pathway, and POR is required for transfer of electrons to CYPs. This results in oscillations in CYP activity. In addition, the expression of some Cyp genes is controlled by DBP, including CYP3A4, CYP2A4, and CYP2A5, or directly by oscillator components (CYP2E1). In all, this results in circadian regulation of Phase I metabolism, which has maximum function during the night, which is when food consumption peaks. Phases II and III metabolism are also influenced by the clock. For example, Phase II processes are scheduled so that glutathione conjugation occurs at the beginning of the fasting period (daytime), followed by glucuronidation towards the end of the light phase, and sulfation occurring at the day-to-night period. The expression of several phase III metabolism transporters are also regulated by the circadian oscillator, including organic cation transporter 1 (OCT1), organic anion transporters 1, 2, and 3 (OAT1–3), multidrug resistance protein 1 (MDR1), and others [49].

The Circadian Clock and the Endocrine System

Since the endocrine system enables communication and coordinated physiological functions across tissues and organs throughout the body, one might expect circadian influence over those processes that occur with regularity. Indeed, the circulating levels of many hormones and other endocrine factors oscillate strongly over the course of the day/night cycle in normal conditions and chronodisruption interferes with their synthesis, release and sometimes their effect within target cells. The disruption of circadian rhythms certainly means a disruption of endocrine systems, which, in turn, contributes to pathologies, including metabolic syndrome, obesity, type-2 diabetes, and cancer.

Melatonin

Melatonin is a pineal hormone with synthesis and secretion occurring during the night regardless of whether activity occurs during the day as in diurnal organisms or the night: melatonin can thus be considered the chemical signal through which physiology recognizes darkness [63]. Light exposure inhibits the production of melatonin according to its intensity, wavelength, and duration. Accordingly, reduced production is observed following shift work or other light-at-night environments. Additionally, production declines with age along with various other outputs of the circadian system. The pineal gland receives direct innervation from the SCN and, in terms of both synthesis and release, melatonin is under direct regulation of the clock. In a positive feedback loop, melatonin also signals to the SCN, resynchronizing clock gene expression and thereby fortifying rhythmicity. It is also one of the most important signals serving to synchronize peripheral clocks, regulating the phase and period of the transcription/translation cycle of peripheral clock genes. This action supports the segregation of daytime appropriate physiology from that of the night; however, its effects in their entirety extend beyond circadian rhythms with important consequences for overall metabolic function and health. Acting on adipocytes, for instance, melatonin inhibits lipogenesis and increases the production of the satiety hormone, leptin [64]; an effect that is potentiated with rhythmicity of melatonin exposure, as would occur endogenously. Additionally, melatonin functions synergistically with insulin, being important for its proper synthesis and secretion as well as improving insulin sensitivity in target cells. Consequently, pinealectomized animals demonstrate glucose intolerance and insulin resistance which can be reverted by melatonin replacement therapy [65].

Melatonin is known to have powerful oncostatic and oncoprotective effects. It can prevent and reverse tumorigenesis in murine models. Furthermore, circulating melatonin levels are inversely correlated with tumor growth rates in cancer patients, and positively associated with survival [66]. Interestingly, blind subjects with total absence of photoreception have reduced cancer rates in comparison to those that retain at least some light sensitivity, possibly due to a decreased susceptibility to alterations in melatonin secretion [22, 67].

Melatonin is thought to influence cancer occurrence and progression both directly and indirectly. Several receptor-dependent and -independent actions of melatonin within tumors oppose various hallmarks of cancer: reducing proliferation, boosting antioxidant defenses, regulating cellular metabolism, and blocking invasion and metastasis. Alternatively, melatonin's impact on the circadian system stimulates robust oscillations and coordinates physiology across the body, supporting healthy circadian and metabolic states which in turn are protective.

Glucocorticoids

Glucocorticoids (GCs) are steroid hormones derived from the adrenal gland which function in a wide array of physiological processes, including inflammation, glucose homeostasis, and cell proliferation [68]. Their anti-inflammatory effects are exploited pharmaceutically to combat pain, allergies, arthritis as well as lymphomas and leukemia. They impinge on various other physiological processes as well, with the outcome depending on the target cell type—even positively or negatively affecting cell proliferation, for instance. Glucocorticoids exhibit strong daily oscillations in plasma concentrations. The SCN drives these fluctuations via the hypothalamic-pituitary-adrenal axis and their rhythmic presence participates in the entrainment of peripheral clocks [69–71]. In jet-lag conditions, both the synthesis and release of GCs are disturbed, which, in turn, contribute to desynchrony of peripheral clocks as well as disrupting various other physiological processes that GCs normally regulate [72, 73].

The role of glucocorticoids in cancer varies. The outcome of GC signaling is tissue-specific and likewise the effect of GC signaling in cancerous cells depends on the type; pro-apoptotic in certain cancers, but pro-survival and promoting resistance to cell death in others [68, 74]. In obesity, GCs also have pleiotropic effects but generally promote fat deposition.

Insulin/Insulin-Like Growth Factor

Insulin is produced in the pancreas and regulates energy metabolism throughout the body; its cousin, insulin-like growth factor (IGF) is secreted from the liver and promotes cell proliferation [75]. Although serving some distinct functions, these proteins are very similar in structure and share many components of their signaling pathways. High levels of insulin are observed in the blood of obese and diabetic patients, and insulin/IGF signaling has been implicated in the risk, incidence, and tumorigenic responses of various human cancers in such individuals [76, 77]. Intriguingly, in human breast cancer patients, a disruption to IGF-1 rhythmicity has been observed whereas other rhythms remain intact; it is possible that the timing of release may be as important as the overall levels [78].

Insulin release is responsive to nutrient availability in the blood; however, it is also under circadian control [79, 80]. Many components of the insulin/IGF signaling pathway are regulated in a circadian manner at the level of transcription, and blood levels of both oscillate over 24 h even in the absence of food [81]. Conversely, insulin stimulates circadian gene expression [82]. Additionally, mice fed a high-fat diet (HFD) become obese and hyperinsulinemic and have disrupted feeding rhythms and free-running period length [83].

The Circadian Clock and Energy Metabolism

The rise in obesity and excess body weight rates constitute one of the greatest modern global health challenges. In the USA alone, ~60 % of the population is considered overweight, with half of those considered clinically obese, 8–11 % suffering from type2 diabetes, and a quarter exhibiting metabolic syndrome. Interestingly, this obesity epidemic has been paralleled by the rise of the 24-h society, which is characterized by the proliferation of, and chronic exposure to, light pollution, and altered activity profiles inherent to modern economies (e.g., rotating shift work, jet lag). Indeed, epidemiological studies have consistently linked circadian dysfunction to metabolic syndrome, weight gain, obesity, and type-2 diabetes [84]. A recent study of more than 113,000 women living in the UK found a strong association between increasing levels of light exposure at night with weight gain and obesity [85]. Similarly, rotating shift work, sleep deprivation, and sleep disruption are positively associated with type-2 diabetes and impairment in glucose metabolism [86–88]. Consistently, human volunteers subjected to circadian misalignment that mimic shift work and jet lag had marked increases in blood glucose and insulin levels, increased arterial pressure, decreased leptin levels, and sleep quality [89, 90], whereas disrupted or insufficient sleep increases the risk of developing T2D [91, 92]. Intriguingly, human circadian genetic variability has been found to impact metabolic function. For example, a number of SNPs that occur in human circadian genes are correlated with propensity for obesity and type-2 diabetes (*Clock* and *Bmal1*), heightened fasting (NPAS2 and PER2) or otherwise correlated with (CRY2 gene variants) blood glucose levels [93–95].

Importantly, experimental data firmly corroborate human population-based studies, and have helped establish the circadian oscillator as a key regulator of energy homeostasis, from the animal as a whole to the cellular level. For example, whole-body *Bmal1* knockout mice present with a number of metabolic phenotypes, including hypoinsulinemia, higher body fat mass content, and fasting hypoglycemia, with the latter also occurring in liver-specific *Bmal1* knockouts [96, 97]. In addition, in both *Bmal1* knockouts and *Clock* mutant mice, circadian rhythms in insulin sensitivity and glucose tolerance are eliminated, and gluconeogenesis is impaired [96, 98]. *Clock* mutant mice, which harbor a dysfunctional allele, are also predisposed to weight gain under both regular and high-fat dietary paradigms and exhibit dyslipidemia, hyperglycemia, and hypoinsulinemia [99]. Cryptochrome-deficient mice exhibit abnormal glucose metabolism. *Cry* double-null mice ($Cry1^{-/-};Cry2^{-/-}$) show heightened postprandial blood glucose levels, are glucose intolerant, and are sensitized to HFD-induced weight gain and hyperinsulinemia [100]. $Cry1^{-/-}$ or $Cry2^{-/-}$ animals also show glucose intolerance, yet *Cry1* but not *Cry2* knockouts are protected against HFD-induced obesity [101]. Similarly, *Per2* mutant mice have lower body fat content and total body weight, and show aberrations in blood glucose regulation, with fasting hypoglycemia, impaired gluconeogenesis and increased insulin sensitivity despite being hyperinsulinemic [102]. $Rev\text{-}erb\alpha^{-/-}$ animals have increased white adipose tissue, elevated blood glucose, decreased blood lipid levels, and abnormal bile acid levels [103, 104]. More recently, mice with a *Per1* gene mutation paralogous to that occurring in humans affected with FASPS were found to be predisposed to HFD-induced obesity [105].

Genomic analyses of tissues such as liver, white and brown adipose tissues, and skeletal muscle have been instrumental in furthering our understanding of how the clock regulates metabolism and why its dysfunction contributes to pathological states. In liver, a key organ in the regulation of systemic energy homeostasis whose function is extensively influenced by the circadian oscillator, many of the direct targets of the hepatic oscillator are essential components of genomic programs and pathways that regulate energy homeostasis. Indeed, circadian clock components (BMAL1, NPAS2, PER1, PER2, CRY, CRY2, NR1D1, and NR1D2) are enriched at the promoters of key regulators of glucose (*G6pc, Pck1, Pcx, Pdk1)*, lipid, and cholesterol homeostasis (*Acaca, Hmgcr, Scap, Insig2, Cyp7a1*) [106, 107]. Additionally, the circadian oscillator influences metabolism indirectly by imposing strong rhythms in the levels of non-clock transcription factors. Notably, out of the 41 hepatic-expressed NHRs, 20 are rhythmically expressed, including the non-clock NHR components PPARα,δ,γ, thyroid hormone receptor, and ERRs (Estrogen-related receptors) [108, 109]. Likewise, the genes that code for the sterol homeostasis regulators SREBP1 and SREBP2– *Srebf1* and *Srebf2*– are expressed with a circadian pattern that, respectively, peaks at the beginning and end of the dark period [110, 111].

White adipose tissue (WAT) is a widely distributed, complex tissue that is a major site of fat storage and de novo fatty acid synthesis. As with liver, the circadian clock involvement in WAT physiology is extensive, affecting both adult tissue function and adipogenesis. In adult WAT, a major role of the circadian clock is to regulate the balance between lipolysis and lipogenesis needed to store energy for use under fasting conditions, yet prevent dyslipidemia and excess fat accumulation. For example, BMAL1 drives the activation of the lipogenic genes *Elovl6, Scd1, Atgl*, and *Hsl* through direct binding to the promoter regions [112, 113]. During adipogenesis, Bmal1 gene expression is enhanced [114]. Consistently, *Bmal1*$^{-/-}$ mice exhibit reduced adiposity and low levels of long-chain polyunsaturated fatty acids. Like BMAL!, REV-ERBα is highly expressed during adipocyte differentiation, although its exact role is this process is not clear; as cell-based experiments show REV-ERBα is required for adipocyte differentiation yet *Rev-Erb* knockout mice show no adipose tissue defects. WAT is also the source of endocrine factors known as adipokines, including adiponectin, leptin, and resistin, all of which are released into the bloodstream in a circadian pattern. In mammals, brown adipose tissue (BAT) is a major site of thermogenesis and a key organ in the regulation of body temperature. In BAT, thermogenesis is achieved via dissipation of the mitochondrial proton gradient through uncoupling protein 1 (UCP1), whose expression has been found to be regulated by *Rev-Erbα* and *Per2* [115, 116]. In skeletal muscle, the circadian clock imposes rhythmicity on the expression of 215 genes, including genes involved in triglyceride hydrolysis (*Ces3, Pnpla3*), fatty acid oxidation (*Pgc1b, Myod1, Ucp3*), and synthesis of fatty acids and cholesterol (*Pank1, Dbt, Dgat2, Acat2, Idh1, S3–12*) [117].

The clock can also fine-tune its control of energy homeostasis through non-transcriptional mechanisms. For instance, cryptochrome proteins can prevent transcription of gluconeogenic genes by preventing activation of the cAMP response element-binding protein (CREB) by glucagon-activated Gsa, as well as by direct repression of the glucocorticoid receptor [118, 119]. Similarly, PER2 can impact glucose metabolism through physical association and modulation of PPARα, REV-ERBα, and possibly HNF4α, and thus regulate their activity [120]. Finally, the clock can influence energy homeostasis through the production of metabolites that activate energetic regulators. The gene for the key enzyme involved in the production of nicotinamide adenine dinucleotide (NAD+), NAD phosphoribosyltransferase

(*Nampt*), harbors E-boxes in its promoter region that are bound by CLOCK, and are expressed in a circadian fashion [121, 122]. In turn, hepatic NAD+ levels oscillate, and this coincides with the activity of SIRT1, an NAD+-dependent histone deacetylase (HDAC) that is one of the central regulators of energy homeostasis [123, 124].

Timing and Quality of Caloric Intake and the Clock

Perhaps the most obvious mechanism by which the circadian oscillator regulates energy metabolism is by orchestrating feeding behavior. Yet, until recently, the role played by the timing of food intake in the maintenance of energy metabolism homeostasis had not been appreciated. The initial cues came from the observation that mice fed a high-fat diet during their inactive period gained more weight than their control counterparts in spite of similar caloric intake and activity levels [125]. This observation was then followed by studies conducted by us and others in which mice were protected from HFD-induced obesity and diabetes by manipulating the feeding time [126, 127]. Specifically, mice in which access to food was limited (temporally restricted feeding; tRF) exclusively during nighttime (active period in mice) did not show the weight gain and metabolic dysfunction characteristic of HFD-fed mice with ad libitum food access, even though the ad libitum and TRF paradigms were isocaloric [126].

Interestingly, the obesoprotective effects of TRF appear to be due to enhancement in circadian oscillator function. Indeed, tRF restored the amplitude in expression of circadian clock components as well as in a number of clock targets and other rhythmic genes. Such improvement is reflective of the intricate relationship between food-derived input and the local circadian oscillator. Under normal conditions, 2997 genes are rhythmically expressed in the livers of wild-type mice, of which only 368 maintain rhythmicity in the absence of food intake [128]. On the other hand, the imposition of a temporally restricted feeding paradigm in cry double-knockout mice, which have a functional circadian clock, restores rhythmicity of 617 genes, whereas such a paradigm in WT mice increases the number of genes with circadian oscillations to 4960.

However, the relationship between metabolism and the circadian clock is bidirectional. As mentioned previously, energy metabolism regulators and signaling pathways impinge on the clock and vice versa. As such, dietary conditions affect the function of the oscillator. Consistently, dietary quality affects the circadian clock. For instance, an HFD and altered behavioral rhythms, including preference for food intake during daytime hours, dampened rhythms in behavior and in the oscillations of circadian clock component gene expression [83]; a high-carbohydrate, high-protein diet phase advances the phase of clock component abundance rhythms and increases the overall levels of BMAL1 and CRY1 [129]; and a high-salt diet similarly results in phase advances of clock component levels [130].

In recent years, the role that metabolic dysfunction plays in tumorigenesis and cancer progression has received increasing attention [131, 132]. Most notably, the

Warburg effect, where cancer cells reprogram their metabolism as to favor glycolysis-derived energy production over mitochondrial respiration, irrespective of oxygen availability, has been intensively studied [133]. Although first postulated by Otto Warburg in 1924 as a possible cause of cancer, the Warburg effect is now considered an acquired trait that enables cancer cells to survive in the otherwise proliferation-limiting hypoxic environment that arises as solid tumors grow. Hypoxia leads to the activation of hypoxia-inducible factors (HIF), amongst whose functions are included the promotion of angiogenesis and, importantly, the increase of glycolysis to compensate for loss of oxygen-dependent mitochondrial respiration [134]. Not surprisingly, HIFs are dysregulated in cancer. Interestingly, HIF proteins, like CLOCK, BMAL1, and PERs, are bHLH-PAS transcription factors; HIFs, CLOCK, and BMAL1 can physically associate to form transcriptionally active complexes (HIF) [135, 136]. In addition, key glycolytic regulators and glucose transporters are dysregulated in cancer, including hexokinase, phosphofructo kinase 1 and 2, and GLUT transporters, all of which are expressed in a circadian pattern in several tissues [137–139].

Conclusions

In all, chronodisruption, metabolic dysfunction, and cancer are so intricately linked that it may be futile to establish a cause–effect relationship between them. Instead, the simplest view that emerges is one where the different processes that underlie these pathologies exist in a dynamic equilibrium and that disrupting any one of them triggers a pathological chain reaction that predisposes organisms to the others. Consider the following scenario: gating of the cell cycle by the circadian clock so that DNA replication and mitosis occurs during the fasting period when food intake and associated metabolically and environmentally derived toxins are at a minimum [140]. Circadian disruption arising from obesity could lead to a temporal spreading of cell division, so that more individual cells are replicating under conditions where genomic insults are more abundant. Simultaneously, this circadian disorganization may result in blunted rhythms and overall lower levels and activities of proteins involved in detoxification mechanisms, exacerbating the increased mutation risk. Conversely, certain cancers, such as insulinomas, trigger metabolic changes that affect circadian and non-circadian mechanisms, again highlighting the extensive interrelatedness of these pathological states. Finally, environmental chronodisruption, such as chronic jet lag or rotating shift work, results in a generalized circadian disorganization that results in hormonal imbalances, including suppression of melatonin, glucose, and lipid metabolic dysregulation, and impaired gating of the cell cycle. As metabolism becomes more perturbed, the circadian oscillator is further disrupted, leading to the aforementioned conditions, thus facilitating oncogenesis.

Acknowledgments KD and LD are supported in part by NIH grant numbers P20RR021940 and P20GM103549. Research in SP's lab is supported by NIH grant numbers DK091618, EY016807, P30 CA014195, and American Federation for Aging Research grant number M14322.

References

1. Feillet C, et al. Phase locking and multiple oscillating attractors for the coupled mammalian clock and cell cycle. Proc Natl Acad Sci U S A. 2014;111:9828–33.
2. Chen R, D'Alessandro M, Lee C. miRNAs are required for generating a time delay critical for the circadian oscillator. Curr Biol. 2013;23:1959–68.
3. Spengler ML, Kuropatwinski KK, Schumer M, Antoch MP. A serine cluster mediates BMAL1-dependent CLOCK phosphorylation and degradation. Cell Cycle. 2009;8:4138–46.
4. Yin L, Wang J, Klein PS, Lazar MA. Nuclear receptor Rev-erbalpha is a critical lithium-sensitive component of the circadian clock. Science. 2006;311:1002–5.
5. Reischl S, Kramer A. Kinases and phosphatases in the mammalian circadian clock. FEBS Lett. 2011;585:1393–9.
6. Lee J, et al. Dual modification of BMAL1 by SUMO2/3 and ubiquitin promotes circadian activation of the CLOCK/BMAL1 complex. Mol Cell Biol. 2008;28:6056–65.
7. Cardone L, et al. Circadian clock control by SUMOylation of BMAL1. Science. 2005;309:1390–4.
8. Asher G, et al. SIRT1 regulates circadian clock gene expression through PER2 deacetylation. Cell. 2008;134:317–28.
9. Gossan NC, et al. The E3 ubiquitin ligase UBE3A is an integral component of the molecular circadian clock through regulating the BMAL1 transcription factor. Nucleic Acids Res. 2014;42:5765–75.
10. Panda S, Hogenesch JB, Kay SA. Circadian light input in plants, flies and mammals. Novartis Found Symp. 2003;253:73–82 (discussion 82–8, 102–9, 281–4).
11. Panda S, Hogenesch JB, Kay SA. Circadian rhythms from flies to human. Nature. 2002;417:329–35.
12. Hastings MH, Reddy AB, Maywood ES. A clockwork web: circadian timing in brain and periphery, in health and disease. Nat Rev Neurosci. 2003;4:649–61.
13. Megdal SP, Kroenke CH, Laden F, Pukkala E, Schernhammer ES. Night work and breast cancer risk: a systematic review and meta-analysis. Eur J Cancer. 2005;41:2023–32.
14. Hansen J, Lassen CF. Nested case-control study of night shift work and breast cancer risk among women in the Danish military. Occup Environ Med. 2012;69:551–6.
15. Viswanathan AN, Hankinson SE, Schernhammer ES. Night shift work and the risk of endometrial cancer. Cancer Res. 2007;67:10618–22.
16. Pauley SM. Lighting for the human circadian clock: recent research indicates that lighting has become a public health issue. Med Hypotheses. 2004;63:588–96.
17. Schernhammer ES, et al. Night-shift work and risk of colorectal cancer in the nurses' health study. J Natl Cancer Inst. 2003;95:825–8.
18. Haus EL, Smolensky MH. Shift work and cancer risk: potential mechanistic roles of circadian disruption, light at night, and sleep deprivation. Sleep Med Rev. 2013;17:273–84.
19. Conlon M, Lightfoot N, Kreiger N. Rotating shift work and risk of prostate cancer. Epidemiology. 2007;18:182–3.
20. Kubo T, et al. Industry-based retrospective cohort study of the risk of prostate cancer among rotating-shift workers. Int J Urol. 2011;18:206–11.
21. Buja A, et al. Cancer incidence among male military and civil pilots and flight attendants: an analysis on published data. Toxicol Ind Health. 2005;21:273–82.
22. Hahn RA. Profound bilateral blindness and the incidence of breast cancer. Epidemiology. 1991;2:208–10.

23. Mormont MC, et al. Marked 24-h rest/activity rhythms are associated with better quality of life, better response, and longer survival in patients with metastatic colorectal cancer and good performance status. Clin Cancer Res. 2000;6:3038–45.
24. Chu LW, et al. Variants in circadian genes and prostate cancer risk: a population-based study in China. Prostate Cancer Prostatic Dis. 2008;11:342–8.
25. Couto P, et al. Association between CLOCK, PER3 and CCRN4 L with nonsmall cell lung cancer in Brazilian patients. Mol Med Rep. 2014;10:435–40.
26. Hoffman AE, et al. The core circadian gene Cryptochrome 2 influences breast cancer risk, possibly by mediating hormone signaling. Cancer Prev Res (Phila). 2010;3:539–48.
27. Zhu Y, et al. Ala394Thr polymorphism in the clock gene NPAS2: a circadian modifier for the risk of non-Hodgkin's lymphoma. Int J Cancer. 2007;120:432–5.
28. Zhao B, et al. A functional polymorphism in PER3 gene is associated with prognosis in hepatocellular carcinoma. Liver Int. 2012;32:1451–9.
29. Zhou F, et al. Functional polymorphisms of circadian positive feedback regulation genes and clinical outcome of Chinese patients with resected colorectal cancer. Cancer. 2012;118:937–46.
30. Kettner NM, Katchy CA, Fu L. Circadian gene variants in cancer. Ann Med. 2014;46:208–20.
31. Hu ML, et al. Deregulated expression of circadian clock genes in gastric cancer. BMC Gastroenterol. 2014;14:67.
32. Yeh CM, et al. Epigenetic silencing of ARNTL, a circadian gene and potential tumor suppressor in ovarian cancer. Int J Oncol. 2014;45:2101–7.
33. Rana S, et al. Deregulated expression of circadian clock and clock-controlled cell cycle genes in chronic lymphocytic leukemia. Mol Biol Rep. 2014;41:95–103.
34. Mazzoccoli G, et al. Clock gene expression levels and relationship with clinical and pathological features in colorectal cancer patients. Chronobiol Int. 2011;28:841–51.
35. Yu H, et al. Cryptochrome 1 overexpression correlates with tumor progression and poor prognosis in patients with colorectal cancer. PLoS One. 2013;8:e61679.
36. Wang L, et al. hClock gene expression in human colorectal carcinoma. Mol Med Rep. 2013;8:1017–22.
37. Antoch MP, et al. Disruption of the circadian clock due to the Clock mutation has discrete effects on aging and carcinogenesis. Cell Cycle. 2008;7:1197–204.
38. Ozturk N, Lee JH, Gaddameedhi S, Sancar A Loss of cryptochrome reduces cancer risk in p53 mutant mice. Proc Natl Acad Sci U S A. 2009;106:2841–6.
39. Jensen LD, Cao Y. Clock controls angiogenesis. Cell Cycle. 2013;12:405–8.
40. Wood PA, et al. Period 2 mutation accelerates ApcMin/+ tumorigenesis. Mol Cancer Res. 2008;6:1786–93.
41. Yang X, et al. The circadian clock gene Per1 suppresses cancer cell proliferation and tumor growth at specific times of day. Chronobiol Int. 2009;26:1323–39.
42. Gu X, et al. The circadian mutation PER2(S662G) is linked to cell cycle progression and tumorigenesis. Cell Death Differ. 2012;19:397–405.
43. Lee S, Donehower LA, Herron AJ, Moore DD, Fu L. Disrupting circadian homeostasis of sympathetic signaling promotes tumor development in mice. PLoS One. 2010;5:e10995.
44. Filipski E, et al. Host circadian clock as a control point in tumor progression. J Natl Cancer Inst. 2002;94:690–7.
45. Filipski E, et al. Effects of light and food schedules on liver and tumor molecular clocks in mice. J Natl Cancer Inst. 2005;97:507–17.
46. Wu M, et al. Experimental chronic jet lag promotes growth and lung metastasis of Lewis lung carcinoma in C57BL/6 mice. Oncol Rep. 2012;27:1417–28.
47. Anisimov VN, Vinogradova IA, Panchenko AV, Popovich IG, Zabezhinski MA. Light-at-night-induced circadian disruption, cancer and aging. Curr Aging Sci. 2012;5:170–7.
48. Lee JH, Sancar A. Regulation of apoptosis by the circadian clock through NF-kappaB signaling. Proc Natl Acad Sci U S A. 2011;108:12036–41.
49. Zmrzljak UP, Rozman D. Circadian regulation of the hepatic endobiotic and xenobitoic detoxification pathways: the time matters. Chem Res Toxicol. 2012;25:811–24.

50. Savvidis C, Koutsilieris M. Circadian rhythm disruption in cancer biology. Mol Med. 2012;18:1249–60.
51. Everett LJ, Lazar MA. Nuclear receptor Rev-erbα: up, down, and all around. Trends Endocrinol Metab. 2014;25(11):586–92.
52. Pilgrim C, Erb W, Maurer W. Diurnal fluctuations in the numbers of DNA synthesizing nuclei in various mouse tissues. Nature. 1963;199:863.
53. Gomes JR, et al. Circadian variation of the cell proliferation in the jejunal epithelium of rats at weaning phase. Cell Prolif. 2005;38:147–52.
54. Scheving LA. Biological clocks and the digestive system. Gastroenterology. 2000;119:536–49
55. Smaaland R. Circadian rhythm of cell division. Prog Cell Cycle Res. 1996;2:241–66.
56. Bjarnason GA, Jordan R. Circadian variation of cell proliferation and cell cycle protein expression in man: clinical implications. Prog Cell Cycle Res. 2000;4:193–206.
57. Matsuo T, et al. Control mechanism of the circadian clock for timing of cell division in vivo. Science. 2003;302:255–9.
58. Gery S, et al. The circadian gene per1 plays an important role in cell growth and DNA damage control in human cancer cells. Mol Cell. 2006;22:375–82.
59. Unsal-Kacmaz K, Mullen TE, Kaufmann WK, Sancar A. Coupling of human circadian and cell cycles by the timeless protein. Mol Cell Biol. 2005;25:3109–16.
60. Mullenders J, Fabius AW, Madiredjo M, Bernards R, Beijersbergen RL. A large scale shRNA barcode screen identifies the circadian clock component ARNTL as putative regulator of the p53 tumor suppressor pathway. PLoS One. 2009;4:e4798.
61. Cotta-Ramusino C, et al. A DNA damage response screen identifies RHINO, a 9-1-1 and TopBP1 interacting protein required for ATR signaling. Science. 2011;332:1313–7.
62. Gachon F, Olela FF, Schaad O, Descombes P, Schibler U. The circadian PAR-domain basic leucine zipper transcription factors DBP, TEF, and HLF modulate basal and inducible xenobiotic detoxification. Cell Metab. 2006;4:25–36.
63. Reiter RJ. Melatonin: the chemical expression of darkness. Mol Cell Endocrinol. 1991;79:C153–8.
64. Ng TB, Wong CM. Effects of pineal indoles and arginine vasotocin on lipolysis and lipogenesis in isolated adipocytes. J Pineal Res. 1986;3:55–66.
65. Cipolla-Neto J, Amaral FG, Afeche SC, Tan DX, Reiter RJ. Melatonin, energy metabolism, and obesity: a review. J Pineal Res. 2014;56:371–81.
66. Cutando A, Lopez-Valverde A, Arias-Santiago S, DE Vicente J, DE Diego RG. Role of melatonin in cancer treatment. Anticancer Res. 2012;32:2747–53.
67. Feychting M, Osterlund B, Ahlbom A. Reduced cancer incidence among the blind. Epidemiology. 1998;9:490–4.
68. Dickmeis T, Foulkes NS. Glucocorticoids and circadian clock control of cell proliferation: at the interface between three dynamic systems. Mol Cell Endocrinol. 2011;331:11–22.
69. Son GH, Chung S, Kim K. The adrenal peripheral clock: glucocorticoid and the circadian timing system. Front Neuroendocrinol. 2011;32:451–65.
70. Kalsbeek A, et al. Circadian rhythms in the hypothalamo-pituitary-adrenal (HPA) axis. Mol Cell Endocrinol. 2012;349:20–9.
71. Kino T, Chrousos GP. Circadian CLOCK-mediated regulation of target-tissue sensitivity to glucocorticoids: implications for cardiometabolic diseases. Endocr Dev. 2011;20:116–26.
72. Greene MW. Circadian rhythms and tumor growth. Cancer Lett. 2012;318:115–23.
73. Spiga F, Walker JJ, Terry JR, Lightman SL. HPA axis-rhythms. Compr Physiol. 2014;4:1273–98.
74. Herr I, Gassler N, Friess H, Buchler MW. Regulation of differential pro- and anti-apoptotic signaling by glucocorticoids. Apoptosis. 2007;12:271–91.
75. Humbel RE. Insulin-like growth factors I and II. Eur J Biochem. 1990;190:445–62.
76. Calle EE, Rodriguez C, Walker-Thurmond K, Thun MJ. Overweight, obesity, and mortality from cancer in a prospectively studied cohort of U.S. adults. N Engl J Med. 2003;348:1625–38.

77. Pollak M. The insulin and insulin-like growth factor receptor family in neoplasia: an update. Nat Rev Cancer. 2012;12:159–69.
78. Haus E, Dumitriu L, Nicolau GY, Bologa S, Sackett-Lundeen L. Circadian rhythms of basic fibroblast growth factor (bFGF), epidermal growth factor (EGF), insulin-like growth factor-1 (IGF-1), insulin-like growth factor binding protein-3 (IGFBP-3), cortisol, and melatonin in women with breast cancer. Chronobiol Int. 2001;18:709–27.
79. Mejean L, et al. Circadian and ultradian rhythms in blood glucose and plasma insulin of healthy adults. Chronobiol Int. 1988;5:227–36.
80. Haus E. Chronobiology in the endocrine system. Adv Drug Deliv Rev. 2007;59:985–1014.
81. Gamble KL, Berry R, Frank SJ, Young ME. Circadian clock control of endocrine factors. Nat Rev Endocrinol. 2014;10:466–75.
82. Balsalobre A, Marcacci L, Schibler U. Multiple signaling pathways elicit circadian gene expression in cultured Rat-1 fibroblasts. Curr Biol. 2000;10:1291–4.
83. Kohsaka A, et al. High-fat diet disrupts behavioral and molecular circadian rhythms in mice. Cell Metab. 2007;6:414–21.
84. Monk TH, Buysse DJ. Exposure to shift work as a risk factor for diabetes. J Biol Rhythms. 2013;28:356–9.
85. McFadden E, Jones ME, Schoemaker MJ, Ashworth A, Swerdlow AJ. The relationship between obesity and exposure to light at night: cross-sectional analyses of over 100,000 women in the breakthrough generations study. Am J Epidemiol. 2014;180:245–50.
86. Kalsbeek A, la Fleur S, Fliers E. Circadian control of glucose metabolism. Mol Metab. 2014;3:372–83.
87. Suwazono Y, et al. A longitudinal study on the effect of shift work on weight gain in male Japanese workers. Obesity (Silver Spring). 2008;16:1887–93.
88. Kroenke CH, et al. Work characteristics and incidence of type 2 diabetes in women. Am J Epidemiol. 2007;165:175–83.
89. Scheer FA, Hilton MF, Mantzoros CS, Shea SA. Adverse metabolic and cardiovascular consequences of circadian misalignment. Proc Natl Acad Sci U S A. 2009;106:4453–8.
90. Gangwisch JE. Epidemiological evidence for the links between sleep, circadian rhythms and metabolism. Obes Rev. 2009;10(Suppl 2):37–45.
91. Meisinger C, Heier M, Loewel H, Study MKAC. Sleep disturbance as a predictor of type 2 diabetes mellitus in men and women from the general population. Diabetologia. 2005;48:235–41.
92. Beihl DA, Liese AD, Haffner SM. Sleep duration as a risk factor for incident type 2 diabetes in a multiethnic cohort. Ann Epidemiol. 2009;19:351–7.
93. Kelly MA, et al. Circadian gene variants and susceptibility to type 2 diabetes: a pilot study. PLoS One. 2012;7:e32670.
94. Ruano EG, Canivell S, Vieira E. REV-ERB ALPHA polymorphism is associated with obesity in the Spanish obese male population. PLoS One. 2014;9:e104065.
95. Woon PY, et al. Aryl hydrocarbon receptor nuclear translocator-like (BMAL1) is associated with susceptibility to hypertension and type 2 diabetes. Proc Natl Acad Sci U S A. 2007;104:14412–7.
96. Marcheva B, et al. Disruption of the clock components CLOCK and BMAL1 leads to hypoinsulinaemia and diabetes. Nature. 2010;466:627–31.
97. Lamia KA, Storch KF, Weitz CJ. Physiological significance of a peripheral tissue circadian clock. Proc Natl Acad Sci U S A. 2008;105:15172–7.
98. Rudic RD, et al. BMAL1 and CLOCK, two essential components of the circadian clock, are involved in glucose homeostasis. PLoS Biol. 2004;2:e377.
99. Turek FW, et al. Obesity and metabolic syndrome in circadian Clock mutant mice. Science. 2005;308:1043–5.

100. Barclay JL, et al. High-fat diet-induced hyperinsulinemia and tissue-specific insulin resistance in Cry-deficient mice. Am J Physiol Endocrinol Metab. 2013;304:E1053–63.
101. Griebel G, Ravinet-Trillou C, Beeske S, Avenet P, Pichat P. Mice deficient in cryptochrome 1 (cry1 (−/−)) exhibit resistance to obesity induced by a high-fat diet. Front Endocrinol (Lausanne). 2014;5:49.
102. Carvas JM, et al. Period2 gene mutant mice show compromised insulin-mediated endothelial nitric oxide release and altered glucose homeostasis. Front Physiol. 2012;3:337.
103. Le Martelot G, et al. REV-ERBalpha participates in circadian SREBP signaling and bile acid homeostasis. PLoS Biol. 2009;7:e1000181.
104. Cho H, et al. Regulation of circadian behaviour and metabolism by REV-ERB-alpha and REV-ERB-beta. Nature. 2012;485:123–7.
105. Liu Z, et al. PER1 phosphorylation specifies feeding rhythm in mice. Cell Rep. 2014;7:1509–20.
106. Hughes ME, et al. Harmonics of circadian gene transcription in mammals. PLoS Genet. 2009;5:e1000442.
107. Panda S, et al. Coordinated transcription of key pathways in the mouse by the circadian clock. Cell. 2002;109:307–20.
108. Yang X, et al. Nuclear receptor expression links the circadian clock to metabolism. Cell. 2006;126:801–10.
109. Yang X, Lamia K, Evans R. Nuclear receptors, metabolism, and the circadian clock. Cold Spring Harb Symp Quant Biol. 2007;72:387–94.
110. Masri S, Zocchi L, Katada S, Mora E, Sassone-Corsi P. The circadian clock transcriptional complex: metabolic feedback intersects with epigenetic control. Ann N Y Acad Sci. 2012;1264:103–9.
111. Masri S, et al. Partitioning circadian transcription by SIRT6 leads to segregated control of cellular metabolism. Cell. 2014;158:659–72.
112. Paschos GK, et al. Obesity in mice with adipocyte-specific deletion of clock component Arntl. Nat Med. 2012;18:1768–77.
113. Shostak A, Meyer-Kovac J, Oster H. Circadian regulation of lipid mobilization in white adipose tissues. Diabetes. 2013;62:2195–203.
114. Shimba S, et al. Brain and muscle Arnt-like protein-1 (BMAL1), a component of the molecular clock, regulates adipogenesis. Proc Natl Acad Sci U S A. 2005;102:12071–6.
115. Gerhart-Hines Z, et al. The nuclear receptor Rev-erbalpha controls circadian thermogenic plasticity. Nature. 2013;503:410–3.
116. Chappuis S, et al. Role of the circadian clock gene Per2 in adaptation to cold temperature. Mol Metab. 2013;2:184–93.
117. McCarthy JJ, et al. Identification of the circadian transcriptome in adult mouse skeletal muscle. Physiol Genomics. 2007;31:86–95.
118. Zhang E, et al. Cryptochrome mediates circadian regulation of cAMP signaling and hepatic gluconeogenesis. Nat Med. 2010;16:1152–6.
119. Lamia KA, et al. Cryptochromes mediate rhythmic repression of the glucocorticoid receptor. Nature. 2011;480:552–6.
120. Asher G, Schibler U. Crosstalk between components of circadian and metabolic cycles in mammals. Cell Metab. 2011;13:125–37.
121. Ramsey KM, et al. Circadian clock feedback cycle through NAMPT-mediated NAD + biosynthesis. Science. 2009;324:651–4.
122. Nakahata Y, Sahar S, Astarita G, Kaluzova M, Sassone-Corsi P. Circadian control of the NAD+ salvage pathway by CLOCK-SIRT1. Science. 2009;324:654–7.
123. Imai S, Guarente L. NAD+ and sirtuins in aging and disease. Trends Cell Biol. 2014;24:464–71.
124. Houtkooper RH, Pirinen E, Auwerx J. Sirtuins as regulators of metabolism and healthspan. Nat Rev Mol Cell Biol. 2012;13:225–38.
125. Arble D, Bass J, Laposky A, Vitaterna M, Turek F. Circadian timing of food intake contributes to weight gain. Obesity (Silver Spring, Md.). 2009;17:2100–2.

126. Hatori M, et al. Time-restricted feeding without reducing caloric intake prevents metabolic diseases in mice fed a high-fat diet. Cell Metab. 2012;15:848–60.
127. Sherman H, et al. Timed high-fat diet resets circadian metabolism and prevents obesity. FASEB J. 2012;26:3493–502.
128. Vollmers C, et al. Time of feeding and the intrinsic circadian clock drive rhythms in hepatic gene expression. Proc Natl Acad Sci U S A. 2009;106:21453–8.
129. Oishi K, Uchida D, Itoh N. Low-carbohydrate, high-protein diet affects rhythmic expression of gluconeogenic regulatory and circadian clock genes in mouse peripheral tissues. Chronobiol Int. 2012;29:799–809.
130. Oike H, Nagai K, Fukushima T, Ishida N, Kobori M. High-salt diet advances molecular circadian rhythms in mouse peripheral tissues. Biochem Biophys Res Commun. 2010;402:7–13.
131. Cairns RA, Harris IS, Mak TW. Regulation of cancer cell metabolism. Nat Rev Cancer. 2011;11:85–95.
132. Hanahan D, Weinberg RA. Hallmarks of cancer: the next generation. Cell. 2011;144:646–74.
133. Koppenol WH, Bounds PL, Dang CV. Otto Warburg's contributions to current concepts of cancer metabolism. Nat Rev Cancer. 2011;11:325–37.
134. Denko NC. Hypoxia, HIF1 and glucose metabolism in the solid tumour. Nat Rev Cancer. 2008;8:705–13.
135. Hogenesch JB, et al. The basic helix-loop-helix-PAS protein MOP9 is a brain-specific heterodimeric partner of circadian and hypoxia factors. J Neurosci. 2000;20:RC83.
136. Hogenesch JB, Gu YZ, Jain S, Bradfield CA. The basic-helix-loop-helix-PAS orphan MOP3 forms transcriptionally active complexes with circadian and hypoxia factors. Proc Natl Acad Sci U S A. 1998;95:5474–9.
137. Szablewski L. Expression of glucose transporters in cancers. Biochim Biophys Acta. 2013;1835:164–9.
138. Salani B, et al. Metformin, cancer and glucose metabolism. Endocr Relat Cancer. 2014;21:R461–71.
139. Ros S, Schulze A. Balancing glycolytic flux: the role of 6-phosphofructo-2-kinase/fructose 2,6-bisphosphatases in cancer metabolism. Cancer Metab. 2013;1:8.
140. Pacha J, Sumova A. Circadian regulation of epithelial functions in the intestine. Acta Physiol (Oxf). 2013;208:11–24.

Chapter 2
Environmental Manipulation and Neuropeptide Effects on Energy Balance and Cancer

Lei Cao

Abstract Social and environmental factors have profound impacts on energy balance and cancer. Yet many experimental studies of the metabolic syndromes and cancer utilize animals in laboratory conditions without adequate social interactions. We recently demonstrate that environments that are more complex and challenging, but not stressful per se, have robust effects on body composition, energy balance, and peripheral cancer progression. One key underlying mechanism is the activation of a specific neuroendocrine brain-adipocyte axis, the hypothalamic-sympathoneural-adipocyte (HSA) axis. The social, physical, and cognitive stimuli provided by the enriched environments induce brain-derived neurotrophic factor (BDNF) in the hypothalamus and the ensuing sympathetic innervation of adipose tissue. The remodeling of the adipose tissue, including the white-to-brown phenotypic switch and the suppression of leptin, leads to antiobesity and anticancer phenotype. This chapter summarizes this work and discusses how environmental enrichment (EE) can serve as a valuable animal model to study eustress (positive or benign stress), metabolism, cancer, and aging.

Keywords Environmental enrichment · BDNF · Adipose tissue · Leptin · Cancer Hypothalamic-sympathoneural-adipocyte (HSA) axis · Aging · Obesity · Stress

---

L. Cao (✉)
College of Medicine, The Ohio State University, 910 BRT, 460 W 12th Avenue, Columbus, OH 43210, USA
e-mail: lei.cao@osumc.edu

N. A. Berger (ed.), *Murine Models, Energy Balance, and Cancer*,
Energy Balance and Cancer, DOI 10.1007/978-3-319-16733-6_2

What Is Environmental Enrichment?

Environmental enrichment (EE) refers to a more complex housing condition (compared with standard laboratory animal housing) that facilitates enhanced sensory, cognitive, motor, and social stimulation [1, 2]. Standard housing conditions most commonly constitute cages with bedding, ad libitum access to food and water, and single-sex housing in groups (typically 3–5 mice per cage). EE protocols vary widely between laboratories and there is no consensus on which EE paradigms are ideal to achieve optimal benefits on brain and behavior. EE housing typically includes a combination of increased physical space, larger groups, and objects for activity or hiding and running wheels. Figure 2.1 shows the EE housing that is established in our lab. Mice are housed in groups (15–20 mice per cage) in large cages of 1.5 m × 1.5 m × 1.0 m supplemented with running wheels, tunnels, igloos, huts, retreats, wood toys, a maze, and nesting material with food and water ad libitum. One key aspect is the provision of environmental complexity with EE objects that provide a range of opportunities for visual, somatosensory, and olfactory stimulation. Environmental novelty is another key aspect that is achieved by weekly changing the objects and the locations of the objects [2]. The larger space and the running wheels encourage free roaming and physical activities. The larger number of animals living in the EE housing increases social stimulation. Hiding places allow animals to retreat to safe havens and thus enhance the sense of self-control. We videotaped the activities of mice living in EE using a night-view camera (in press JoVE). Figure 2.2 was snapshots that showed different activities of individual mouse in EE: free roaming, hiding in huts, running on wheels, eating/drinking, exploring maze, etc. Interestingly, one mouse was interested in getting on the wheel that was being used by two other mice. This moment captured the rich social contact in this complex environment.

EE is associated with many adaptive changes in the brain and behavior, including increased hippocampal neurogenesis, enhanced motor and cognitive performance,

Fig. 2.1 EE housing

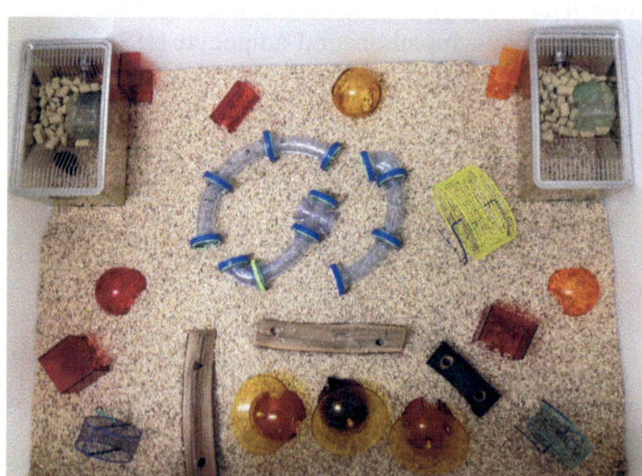

Fig. 2.2 Snapshots of EE

greater resistance and improved recovery to external insults and ischemia, reduced apoptosis, and milder phenotype in models of neurological disease. The robust and powerful influences of EE on brain function and the underlying mechanisms have been reviewed elsewhere [2]. This chapter summarizes recent work showing the impact of EE on energy balance and cancer, and discusses the use of EE as a unique and valuable model for research in metabolism, cancer, stress, and aging.

Macroenvironments and Cancer

There is mounting evidence that environmental factors and lifestyle have profound effects in the initiation, promotion, and progression of cancer [3, 4]. The growth of cancer is dependent, in part, on its microenvironment constructed by several types of cells, including pericytes, cancer-associated fibroblasts, endothelial cells, local and bone marrow-derived stromal stem and progenitor cells, and immune cells [5, 6]. In case of solid tumors, the microenvironment provides a tumor ecosystem for bidirectional communications between cancer cells and the tumor-associated stroma to support the full malignancy [6, 7]. This local microenvironment is influenced by systemic factors, and the cancer itself induces both local and distant changes through paracrine signaling and interactions with the immune and nervous systems [8, 9]. The past decade has seen more appreciation of the macrophysiological milieus that can shape individual variability in the natural course of cancer and responsiveness to therapies [6, 10]. However, the effect and mechanisms of the macroenvironment on systemic cancer, specifically an individual's interaction with its physical living and social environment, are much less well defined.

EE Suppresses Cancer Progression via Activating the HSA Axis

Our long-term interest is to understand how environmental stimuli to the central nervous system (CNS) shape biological processes and disease pathology, and how this knowledge can be harnessed to improve health and treat diseases. EE is a remarkable environmental model demonstrating robust impact on brain structure and function. We previously published that EE leads to the changes in expression of growth factor, enhancement of neurogenesis, and survival of cells within the CNS [11, 12]. Moreover, EE has considerable impact on the phenotype of a variety of toxin- and genetically induced models of human neurological disease [2]. This intrigued us with the hypothesis that an EE, one that improved brain function and resiliency, can also affect the body's overall state of health and lead to an anticancer phenotype. We tested this hypothesis in both implantation of tumor cells and spontaneous cancer model [13]. Living in EE for 6 weeks prior to the implantation of syngeneic B16 melanoma resulted in a remarkable suppression of tumor growth, approximately 80 % decrease in tumor mass. Furthermore, ~17 % mice in EE showed no palpable tumor at the end of the experiment while all mice in standard housing developed tumor. In a second syngeneic implantation model, we delayed the EE until the MC38 colon cancer became visible. EE reduced the growth rate of established MC38 colon cancer. Moreover, EE was highly effective in $APC^{min/+}$ mice, a spontaneous model with a germline mutation in APC similar to humans with familial adenomatous polyposis, and a gene in which somatic mutations occur in 80 % of human colon cancer. EE decreased the total number of polyps in the small intestine by approximately 50 % and substantially reduced the size of polyps [13]. In a comprehensive set of experiments using transgenic animals, somatic gene transfer, controlled release liposomes, and osmotic minipump drug infusion, we dissected out a key mechanism underlying the EE-induced anticancer phenotype and coined the term "hypothalamic-sympathoneural-adipocyte" (HSA) axis to describe this brain-adipocyte axis. The HSA axis refers to the specific neuroendocrine pathway linking the hypothalamus to white adipose tissue (WAT) [1]. The hypothalamic component is mediated via brain-derived neurotrophic factor (BDNF), which is highly expressed in the ventromedial (VMH), dorsomedial (DMH), and arcuate nuclei. In response to the stimuli provided by EE, hypothalamic BDNF is induced as an effector immediate early gene and leads to the activation of a component of the sympathetic nervous system (SNS), preferentially the sympathoneural innervation of WAT. The activation of the axis therefore involves an increase in hypothalamic BDNF, WAT norepinephrine, and suppression of leptin expression and release via β-adrenergic receptors (ARs) on the adipocytes. This marked drop of leptin in circulation is associated with an anticancer phenotype (Fig. 2.3).

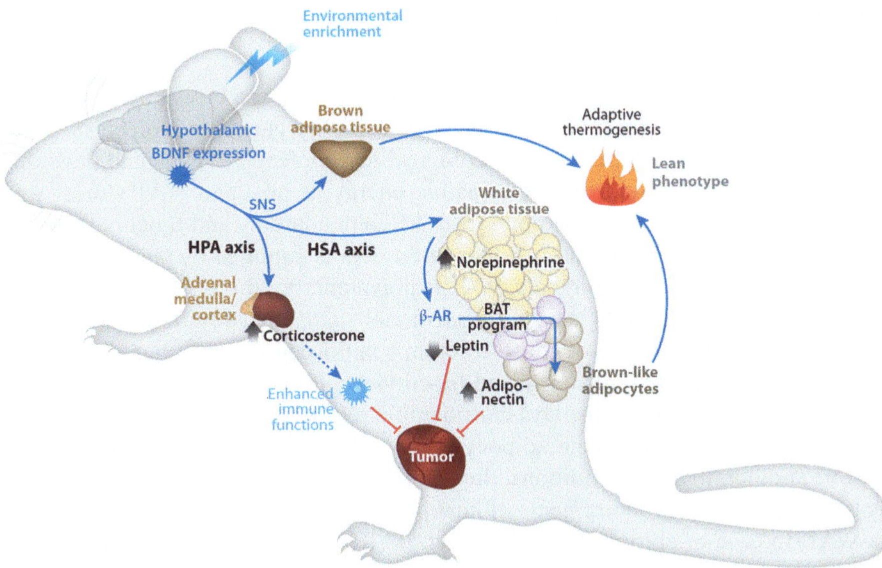

Fig. 2.3 Mechanisms of EE-induced anticancer and antiobesity phenotypes (With permission from the Annual Review of Neuroscience, volume 35 © 2012 by Annual Reviews, http:www.annualreviews.org)

Environment, Stress, and Obesity

Obesity is a complex acquired disease with both environmental and genetic factors. There is strong evidence that social and environmental factors have profound effects on weight and the development of obesity and associated metabolic syndromes [14, 15]. From an evolutionary viewpoint, obesity and metabolic syndromes can be considered as a maladaptive consequence of an initially successful adaptation to high environmental demands. In current human society, when calorie-rich (and usually palatable) food is abundant while the energy-demanding actions (i.e., fight, flight, and hunt) are no longer required, the biological mechanisms adaptive to prehistoric times can easily become maladaptive, and trigger obesity and associated metabolic disorders [15]. Psychological factors also play a role in the development and propagation of obesity and metabolic syndromes that are found most prevalent among individuals under psychosocial pressure or chronic stress [16, 17]. However, many studies of the etiology of obesity and metabolic syndromes utilize the genetically modified rodents in laboratory conditions without adequate social interactions. Because the external environment impacts the internal environment in profound ways that can ultimately alter body weight and body composition, physical and social environments of laboratory animals should be taken into consideration in studies to identify potential molecular therapeutic candidates and regulation pathways for obesity treatment.

Adipose Organ and Plasticity

Two types of adipose tissues are found in mammals, WAT and brown adipose tissue (BAT). Although both are involved in the use of highly energetic molecules, WAT and BAT are widely different at morphological and molecular levels and have different functional roles with BAT dissipating energy as heat and WAT storing and distributing excessive energy. Despite their different anatomy and functions, brown and white adipocytes are found together in fat depots supplied by specific vessels and nerves. In most adult mammals, WAT is prevalent, but the WAT/BAT ratio varies with genetic background, sex, age, nutritional status, and environmental conditions supporting the concept of adipose organ, a multi-depot organ characterized by two tissues with different anatomy and functional roles (reviewed by Frontini and Cinti [18]). Adipose organ displays considerable plasticity. Accumulating evidence supports the concept that some adipocytes in each population can reversibly turn into one another. In basal conditions, the adipose organ uses the two tissues to meet the two physiological requirements of heat production via BAT and energy storage via WAT. In case of chronic energy surplus, BAT is able to transform into WAT to store more energy molecules, whereas, in the event of sustained heat requirement (i.e., chronic cold exposure), WAT can convert to BAT [18]. The nature of this white-to-brown conversion is not clear. Several data support the notion that this process occurs through direct transformation of adult cells, i.e., via physiological reversible transdifferentiation [19, 20] accompanied by tissue reorganization with changes in the density of capillaries and parenchymal nerve fibers [21]. However, other data suggest that the brown adipocytes emerging in rodents in classic WAT depots in response to β-adrenergic stimulation derive from de novo differentiation of stem cells or committed brown preadipocytes [22, 23]. The third source is the activation of “brite” (brown-in-white) adipocyte which is thermogenically competent but developmentally and molecularly distinct from classical brown adipocyte [24]. Many of the anatomo-physiological features of murine fat depots can apply to humans.

Brown Fat and Obesity Prevention and Treatment

BAT has recently become a potential target for pharmacological and genetic manipulation to treat human obesity, because positron emission tomography has provided evidence that adult humans retain considerable amounts of metabolically active BAT depots which can be induced in response to cold and SNS activation [25–27]. BAT dissipates energy directly as heat through uncoupling fatty acid oxidation from ATP production by uncoupling protein 1 (UCP1) and is vital for the regulation of body temperature and is also involved in the control of body weight [28, 29]. Increasing energy expenditure is always attractive to counteract obesity. But is the small amount of BAT, about 0.05 ~ 1 % of body weight in adult humans in contrast to 5 ~ 10 % in mouse, enough to make a metabolic difference? It has been estimated that 40–50 g of maximally stimulated BAT in man could correspond to as much as 20 % of energy expenditure, equivalent to 20 kg of body weight, over a year [30], which could be highly relevant in terms of human energy expenditure. Moreover, even a small but consistent imbalance in the energy input/output equation can ultimately lead to obesity/leanness. A recent review by Cannon and Nedergaard concludes that modifying thermogenesis via cold exposure or manipulating UCP1 leads to changes in energy balance and body weight not fully compensated by changes in food intake [31]. Indeed, several studies demonstrate an inverse correlation between BAT activity in adult humans and body fat [26, 27, 32–35]. A recent study in morbid obese subjects shows that in an extremely wide range of body composition, BAT activity is highly correlated with body mass index (BMI) and body fat percentage [36]. Furthermore, several histological studies have reported that brown adipocytes dispersed among white fat in 24 % of adult humans biopsies and reaching 50 % of cases with exclusion of patients over 50 years old [33]. In addition, studies of human fat cell dynamics show a high turnover rate of adipocytes with approximately 10 % of fat cells being renewed annually at all adult ages and levels of BMI [37]. Human white adipocytes from subcutaneous fat tissues can be manipulated in vitro to develop "brown" characteristics by forced expression of transcription coactivator PGC-1α [38, 39]. These findings further suggest the therapeutic potential of BAT-oriented strategies to enhance energy expenditure by facilitating brown adipocyte maintenance, stimulating preexisting brown precursors, and inducing white-to-brown transformation [18, 40, 41]. However, as Nedergaard and Cannon point out in a recent review, UCP1 in intact cells is constantly inhibited by purine nucleotides in the cytosol. Therefore, the strategies solely based on increasing the total amount of BAT or UCP1 may be flawed because the UCP1 is not automatically active and its inhibition has to be overcome through a process initiated physiologically by SNS stimulation [42]. There are no indications that the amount of UCP1 (i.e., the capacity for thermogenesis) influences basal brown adipocyte metabolic rate or whole body metabolic rate. Indeed, large amounts of BAT emerge in rodents when adapted to chronic cold. But immediately after mice are transferred to warm conditions, BAT activity is turned off within seconds [42]. Thus, an individual constantly decides how much combustion is needed to meet environmental demand and that is independent of the capacity for thermogenesis. It is plausible that the brain plays the commanding role. Our recent work on EE has revealed a novel avenue to induce white-to-brown conversion and continuous but preferential SNS activation in WAT.

EE Induces White Fat Browning and Leanness via the HSA Axis

Our recent studies on EE have revealed a new type of thermogenesis that contributes to the profound impact on body composition and metabolism in mice living in EE (Fig. 2.3). EE induces a molecular and functional switch from WAT to BAT in the absence of chronic cold exposure or prolonged pharmacological β-adrenergic stimulation [43]. The molecular characteristics of EE-induced WAT "browning" include the induction of BAT specific markers; suppression of WAT markers; and no changes in adipocyte markers shared by both white and brown adipocytes. The induction of BAT genetic program by EE is remarkable when compared to the most effective physiological and pharmacological approaches to induce browning up to date. The emergence of "brite" or "beige" cells is under genetic control with large differences among inbred strains of mice [44]. C57BL/6 mice, the strain we used in our experiments, respond poorly to SNS activation probably due to the low levels of β-AR expression [44, 45]. Cold exposure at 5 °C for 7 days failed to induce brown-like cells in retroperitoneal white fat pad (RWAT) with only low induction of *Ucp1* expression [44]. In contrast, 4-week EE led to a 27-fold induction of *Ucp1* which is more effective compared to the 9-fold induction observed in the same strain of mice acclimated to 5 °C cold for 4 weeks or the 23-fold induction after injection of β-agonist CL316243 for 11–12 days [46]. Of interest, EE significantly increased both β2 and β3-AR gene expression and protein levels in RWAT suggesting the absence of desensitization of β-ARs after extended exposure to β-agonist [47]. Consistent with the brown gene program induced in WAT, cells with the multilocular morphology characteristic of brown adipocytes were observed in WAT and mitochondrial DNA content was increased by twofold indicating enhanced mitochondrial biogenesis. Long-term EE led to stronger WAT browning. After 12-week EE, clear macroscopic changes in fat pads became visible to the eye with white fat turning brown and brown fat going even darker. Widespread UCP1 staining was observed associated with more robust induction of brown gene program. Our mechanistic studies demonstrate that the activation of the HSA axis mediates the EE-induced WAT browning, reduction of adiposity via increase in energy expenditure, and resistance to diet-induced obesity (DIO) [43]. We recently identified adipose vascular endothelial growth factor (VEGF) as a key component of the HSA axis underlying the browning effect of EE (Endocrinology in press). VEGF is the only bona fide endothelial cell growth factor and its presence is essential for initiation of the angiogenic program [48, 49]. EE stimulated the VEGF expression in a fat depot-specific manner prior to the emergence of beige cells. The VEGF upregulation was independent of hypoxia but required intact sympathetic tone to the adipose tissue. Targeted adipose overexpression of VEGF by a novel recombinant adeno-associated virus (rAAV) reproduced the angiogenic and browning effect of EE. Adipose-specific VEGF knockout or pharmacological VEGF blockade by antibody abolished the beige cell recruitment induced by EE. Furthermore, VEGF may have actions independent of angiogenesis. One of the key findings of this study is that VEGF integrates multiple upstream stimulations to a common pathway that is essential to the emergence of beige cells. In addition to EE, VEGF blockade could substantially block the browning induced by the β3-adrenergic agonist CL-316,243, the PPARγ ligand rosiglitazone, and voluntary running. UCP1 was particularly sensitive to the anti-VEGF antibody whose upregulation by EE, running, or rosiglitazone was completely abolished. Thus, VEGF signaling is likely a downstream pathway shared by diverse upstream mechanisms that all lead to the induction of beige cells suggesting coordination between angiogenesis and the white-to-brown switch. Targeting this common pathway may have therapeutic potential.

Obesity Increases Cancer Risk and Mortality

The worldwide epidemic of obesity and global incidence of cancer are both rising [50, 51]. The World Health Organization estimates that in 2008 more than 1.4 billion adults were overweight (BMI≥25 kg/m^2). Of these overweight adults, over 200 million men and nearly 300 million women were obese (BMI≥30 kg/m^2). More alarmingly, more than 40 million children under the age of 5 were overweight or obese in 2012. Overweight and obesity are leading risks for global deaths. In 2010, overweight and obesity were estimated to cause around 3.4 million deaths in adults worldwide (www.who.int/mediacentre/factsheets/fs311/en/). Obesity increases the risk of certain types of cancer including postmenopausal breast, renal, ovarian, esophagus, pancreas, prostate, hepatobiliary, colorectal, and melanoma [52–54]. Between 7 and 41 % of certain cancer burdens are attributable to overweight and obesity. Obesity may account for 14 % of all deaths from cancer, and 20 % of those occur in women [55]. Although we know that weight gain can result in higher cancer rates, the converse finding of weight loss and lower cancer rates was much harder to achieve due to the lack of successful weight-loss modalities [56, 57]. Both randomized and prospective studies reveal that the long-term weight loss with intensive lifestyle intervention in diabetic patients (BMI of 30) is ~2 % at 10 years [58] and is even less in patients with BMI above 35 [59]. Metabolic or bariatric surgery can provide sustained weight loss [60]. Recent longitudinal studies on metabolic surgery demonstrate that successful weight loss leads to lower cancer rates and decreased cancer mortality and therefore help to establish a causal association between obesity and cancer [59, 61, 62]. The mechanisms for the obesity-cancer association are multifactorial and include insulin resistance, increased growth factors and anabolic hormones, altered balance of adipokines, oxidative stress, inflammation, increased bioavailable sex hormones, deterioration in immunosurveilance, and altered cellular energetic [63, 64] (Fig. 2.4).

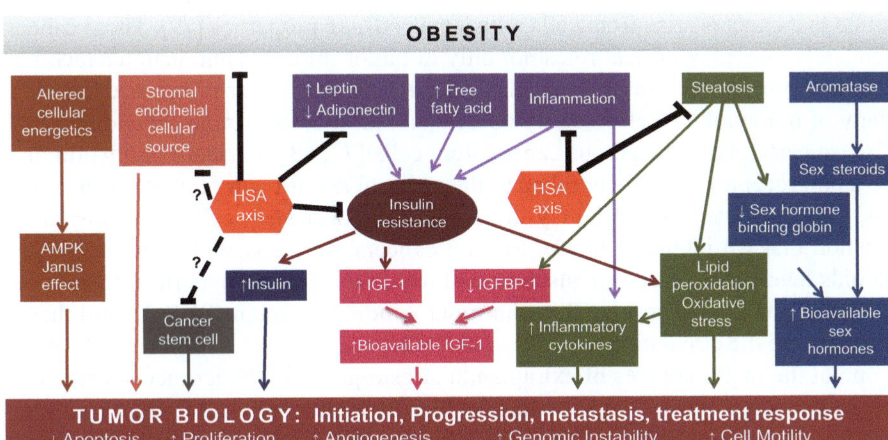

Fig. 2.4 Mechanisms of cancer inhibition by activating the HSA axis

HSA Activation Targets Both Obesity and Cancer

EE induces antiobesity and anticancer phenotypes via the HSA axis [13, 43]. Although likely just one component to the body's response to eustress, the HSA axis links physical and social environment to the regulation of fat phenotype and energy balance and thereby can regulate multiple organ systems and ultimately affect cancer progression. Given the robust anticancer effects in animals of normal weight and the potent modulation of fat in various models of obesity, one can predict that the anticancer effect of HSA activation is likely to be even more robust in obese individuals. Indeed, 3 weeks of EE led to approximate 70 % decrease of melanoma mass in DIO mice, compared with a 50 % decrease in mice of normal weight [13]. The adipokines, in particular, leptin and adiponectin, are recognized for their influence on cancer risk and cancer biology [65–67]. Both animal and clinical data suggest that the balance of adiponectin to leptin (adiponectin/leptin ratio) may be important in the development of cancer [68, 69]. Activation of the HSA axis leads to a sharp drop of leptin and a significant increase of adiponectin level and therefore resulting in a favorable shift of leptin/adiponectin ratio. Adipokine-independent mechanisms may also exist and are likely to be revealed in obesity models. The chronically increased insulin levels have been associated with various cancers [70–72]. Both obesity and diabetes are risk factors for hepatocellular carcinoma [73]. Our data show that genetically activating HSA axis by hypothalamic BDNF gene transfer potently alleviates obesity-associated insulin resistance, hyperinsulinemia, dislipidemia, and liver steatosis [43, 74]. In addition, an association between inflammation and cancer has been established [75] and obese adipose tissue is a mediator of low-grade inflammation and a source of local and circulating proinflammatory cytokines [76, 77]. We recently published that hypothalamic gene transfer of BDNF to obese middle-aged female mice substantially alleviated obesity, decreased mammary tumor progression, and prevented metastasis. Interestingly, the BDNF treatment robustly suppressed the inflammatory markers in the circulation, the adipose tissue, the mammary tumor, and the hypothalamus [78]. Accumulating evidence supports the notion that cancers arise from cells with stem cell-like properties [79]. These cancer stem cells may play critical roles not only in tumor initiation and maintenance but also in tumor invasion, metastasis, drug resistance, and cancer recurrence [79–82]. Many of the cytokines elevated by obesity can promote self-renewal and survival of these particularly dangerous cancer stem cells [77, 83]. Furthermore, leptin and the leptin receptor have been linked to the survival and growth of cancer stem cells [84, 85]. How the HSA axis may influence the cancer stem cells is an intriguing question that we intend to tackle. Several cancers are associated with altered sex steroids levels [86]. Insulin resistance and the compensatory hyperinsulinemia inhibit the hepatic production of sex hormone-binding globulin (SHBG), and therefore increase the circulating concentrations of bioavailable sex hormones [87]. WAT is one of the major sources of extraglandular estrogen [88]. Other mechanisms implicated in obesity-cancer link include cellular energetics [89] and stromal contribution from cells originating within WAT [90]. How the HSA axis may affect cellular energetics and adipose stromal/vascular cell trafficking are fascinating topics and warrant further investigations. Hanahan and Weinberg recently updated the concept of cancer hallmarks and noted that the clinical responses to the hallmark-targeting cancer drugs were generally transitory followed by almost-inevitable relapses. One interpretation is that some cancer cells may survive a therapeutic agent targeting one key pathway by relying on other hallmark capabilities. Thus, the combination of mechanism-guided therapies co-targeting multiple hallmarks is likely to be more effective and durable [91]. The HSA axis has the capacity to affect multiple pathways implicated in obesity-cancer association (Fig. 2.4) and therefore represents an attractive target for combination therapy.

EE as a Eustress Model

The stress response involves the brain's ability to perceive a threat resulting in activation of the sympathetic-adrenal-medullary (SAM) axis and the hypothalamic-pituitary-adrenal (HPA) axis, thereby causing the release of catecholamines (mainly epinephrine and norepinephrine), glucocorticoids (cortisol in humans and corticosterone in rodents), and other stress hormones from the adrenal gland, brain, and sympathetic nerve terminals. These factors can modulate the activity of various components of the tumor microenvironment. Clinical and epidemiological studies have recognized that specific psychosocial factors, such as stress, chronic depression and lack of social support, are risk factors for the development and progression of cancer [92, 93]. A recent meta-analysis reveals that stress-related psychosocial factors are associated with a higher cancer incidence in premorbid healthy people, poorer survival in cancer patients, and higher cancer mortality [94]. Psychosocial factors, such as hopelessness, denial, suppression of negative emotions, and lack of social support, can predict progression of already diagnosed cancers [93]. Experimental studies parallel these clinical data demonstrating that experimentally imposed distress can modulate cancer progression [95]. Recent mechanistic studies have started to identify the specific pathways that could mediate such effects. The majority of studies on stress-cancer relationship support the notion that psychosocial/behavioral risk factors initiate a cascade of information-processing pathways in the CNS triggering the SAM axis and/or the HPA axis [96, 97]. Downstream activation of the SAM and HPA axes exerts selective physiologic pressures that initiate molecular signaling pathways involved in DNA repair, cell survival, angiogenesis, inflammation, invasion, metastasis, and drug resistance [6, 98, 99]. However, the vast majority of mechanistic research on the stress-cancer relationship used experimentally imposed distress such as restraint stress [100–102] and social isolation (SI) [103–105] to model adversity. In contrast, scarce attention has been paid to the positive functions of a more mild stress or challenging environment (eustress/hormesis). Hans Selye who first brought the concept of stress into experimental medicine noted that departures from homeostasis could be "eustressful" or "distressful" and that health effects would vary accordingly [106, 107]. Eustress (positive stress) is associated with adaptive responses and benign or beneficial effects on health whereas maladaptive distress (negative stress) is associated with exposure to more severe aversive or hostile environments resulting in ill health [106, 107]. Failure to note the distinction between eustress and distress misunderstands the nature of pathways linking environmental conditions to cancer and hampers our ability to develop effective treatments.

Our research on eustress and cancer uses EE as an experimental paradigm to explore effects of a more complex environment devoid of elements traditionally thought of as being stressful. The EE-induced anticancer and antiobesity phenotypes could not be accounted for by physical activity alone but could be largely reproduced by overexpression of BDNF in the hypothalamus [13, 43]. Experimental animal studies have almost invariably shown that EE has beneficial effects on an animal's well-being. However, EE also moderately, but significantly, increases corticosterone level in serum and norepinepherine level in WAT [13, 43]. Both corticosterone and norepinepherine are signatures of stress. It may appear paradoxical that chronic mild stress could be associated with an anticancer phenotype. However, both hormones are metabolic regulators catabolizing energy stores to meet immediate metabolic demands. EE increases energy utilization, so it is not surprising that both the HPA axis and the SNS are activated and WAT is mobilized to meet elevated energy expenditure [1]. We believe EE can be used as a valuable model of eustress.

What Are the Mechanisms Underlying Opposite Effects of Eustress Versus Distress on Cancer?

In contrast to EE, SI of naturally gregarious rodents is a widely used model for social deprivation in humans that can produce long-term changes including neophobia, aggression, and cognitive rigidity [108, 109]. SI is associated with increased tumor progression in models of breast cancer [110, 111] and Ehrlich tumor [112]. Hasegawa and Saiki report that SI increases B16 melanoma growth in male C57BL/6 mice [104], and oral administration of β blocker propranolol abrogates the stress-accelerated B16 melanoma growth [104]. Interestingly, we used the same B16 melanoma transplant to the same strain C57BL/6 mice and showed remarkable tumor suppression induced by EE. And the EE-induced leptin drop and inhibition of tumor growth could be completely blocked by oral propranolol [13]. How to reconcile the contradiction? We have proposed a hypothesis in a review article [1]. This seemingly paradoxical phenomenon reflects the distinct physiological responses to eustress versus distress that may involve three axes: the well-known HPA axis, the SAM axis, and the lesser known HSA axis (Fig. 2.5). The HPA axis is activated in response to both EE and SI but plays a relatively minor role through its modulation of immune function. SAM activation resulting in elevated norepinephrine levels in the circulation and possibly a local increase in the solid tumor itself may facilitate melanoma growth in SI by stimulating proangiogenic factors, including VEGF, similar to the observations reported in other cancer models [105]. The HSA response in SI remains to be investigated. Our preliminary data showed that SI increased fat mass and circulating leptin level, and downregulated BDNF expression in hypothalamus (Cao unpublished data), suggesting a likely suppression of the HSA axis in SI. In contrast, a robust activation of HSA is associated with and responsible in part for the EE-induced melanoma inhibition. Of note, EE resulted in a significant increase of norepinephrine specifically in WAT but not in muscle and caused no change in circulation when mice were on normal chow diet, suggesting minimal activation of the SAM [43] (Fig. 2.5). We are currently testing the hypothesis that EE inhibits melanoma mainly through activating HSA while SI promotes melanoma growth mainly through activating SAM (possibly also suppressing HSA, Fig. 2.5). Accomplishing this research will have significant implications in individualized treatment. For example, several preclinical studies have shown that β blockers abrogate the deleterious effect of distress on cancer growth [102, 105]. Recent epidemiological and clinical studies support the link between β blockers and reduced cancer risk [113, 114]. These data have led to the proposal of using β blockers as a therapeutic intervention for cancer [115–117]. However, the mediators of stress operate in a nonlinear network and the interactions are very complex. When any one mediator is changed, there are compensatory changes in the other mediators that depend on the time course and the magnitude of changes of each of the mediators [118]. Therefore, the effects of environmental and psychosocial factors on cancer should be analyzed in the context of the relevant environmental circumstance, individual adaptation and coping strategy, and outcome on the specific cancer microenvironment biology. When individuals are experiencing distress and inability to cope when their SAM axis is overactive, β blockers may be considered as a treatment for cancer. However, when the HSA axis activation is the major force, for example, promoting eustress as a prevention strategy, the use of β blockers might attenuate eustress-induced anticancer effects. The complex interplay between the CNS and cancer often complicates the interpretation of human studies because of the difficulties of teasing apart pathways in a clinical setting [1]. Animal studies in which environmental and psychosocial conditions can be well controlled make it possible to distinguish the pathways and elucidate mechanisms, and to identify molecular mediators in the brain that may reveal potential therapeutic targets for cancer treatment.

Potential Effects of EE on Aging and Lifespan

As life expectancy increases, the incidence of diseases such as cancer, cardiovascular disorders, and neurodegeneration rises. It is vital, therefore, that we understand more about the dynamics of aging, how they interact with various environmental and lifestyle factors, and the connections between disease processes and aging in order to develop more effective ways to prevent, diagnose, and treat age-related diseases. Research on model organisms has demonstrated an interaction between genes and environment in determining healthy lifespan [119, 120]. Calorie restriction (CR) without malnutrition remains the most robust and reproducible manipula-

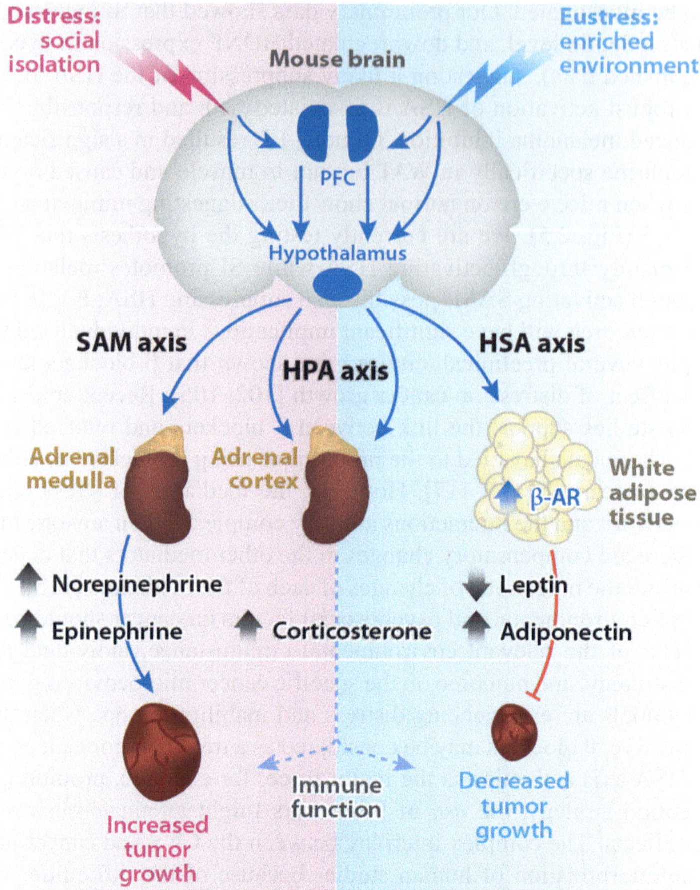

Fig. 2.5 Mechanisms underlying the opposite effects of eustress and distress on cancer. The prefrontal cortex (*PFC*) may mediate the appraisal of environmental stimuli as eustressful (positive stress) or distressful (negative stress). The PFC, sitting at the top of the brain hierarchy modulating stress, is connected through multiple pathways to the hypothalamus, the gate to the periphery, which regulates the activities of three axes: hypothalamic-pituitary-adrenal (*HPA*), sympathetic-adrenal-medullary (*SAM*), and hypothalamic-sympathoneural-adipocyte (*HSA*). The differential activation of the three axes responding to eustressful or distressful events may lead to distinctive health outcomes (eustress associated with improved health, distress linked with impaired health) depending on the magnitude of activation of each axis, the cross talk among them, their preferential target tissues in the periphery, and their influences on the intracellular context at the target tissues. The eustressful environmental enrichment (*EE*) stimulates brain-derived neurotrophic factor (*BDNF*) in the hypothalamus, leading to HSA activation of white adipose tissue. The preferential increase of SNS tone to white fat via beta-adrenergic receptors shuts down leptin and boosts adiponectin production, resulting in tumor suppression. In contrast, the distressful social isolation may stimulate primarily the SAM axis, resulting in high-level norepinephrine in the circulation and possibly increased release in the tumor. This action would facilitate cancer growth by inducing proangiogenic factors such as vascular endothelial growth factor (*VEGF*). The HSA and SAM activations are interactive and could synergistically or counteractively influence tumor growth. In the EE model, the predominant HSA may override the possible weak tumor stimulation brought by mild SAM axis activation. In the SI model, the strong activation of SAM may enhance tumor growth synergistically with the weakened HSA via increased leptin production (With permission from the Annual Review of Neuroscience, volume 35 © 2012 by Annual Reviews, http:www.annualreviews.org)

Table 2.1 Comparison of phenotypic characteristics of models of HSA axis activation (EE/BDNF overexpression) to models for extended longevity (CR, Pten, and FIRKO mice)

Parameter	EE/BDNF	CR	$Pten^{tg}$	FIRKO
Body mass index	↓	↓	↓	↓
Body fat content	↓	↓	↓	↓
Insulin sensitivity	↑	↑	↑	↑
Food intake	↑(→)	↓	↑	↑
IGF-1	↓	↓	↓	↓
Leptin	↓	↓	↓	↑
Adiponectin	↑	↑	↑	↑
Corticosterone	↑	↑	NA	NA
BDNF	↑	↑	NA	NA
Immune functions	↑	↑	NA	NA
Cancer	↓	↓	↓	NA
Energy expenditure	↑	↓	↑	↑
Mitochondrial activity	↑	→	↑	↑
Brown fat function	↑	NA	↑	NA
Longevity	?	↑	↑	↑

NA not applicable

tion able to extend lifespan and delay the onset of age-related disorders in a wide range of model organisms from yeast to monkeys (reviewed by Fontana [119]). Intermittent fasting (IF; a diet with reduced meal frequency such as alternative-day feeding) also increases resistance to toxicity and stress, and extends lifespan [121]. The metabolic and physiological effects of CR that may contribute to its antiaging ability include a reduction in adiposity, higher insulin sensitivity, improved lipid profiles, and reduced inflammation and oxidative stress [122–126]. Our studies on the HSA axis suggest beneficial effects on health and some of the features overlap with CR and/or Pten transgenic mice ($Pten^{tg}$), or fat-specific insulin receptor knock-out (FIRKO) mice with prolonged lifespan [127–129] (Table 2.1). The following characteristics suggest that the HSA axis may play an important role in regulating aging and age-related diseases. (1) HSA axis activation leads to a robust reduction in fat mass with little change in body weight [13, 74]. Studies in several genetically modified mouse models have linked reduction in adiposity to longevity, including the translational inhibitor 4E-BP1 (Eif4ebp1−/−) knockout [130], C/EBPβ knockin (β/β) [131], c-Cbl knockout [132], and FIRKO [128, 133]. The FIRKO mice with fat-specific disruption of the insulin receptor gene have 50 % reduced fat mass, improved whole body insulin sensitivity, and an extended lifespan (18 %) [128, 133]. EE initiated in young mice resulted in over 60 % reduction of abdominal fat when mice were maintained on normal diet whereas body weight was identical to the mice with standard housing [43]. Muscle mass was increased in EE and BDNF overexpressing mice. These data suggest that the HSA axis stimulation is particularly efficient in decreasing adiposity, allowing the dissociation of fat loss from weight loss which is difficult to achieve with other interventions [134]. This specific fat loss induced by HSA axis can provide a new model to clarify the controversy between weight loss and mortality [135]. (2) HSA axis activation is associated with increased whole body metabolic rate, increased oxygen consumption in fat, increased mitochondrial content, and upregulation of genes involved in mitochondrial biogenesis and activity such as PGC-1α [43], resembling FIRKO and Pten^{tg} mice. (3) HSA axis activation alleviates obesity-associated insulin resistance, hyperglycemia, and dyslipidemia [43, 74], mimicking CR. (4) HSA axis activation leads to reduced serum IGF-1 levels, also similar to CR. The IGF-1/growth hormone pathway is one of the most conserved pathways implicated in aging [136]. Both spontaneous and genetically engineered IGF-1 deficiency leads to smaller body size, delayed age-related pathology, and an extended lifespan [136]. (5) HSA axis activation alters adipokine levels with higher adiponectin and lower leptin expression in adipose tissue as well as in the circulation [13, 74], again mimicking CR. Adiponectin, an abundant protein secreted from fat, has insulin-sensitizing, anti-inflammatory, and anti-atherogenic properties in both rodents and humans. Transgenic expression of human adiponectin was shown to inhibit DIO and reduce the morbidity and mortality in DIO models [137]. Adiponectin may also suppress carcinogenesis and inhibit angiogenesis in cancer models [138]. (6) HSA axis activation suppresses tumor growth [13]. Epidemiological studies have revealed that lifestyle and environmental factors can influence cancer initiation, promotion, and progression, suggesting that many cancers are preventable. Among the key players are excessive adiposity, decreased physical activity, and unhealthy diet [139, 140]. CR has been shown to be broadly effective in cancer prevention in rodents and monkeys and the metabolic adaptations to CR are thought to be responsible for CR-mediated anticancer phenotype [63, 123, 141]. Our studies have shown that EE and BDNF overexpression markedly suppresses tumor growth [13]. (7) HSA axis activation is associated with enhanced immunocompetence. Immune dysfunction associated with aging exerts a strong influence on age-related morbidity and mortality [142]. Several age-related changes in immunity have been correlated with increased mortality, including low lymphoproliferative response to mitogens and low natural killer (NK) cytotoxicity [143–145]. We and others have shown that EE and BDNF overexpression in young mice increases lymphocyte proliferation in response to the mitogen Concavalin A, enhanced NK cell activity [13, 146], and increased cytotoxicity of T cells after tumor implantation [13]. (8) EE leads to a modest increase in serum corticosterone consistent with a mild stress. The exposure to the mild, nonaversive challenges of EE may lead to a more adaptive stress system and may therefore buffer the reaction to subsequent major external stressors [13, 146, 147]. Indeed, animals maintained on dietary restriction regimens exhibit increased glucocorticoid levels and yet their health is improved and lifespans are extended [148]. Moreover, both regular physical exercise and cognitive stimulation are beneficial on health and they too increase cortisol levels [149, 150], suggesting that eustress may increase resistance to disease and be beneficial for health and longevity. To date, the majority of studies on EE and aging focus on the effects on cognitive decline and neurodegenerative diseases and a growing body of data shows that EE could reverse age-related neural, cognitive, and behavioral impairments [151–154]. However, scarce evidence is available on the effects of EE on peripheral systems and healthspan or lifespan. We are currently investigating the effects of EE on healthy aging and lifespan from a unique perspective of the recently characterized HSA axis.

Development of a Novel Molecular Therapy of Obesity

EE has been demonstrated to induce beneficial effects in animal models of a wide variety of brain disorders and recently also in metabolic syndromes and cancer. Epidemiological studies suggest that EE has a direct clinical relevance to a range of neurological and psychiatric diseases [155]. A recent study applies EE in the form of increased sensorimotor stimulation to patients of autism spectrum disorder and shows improvements in cognitive performance and autism severity [156]. This study with a relatively small cohort provides a promising example of translational potential of EE. In addition to direct translation, such as environmental stimulations and behavioral interventions, the insights gained in animal studies can help to develop enviromimetics: drugs or treatments that mimic or enhance EE-induced therapeutic effects [155]. Our research on EE and the HSA axis has promoted us to develop a molecular therapy for obesity. Obesity is a complex acquired disease with both environmental and genetic factors. In fact, obesity has one of the strongest genetic components among all the chronic disorders of humans with inheritance accounting for 40–70% of an individual's predisposition to obesity [157]. Although this genetic component stems from a small effect of a large number of genes [158], there are examples of monogenic obesity. Melanocortin-4 receptor (MC4R) mutations have been found in up to 6% of severe early-onset obesity [159]. In the USA, the two largest population studies have found that ~2.5% of grade 2 and grade 3 obese patients have a functional MC4R mutation [160, 161]. MC4R obesity is possibly the most common human monogenic disease [162]. Furthermore, it is suggested that MC4R obese individuals respond poorly to lifestyle intervention, drugs and have much less success with bariatric surgery [163, 164]. The MC4R neurons are considered the final common pathway responsible for the regulation of energy balance [165]. Immediately downstream of the MC4R in this key metabolic regulatory pathway is BDNF [166, 167]. Recent genome-wide association studies have found that BDNF is one of the 18 genetic loci associated with BMI [168, 169]. Haploinsufficiency for BDNF or its receptor TrkB is associated with obesity and hyperphagia [170–172]. BDNF deficiency is also associated with the obesity seen in the Prader–Willi syndrome and the Wilms' tumor, aniridia, genitourinary anomalies, and mental retardation (WAGR) syndrome [171, 173]. Our preclinical data show that rAAV vector-mediated BDNF gene transfer in the hypothalamus efficiently alleviates obesity and diabetes in both DIO and genetic models [74]. To facilitate clinical translation we developed a built-in autoregulatory system to control therapeutic gene expression mimicking the body's natural feedback systems. This autoregulatory system involved a single AAV vector harboring two expression cassettes, one constitutively driving BDNF and the other driving a specific microRNA targeting BDNF under the control of a promoter from the agouti-related protein

(AGRP). As body weight decreased and AGRP was induced, microRNA expression was activated to inhibit transgene expression [74]. This autoregulatory approach led to a sustainable plateau of body weight after substantial weight loss was achieved. We believe BDNF gene therapy using this autoregulatory strategy may provide a targeted molecular therapy for the MC4R obesity, a population particularly recalcitrant to all existing therapies, as well as WAGR and Prader–Willi syndrome. We are currently refining the therapeutic vector and characterizing this approach in relevant animal models to support an investigational new drug (IND) application.

Acknowledgments I am grateful to the past and present lab members for their data and discussions that have helped me to form my ideas of this chapter. I am currently funded by grants from the National Institutes of Health (NCI R01-CA166590, NCI R01-CA163640, NCI R21-CA178227, and NIA R01-AG041250).

References

1. Cao L, During MJ. What is the brain-cancer connection? Ann Rev Neurosci. 2012;35:331–45.
2. Nithianantharajah J, Hannan AJ. Enriched environments, experience-dependent plasticity and disorders of the nervous system. Nat Rev Neurosci. 2006;7(9):697–709.
3. Kolonel LN, Altshuler D, Henderson BE. The multiethnic cohort study: exploring genes, lifestyle and cancer risk. Nat Rev Cancer. 2004;4(7):519–27.
4. Baade PD, Youlden DR, Krnjacki LJ. International epidemiology of prostate cancer: geographical distribution and secular trends. Mol Nutr Food Res. 2009;53(2):171–84.
5. McAllister SS, Weinberg RA. Tumor-host interactions: a far-reaching relationship. J Clin Oncol. 2010;28(26):4022–8.
6. Green McDonald P, O'Connell M, Lutgendorf SK. Psychoneuroimmunology and cancer: a decade of discovery, paradigm shifts, and methodological innovations. Brain Behav Immun. 2013;30 Suppl:S1–9.
7. Hanahan D, Weinberg RA. The hallmarks of cancer. Cell. 2000;100(1):57–70.
8. Aaronson SA. Growth factors and cancer. Science. 1991;254(5035):1146–53.
9. Darnell RB, Posner JB. Paraneoplastic syndromes affecting the nervous system. Semin Oncol. 2006;33(3):270–98.
10. Castano Z, Tracy K, McAllister SS. The tumor macroenvironment and systemic regulation of breast cancer progression. Int J Dev Biol. 2011;55(7–9):889–97.
11. Young D, Lawlor PA, Leone P, Dragunow M, During MJ. Environmental enrichment inhibits spontaneous apoptosis, prevents seizures and is neuroprotective. Nat Med. 1999;5(4):448–53.
12. Cao L, et al. VEGF links hippocampal activity with neurogenesis, learning and memory. Nat Genet. 2004;36(8):827–35.
13. Cao L, et al. Environmental and genetic activation of a brain-adipocyte BDNF/leptin axis causes cancer remission and inhibition. Cell. 2010;142(1):52–64.
14. Bessesen DH. Regulation of body weight: what is the regulated parameter? Physiol Behav. 2011;104:599–607.
15. van Dijk G, Buwalda B. Neurobiology of the metabolic syndrome: an allostatic perspective. Eur J Pharmacol. 2008;585(1):137–46.
16. Bjorntorp P. Abdominal obesity and the metabolic syndrome. Ann Med. 1992;24(6):465–8.
17. Brunner EJ, et al. Adrenocortical, autonomic, and inflammatory causes of the metabolic syndrome: nested case-control study. Circulation. 2002;106(21):2659–65.
18. Frontini A, Cinti S. Distribution and development of brown adipocytes in the murine and human adipose organ. Cell Metab. 2010;11(4):253–6.

19. Himms-Hagen J, et al. Multilocular fat cells in WAT of CL-316243-treated rats derive directly from white adipocytes. Am J Physiol Cell Physiol. 2000;279(3):C670–81.
20. Granneman JG, Li P, Zhu Z, Lu Y. Metabolic and cellular plasticity in white adipose tissue I: effects of beta3-adrenergic receptor activation. Am J Physiol Endocrinol Metab. 2005;289(4):E608–16.
21. Murano I, Barbatelli G, Giordano A, Cinti S. Noradrenergic parenchymal nerve fiber branching after cold acclimatisation correlates with brown adipocyte density in mouse adipose organ. J Anat. 2009;214(1):171–8.
22. Ravussin E, Kozak LP. Have we entered the brown adipose tissue renaissance? Obes Rev. 2009;10(3):265–8.
23. Gesta S, Tseng YH, Kahn CR. Developmental origin of fat: tracking obesity to its source. Cell. 2007;131(2):242–56.
24. Petrovic N, et al. Chronic peroxisome proliferator-activated receptor gamma (PPARgamma) activation of epididymally derived white adipocyte cultures reveals a population of thermogenically competent, UCP1-containing adipocytes molecularly distinct from classic brown adipocytes. J Biol Chem. 2010;285(10):7153–64.
25. Cypess AM, et al. Identification and importance of brown adipose tissue in adult humans. N Engl J Med. 2009;360(15):1509–17.
26. van Marken Lichtenbelt WD, et al. Cold-activated brown adipose tissue in healthy men. N Engl J Med. 2009;360(15):1500–8.
27. Virtanen KA, et al. Functional brown adipose tissue in healthy adults. N Engl J Med. 2009;360(15):1518–25.
28. Enerback S, et al. Mice lacking mitochondrial uncoupling protein are cold-sensitive but not obese. Nature. 1997;387(6628):90–4.
29. Feldmann HM, Golozoubova V, Cannon B, Nedergaard J. UCP1 ablation induces obesity and abolishes diet-induced thermogenesis in mice exempt from thermal stress by living at thermoneutrality. Cell Metab. 2009;9(2):203–9.
30. Rothwell NJ, Stock MJ. Luxuskonsumption, diet-induced thermogenesis and brown fat: the case in favour. Clin Sci (Lond). 1983;64(1):19–23.
31. Cannon B, Nedergaard J. Thermogenesis challenges the adipostat hypothesis for body-weight control. Proc Nutr Soc. 2009;68(4):401–7.
32. Saito M, et al. High incidence of metabolically active brown adipose tissue in healthy adult humans: effects of cold exposure and adiposity. Diabetes. 2009;58(7):1526–31.
33. Nedergaard J, Bengtsson T, Cannon B. Unexpected evidence for active brown adipose tissue in adult humans. Am J Physiol Endocrinol Metab. 2007;293(2):E444–52.
34. Zingaretti MC, et al. The presence of UCP1 demonstrates that metabolically active adipose tissue in the neck of adult humans truly represents brown adipose tissue. Faseb J. 2009;23(9):3113–20.
35. Yoneshiro T, et al. Brown adipose tissue, whole-body energy expenditure, and thermogenesis in healthy adult men. Obesity (Silver Spring). 2011;19(1):13–6.
36. Vijgen GH, et al. Brown adipose tissue in morbidly obese subjects. PLoS ONE. 2011;6(2):e17247.
37. Spalding KL, et al. Dynamics of fat cell turnover in humans. Nature. 2008;453(7196):783–7.
38. Tiraby C, et al. Acquirement of brown fat cell features by human white adipocytes. J Biol Chem. 2003;278(35):33370–6.
39. Mazzucotelli A, et al. The transcriptional coactivator peroxisome proliferator activated receptor (PPAR)gamma coactivator-1 alpha and the nuclear receptor PPAR alpha control the expression of glycerol kinase and metabolism genes independently of PPAR gamma activation in human white adipocytes. Diabetes. 2007;56(10):2467–75.
40. Enerback S. Brown adipose tissue in humans. Int J Obes. 2010;34 Suppl 1:S43–6.
41. Kajimura S, Seale P, Spiegelman BM. Transcriptional control of brown fat development. Cell Metab. 2010;11(4):257–62.
42. Nedergaard J, Cannon B. The changed metabolic world with human brown adipose tissue: therapeutic visions. Cell Metab. 2010;11(4):268–72.

43. Cao L, et al. White to brown fat phenotypic switch induced by genetic and environmental activation of a hypothalamic-adipocyte axis. Cell Metab. 2011;14(3):324–38.
44. Guerra C, Koza RA, Yamashita H, Walsh K, Kozak LP. Emergence of brown adipocytes in white fat in mice is under genetic control. Effects on body weight and adiposity. J Clin Invest. 1998;102(2):412–20.
45. Xue B, et al. Genetic variability affects the development of brown adipocytes in white fat but not in interscapular brown fat. J Lipid Res. 2007;48(1):41–51.
46. Vegiopoulos A, et al. Cyclooxygenase-2 controls energy homeostasis in mice by de novo recruitment of brown adipocytes. Science. 2010;328(5982):1158–61.
47. Arner P. Adrenergic receptor function in fat cells. Am J Clin Nutr. 1992;55(1 Suppl):228S–36S.
48. Papetti M, Herman IM. Mechanisms of normal and tumor-derived angiogenesis. Am J Physiol Cell Physiol. 2002;282(5):C947–70.
49. Zhang QX, et al. Vascular endothelial growth factor is the major angiogenic factor in omentum: mechanism of the omentum-mediated angiogenesis. J Surg Res. 1997;67(2):147–54.
50. World Health Organization. Obesity: preventing and managing the global epidemic. 2000; ISBN: 92 4 120894 5.
51. World Health Organization International Agency for Research on Cancer. 2013; www.iarc.fr/en/media-centre/pr2013/pdfs/pr223_E.pdf.
52. Reeves GK, et al. Cancer incidence and mortality in relation to body mass index in the Million Women Study: cohort study. BMJ. 2007;335(7630):1134.
53. Renehan AG, Tyson M, Egger M, Heller RF, Zwahlen M. Body-mass index and incidence of cancer: a systematic review and meta-analysis of prospective observational studies. Lancet. 2008;371(9612):569–78.
54. Whitlock G, et al. Body-mass index and cause-specific mortality in 900 000 adults: collaborative analyses of 57 prospective studies. Lancet. 2009;373(9669):1083–96.
55. Calle EE, Rodriguez C, Walker-Thurmond K, Thun MJ. Overweight, obesity, and mortality from cancer in a prospectively studied cohort of U.S. adults. N Engl J Med. 2003;348(17):1625–38.
56. Ahn J, et al. Adiposity, adult weight change, and postmenopausal breast cancer risk. Arch Intern Med. 2007;167(19):2091–102.
57. Parker ED, Folsom AR. Intentional weight loss and incidence of obesity-related cancers: the Iowa Women's Health Study. Int J Obes Relat Metab Disord. 2003;27(12):1447–52.
58. Knowler WC, et al. 10-year follow-up of diabetes incidence and weight loss in the Diabetes Prevention Program Outcomes Study. Lancet. 2009;374(9702):1677–86.
59. Sjostrom L, et al. Effects of bariatric surgery on mortality in Swedish obese subjects. N Engl J Med. 2007;357(8):741–52.
60. Buchwald H, et al. Bariatric surgery: a systematic review and meta-analysis. JAMA. 2004;292(14):1724–37.
61. Adams TD, et al. Long-term mortality after gastric bypass surgery. N Engl J Med. 2007;357(8):753–61.
62. Renehan AG. Bariatric surgery, weight reduction, and cancer prevention. Lancet Oncol. 2009;10(7):640–1.
63. Longo VD, Fontana L. Calorie restriction and cancer prevention: metabolic and molecular mechanisms. Trends Pharmacol Sci. 2010;31(2):89–98.
64. Ashrafian H, et al. Metabolic surgery and cancer: protective effects of bariatric procedures. Cancer. 2011;117(9):1788–99.
65. Vona-Davis L, Rose DP. Adipokines as endocrine, paracrine, and autocrine factors in breast cancer risk and progression. Endocr Relat Cancer. 2007;14(2):189–206.
66. Maccio A, et al. Correlation of body mass index and leptin with tumor size and stage of disease in hormone-dependent postmenopausal breast cancer: preliminary results and therapeutic implications. J Mol Med. 2010;88(7):677–86.
67. Paz-Filho G, Lim EL, Wong ML, Licinio J. Associations between adipokines and obesity-related cancer. Front Biosci. 2011;16:1634–50.

68. Nkhata KJ, Ray A, Dogan S, Grande JP, Cleary MP. Mammary tumor development from T47-D human breast cancer cells in obese ovariectomized mice with and without estradiol supplements. Breast Cancer Res Treat. 2009;114(1):71–83.
69. Chen DC, et al. Serum adiponectin and leptin levels in Taiwanese breast cancer patients. Cancer Lett. 2006;237(1):109–14.
70. Gunter MJ, et al. Insulin, insulin-like growth factor-I, and risk of breast cancer in postmenopausal women. J Natl Cancer Inst. 2009;101(1):48–60.
71. Trevisan M, et al. Markers of insulin resistance and colorectal cancer mortality. Cancer Epidemiol Biomarkers Prev. 2001;10(9):937–41.
72. Jee SH, et al. Fasting serum glucose level and cancer risk in Korean men and women. JAMA. 2005;293(2):194–202.
73. Rossi M, et al. Dietary glycemic load and hepatocellular carcinoma with or without chronic hepatitis infection. Ann Oncol. 2009;20(10):1736–40.
74. Cao L, et al. Molecular therapy of obesity and diabetes by a physiological autoregulatory approach. Nat Med. 2009;15(4):447–54.
75. Coussens LM, Werb Z. Inflammation and cancer. Nature. 2002;420(6917):860–7.
76. Olefsky JM. IKKepsilon: a bridge between obesity and inflammation. Cell. 2009;138(5):834–6.
77. Gilbert CA, Slingerland JM. Cytokines, obesity, and cancer: new insights on mechanisms linking obesity to cancer risk and progression. Annu Rev Med. 2013;64:45–57.
78. Liu X, et al. Hypothalamic gene transfer of BDNF inhibits breast cancer progression and metastasis in middle age obese mice. Mol Ther. 2014;22:1275–84.
79. Magee JA, Piskounova E, Morrison SJ. Cancer stem cells: impact, heterogeneity, and uncertainty. Cancer Cell. 2012;21(3):283–96.
80. Li X, et al. Intrinsic resistance of tumorigenic breast cancer cells to chemotherapy. J Natl Cancer Inst. 2008;100(9):672–9.
81. Shafee N, et al. Cancer stem cells contribute to cisplatin resistance in Brca1/p53-mediated mouse mammary tumors. Cancer Res. 2008;68(9):3243–50.
82. Balic M, et al. Most early disseminated cancer cells detected in bone marrow of breast cancer patients have a putative breast cancer stem cell phenotype. Clin Cancer Res. 2006;12(19):5615–21.
83. Korkaya H, Liu S, Wicha MS. Breast cancer stem cells, cytokine networks, and the tumor microenvironment. J Clin Invest. 2011;121(10):3804–09.
84. Zheng Q, et al. Leptin deficiency suppresses MMTV-Wnt-1 mammary tumor growth in obese mice and abrogates tumor initiating cell survival. Endocr Related Cancer. 2011;18(4):491–503.
85. Feldman DE, Chen C, Punj V, Tsukamoto H, Machida K. Pluripotency factor-mediated expression of the leptin receptor (OB-R) links obesity to oncogenesis through tumor-initiating stem cells. Proc Natl Acad Sci U S A. 2012;109(3):829–34.
86. Eliassen AH, Hankinson SE. Endogenous hormone levels and risk of breast, endometrial and ovarian cancers: prospective studies. Adv Exp Med Biol. 2008;630:148–65.
87. Pugeat M, et al. Pathophysiology of sex hormone binding globulin (SHBG): relation to insulin. J Steroid Biochem Mol Biol. 1991;40(4–6):841–9.
88. Flototto T, et al. Hormones and hormone antagonists: mechanisms of action in carcinogenesis of endometrial and breast cancer. Horm Metab Res. 2001;33(8):451–7.
89. Ashrafian H. Cancer's sweet tooth: the Janus effect of glucose metabolism in tumorigenesis. Lancet. 2006;367(9510):618–21.
90. Zhang Y, et al. White adipose tissue cells are recruited by experimental tumors and promote cancer progression in mouse models. Cancer Res. 2009;69(12):5259–66.
91. Hanahan D, Weinberg RA. Hallmarks of cancer: the next generation. Cell. 2011;144(5):646–74.
92. Reiche EM, Nunes SO, Morimoto HK. Stress, depression, the immune system, and cancer. Lancet Oncol. 2004;5(10):617–25.
93. Armaiz-Pena GN, Lutgendorf SK, Cole SW, Sood AK. Neuroendocrine modulation of cancer progression. Brain Behav Immun. 2009;23(1):10–5.

94. Chida Y, Hamer M, Wardle J, Steptoe A. Do stress-related psychosocial factors contribute to cancer incidence and survival? Nat Clin Pract Oncol. 2008;5(8):466–75.
95. Thaker PH, Sood AK. Neuroendocrine influences on cancer biology. Semin Cancer Biol. 2008;18(3):164–70.
96. Glaser R, Kiecolt-Glaser JK. Stress-induced immune dysfunction: implications for health. Nat Rev Immunol. 2005;5(3):243–51.
97. Antoni MH, et al. The influence of bio-behavioural factors on tumour biology: pathways and mechanisms. Nat Rev Cancer. 2006;6(3):240–8.
98. Armaiz-Pena GN, Cole SW, Lutgendorf SK, Sood AK. Neuroendocrine influences on cancer progression. Brain Behav Immun. 2013;30 Suppl:S19–25.
99. Cole SW. Nervous system regulation of the cancer genome. Brain Behav Immun. 2013;30 Suppl:S10–18.
100. Steplewski Z, Vogel WH, Ehya H, Poropatich C, Smith JM. Effects of restraint stress on inoculated tumor growth and immune response in rats. Cancer Res. 1985;45(10):5128–33.
101. Saul AN, et al. Chronic stress and susceptibility to skin cancer. J Natl Cancer Inst. 2005;97(23):1760–7.
102. Sloan EK, et al. The sympathetic nervous system induces a metastatic switch in primary breast cancer. Cancer Res. 2010;70(18):7042–52.
103. Palermo-Neto J, de Oliveira Massoco C, Robespierre de Souza W. Effects of physical and psychological stressors on behavior, macrophage activity, and Ehrlich tumor growth. Brain Behav Immun. 2003;17(1):43–54.
104. Hasegawa H, Saiki I. Psychosocial stress augments tumor development through beta-adrenergic activation in mice. Jpn J Cancer Res. 2002;93(7):729–35.
105. Thaker PH, et al. Chronic stress promotes tumor growth and angiogenesis in a mouse model of ovarian carcinoma. Nat Med. 2006;12(8):939–44.
106. Selye H. Stress without distress. Toronto: McClelland and Stewart, Ltd.; 1974.
107. Milsum JH. A model of the eustress system for health/illness. Behav Sci. 1985;30(4):179–86.
108. Lapiz MD, et al. Influence of postweaning social isolation in the rat on brain development, conditioned behavior, and neurotransmission. Neurosci Behav Physiol. 2003;33(1):13–29.
109. Fone KC, Porkess MV. Behavioural and neurochemical effects of post-weaning social isolation in rodents-relevance to developmental neuropsychiatric disorders. Neurosci Biobehav Rev. 2008;32(6):1087–102.
110. Hasen NS, O'Leary KA, Auger AP, Schuler LA. Social isolation reduces mammary development, tumor incidence, and expression of epigenetic regulators in wild-type and p53-heterozygotic mice. Cancer Prev Res (Phila). 2010;3(5):620–9.
111. Williams JB, et al. A model of gene-environment interaction reveals altered mammary gland gene expression and increased tumor growth following social isolation. Cancer Prev Res (Phila). 2009;2(10):850–61.
112. Palermo-Neto J, Fonseca ES, Quinteiro-Filho WM, Correia CS, Sakai M. Effects of individual housing on behavior and resistance to Ehrlich tumor growth in mice. Physiol Behav. 2008;95(3):435–40.
113. Algazi M, Plu-Bureau G, Flahault A, Dondon MG, Le MG. [Could treatments with beta-blockers be associated with a reduction in cancer risk?]. Rev Epidemiol Sante Publique. 2004;52(1):53–65.
114. Perron L, Bairati I, Harel F, Meyer F. Antihypertensive drug use and the risk of prostate cancer (Canada). Cancer Causes Control. 2004;15(6):535–41.
115. Melamed R, et al. Marginating pulmonary-NK activity and resistance to experimental tumor metastasis: suppression by surgery and the prophylactic use of a beta-adrenergic antagonist and a prostaglandin synthesis inhibitor. Brain Behav Immun. 2005;19(2):114–26.
116. De Giorgi V, et al. Treatment with beta-blockers and reduced disease progression in patients with thick melanoma. Arch Intern Med. 2011;171(8):779–81.
117. Lemeshow S, et al. β-Blockers and survival among Danish patients with malignant melanoma: a population-based cohort study. Cancer Epidemiol Biomarkers Prev. 2011;20(10):2273–9.

118. McEwen BS, Gianaros PJ. Central role of the brain in stress and adaptation: links to socioeconomic status, health, and disease. Ann N Y Acad Sci. 2010;1186:190–222.
119. Fontana L, Partridge L, Longo VD. Extending healthy life span—from yeast to humans. Science. 2010;328(5976):321–6.
120. Smith DL, Jr, Nagy TR, Allison DB. Calorie restriction: what recent results suggest for the future of ageing research. Eur J Clin Invest. 2010;40(5):440–50.
121. Mattson MP. Energy intake, meal frequency, and health: a neurobiological perspective. Annu Rev Nutr. 2005;25:237–60.
122. Anderson RM, Shanmuganayagam D, Weindruch R. Caloric restriction and aging: studies in mice and monkeys. Toxicol Pathol. 2009;37(1):47–51.
123. Colman RJ, et al. Caloric restriction delays disease onset and mortality in rhesus monkeys. Science. 2009;325(5937):201–4.
124. Fontana L, Klein S. Aging, adiposity, and calorie restriction. JAMA. 2007;297(9):986–94.
125. Fontana L, Meyer TE, Klein S, Holloszy JO. Long-term calorie restriction is highly effective in reducing the risk for atherosclerosis in humans. Proc Natl Acad Sci U S A. 2004;101(17):6659–63.
126. Heilbronn LK, et al. Effect of 6-month calorie restriction on biomarkers of longevity, metabolic adaptation, and oxidative stress in overweight individuals: a randomized controlled trial. JAMA. 2006;295(13):1539–48.
127. Katic M, et al. Mitochondrial gene expression and increased oxidative metabolism: role in increased lifespan of fat-specific insulin receptor knock-out mice. Aging Cell. 2007;6(6):827–39.
128. Bluher M, et al. Adipose tissue selective insulin receptor knockout protects against obesity and obesity-related glucose intolerance. Dev Cell. 2002;3(1):25–38.
129. Ortega-Molina A, et al. Pten positively regulates brown adipose function, energy expenditure, and longevity. Cell Metab. 2012;15(3):382–94.
130. Tsukiyama-Kohara K, et al. Adipose tissue reduction in mice lacking the translational inhibitor 4E-BP1. Nat Med. 2001;7(10):1128–32.
131. Chiu CH, Lin WD, Huang SY, Lee YH. Effect of a C/EBP gene replacement on mitochondrial biogenesis in fat cells. Genes Dev. 2004;18(16):1970–5.
132. Molero JC, et al. c-Cbl-deficient mice have reduced adiposity, higher energy expenditure, and improved peripheral insulin action. J Clin Invest. 2004;114(9):1326–33.
133. Bluher M, Kahn BB, Kahn CR. Extended longevity in mice lacking the insulin receptor in adipose tissue. Science. 2003;299(5606):572–4.
134. Heitmann BL, Garby L. Composition (lean and fat tissue) of weight changes in adult Danes. Am J Clin Nutr. 2002;75(5):840–7.
135. Allison DB, et al. Weight loss increases and fat loss decreases all-cause mortality rate: results from two independent cohort studies. Int J Obes Relat Metab Disord. 1999;23(6):603–11.
136. Brown-Borg HM. Hormonal control of aging in rodents: the somatotropic axis. Mol Cell Endocrinol. 2009;299(1):64–71.
137. Otabe S, et al. Overexpression of human adiponectin in transgenic mice results in suppression of fat accumulation and prevention of premature death by high-calorie diet. Am J Physiol Endocrinol Metab. 2007;293(1):E210–8.
138. Fujisawa T, et al. Adiponectin suppresses colorectal carcinogenesis under the high-fat diet condition. Gut. 2008;57(11):1531–8.
139. Calle EE, Kaaks R. Overweight, obesity and cancer: epidemiological evidence and proposed mechanisms. Nat Rev Cancer. 2004;4(8):579–91.
140. Hursting SD, Slaga TJ, Fischer SM, DiGiovanni J, Phang JM. Mechanism-based cancer prevention approaches: targets, examples, and the use of transgenic mice. J Natl Cancer Inst. 1999;91(3):215–25.
141. Albanes D. Total calories, body weight, and tumor incidence in mice. Cancer Res. 1987;47(8):1987–92.
142. Wayne SJ, Rhyne RL, Garry PJ, Goodwin JS. Cell-mediated immunity as a predictor of morbidity and mortality in subjects over 60. J Gerontol. 1990;45(2):M45–8.

143. DelaRosa O, et al. Immunological biomarkers of ageing in man: changes in both innate and adaptive immunity are associated with health and longevity. Biogerontology. 2006;7(5–6):471–81.
144. De la Fuente M, Miquel J. An update of the oxidation-inflammation theory of aging: the involvement of the immune system in oxi-inflamm-aging. Curr Pharm Des. 2009;15(26):3003–26.
145. Guayerbas N, De La Fuente M. An impairment of phagocytic function is linked to a shorter life span in two strains of prematurely aging mice. Dev Comp Immunol. 2003;27(4):339–50.
146. Benaroya-Milshtein N, et al. Environmental enrichment in mice decreases anxiety, attenuates stress responses and enhances natural killer cell activity. Eur J Neurosci. 2004;20(5):1341–7.
147. Larsson F, Winblad B, Mohammed AH. Psychological stress and environmental adaptation in enriched vs. impoverished housed rats. Pharmacol Biochem Behav. 2002;73(1):193–207.
148. Patel NV, Finch CE. The glucocorticoid paradox of caloric restriction in slowing brain aging. Neurobiol Aging. 2002;23(5):707–17.
149. Gotthardt U, et al. Cortisol, ACTH, and cardiovascular response to a cognitive challenge paradigm in aging and depression. Am J Physiol. 1995;268(4 Pt 2):R865–73.
150. Witt KA, Snook JT, O'Dorisio TM, Zivony D, Malarkey WB. Exercise training and dietary carbohydrate: effects on selected hormones and the thermic effect of feeding. Int J Sport Nutr. 1993;3(3):272–89.
151. Goldberg NR, Haack AK, Meshul CK. Enriched environment promotes similar neuronal and behavioral recovery in a young and aged mouse model of Parkinson's disease. Neuroscience. 2011;172:443–52.
152. Harburger LL, Lambert TJ, Frick KM. Age-dependent effects of environmental enrichment on spatial reference memory in male mice. Behav Brain Res. 2007;185(1):43–8.
153. Segovia G, del Arco A, Mora F. Environmental enrichment, prefrontal cortex, stress, and aging of the brain. J Neural Transm. 2009;116(8):1007–16.
154. Mattson MP, Duan W, Lee J, Guo Z. Suppression of brain aging and neurodegenerative disorders by dietary restriction and environmental enrichment: molecular mechanisms. Mech Ageing Dev. 2001;122(7):757–78.
155. Hannan AJ. Environmental enrichment and brain repair: harnessing the therapeutic effects of cognitive stimulation and physical activity to enhance experience-dependent plasticity. Neuropathol Appl Neurobiol. 2014;40(1):13–25.
156. Woo CC, Leon M. Environmental enrichment as an effective treatment for autism: a randomized controlled trial. Behav Neurosci. 2013;127(4):487–97.
157. Herrera BM, Lindgren CM. The genetics of obesity. Curr Diab Rep. 2010;10(6):498–505.
158. Speliotes EK, et al. Association analyses of 249,796 individuals reveal 18 new loci associated with body mass index. Nat Genet. 2010;42(11):937–48.
159. Farooqi IS, et al. Clinical spectrum of obesity and mutations in the melanocortin 4 receptor gene. N Engl J Med. 2003;348(12):1085–95.
160. Calton MA, et al. Association of functionally significant Melanocortin-4 but not Melanocortin-3 receptor mutations with severe adult obesity in a large North American case-control study. Hum Mol Genet. 2009;18(6):1140–7.
161. Lubrano-Berthelier C, et al. Melanocortin 4 receptor mutations in a large cohort of severely obese adults: prevalence, functional classification, genotype-phenotype relationship, and lack of association with binge eating. J Clin Endocrinol Metab. 2006;91(5):1811–8.
162. Farooqi S, O'Rahilly S. Genetics of obesity in humans. Endocr Rev. 2006;27(7):710–8.
163. Reinehr T, et al. Lifestyle intervention in obese children with variations in the melanocortin 4 receptor gene. Obesity (Silver Spring). 2009;17(2):382–9.
164. Potoczna N, et al. Gene variants and binge eating as predictors of comorbidity and outcome of treatment in severe obesity. J Gastrointest Surg. 2004;8(8):971–81 (discussion 981–972).
165. Flier JS. Obesity wars: molecular progress confronts an expanding epidemic. Cell. 2004;116(2):337–50.

166. Xu B, et al. Brain-derived neurotrophic factor regulates energy balance downstream of melanocortin-4 receptor. Nat Neurosci. 2003;6(7):736–42.
167. Nicholson JR, Peter JC, Lecourt AC, Barde YA, Hofbauer KG. Melanocortin-4 receptor activation stimulates hypothalamic brain-derived neurotrophic factor release to regulate food intake, body temperature and cardiovascular function. J Neuroendocrinol. 2007;19(12):974–82.
168. Speliotes EK, et al. Association analyses of 249,796 individuals reveal 18 new loci associated with body mass index. Nat Genet. 2010;42(11):937–48.
169. Jiao H, et al. Genome wide association study identifies KCNMA1 contributing to human obesity. BMC Med Genomics. 2011;4:51.
170. Gray J, et al. Hyperphagia, severe obesity, impaired cognitive function, and hyperactivity associated with functional loss of one copy of the brain-derived neurotrophic factor (BDNF) gene. Diabetes. 2006;55(12):3366–71.
171. Han JC, et al. Brain-derived neurotrophic factor and obesity in the WAGR syndrome. New Engl J Med. 2008;359(9):918–27.
172. Yeo GS, et al. A de novo mutation affecting human TrkB associated with severe obesity and developmental delay. Nat Neurosci. 2004;7(11):1187–9.
173. Han JC, Muehlbauer MJ, Cui HN, Newgard CB, Haqq AM. Lower brain-derived neurotrophic factor in patients with prader-willi syndrome compared to obese and lean control subjects. J Clin Endocrinol Metab. 2010;95(7):3532–6.

Chapter 3
The MRL Mouse: A Model of Regeneration and Cancer

Ellen Heber-Katz and Robert K. Naviaux

Abstract The intersection of regeneration and cancer has long been hypothesized to be due to the well-established inverse relationship between the capacity to heal and the propensity for cancer.

The MRL mouse, a mammalian model of regeneration, provides an opportunity to experimentally explore this hypothesis. We present data relating to metabolism, inflammation, genetics, and diet, which not only support the hypothesis but also provide mechanistic insight linking these pro-carcinogenic traits.

In particular, this mouse has many elements of the tumor microenvironment such as a pro-inflammatory response, enhanced regenerative healing under a high-fat diet (HFD), an environment supportive of tumor growth, the use of a basal metabolism with aerobic glycolysis reminiscent of the Warburg effect, and although it is permissive for tumor transplantation, it is resistant to cancer induction. Genetic studies on the role of pro-inflammatory HFDs and regenerative capacity provide links to dissecting these relationships.

Keywords Aerobic glycolysis · High fat diet · Inflammation · OXPHOS · Mitochondria, MRL mouse, Warburg effect · Quantitative trait loci

E. Heber-Katz (✉)
The Wistar Institute, 3601 Spruce Street, Philadelphia, PA 19104, USA
e-mail: heberkatz@wistar.org

R. K. Naviaux
UCSD School of Medicine, 214 Dickinson St.,
Room C-102, San Diego, CA 92103-8467, USA
e-mail: naviaux@ucsd.edu

N. A. Berger (ed.), *Murine Models, Energy Balance, and Cancer*,
Energy Balance and Cancer, DOI 10.1007/978-3-319-16733-6_3

The Murphy Roths Large (MRL) Mouse as a Model of Mammalian Regeneration

The Murphy Roths Large (MRL) mouse has been used for nearly 40 years as a model to study the autoimmune disease systemic lupus erythematosus [1–3]. This inbred mouse strain actually came about from the crossing of four different strains of mice (AKR, C3H, LG/J, and C57BL/6), and it was generated to be a means of maintaining the achondroplasia *cn* or *Npr2* gene. However, some of these mice developed enlarged lymph nodes due to a mutation of the *fas* gene, which was called the *lpr (lympho-proliferative)* gene mutation [4–6]. The proliferation of B cells and the overproduction of antibodies in these MRL/*lpr* mice led to increased glomerulonephritis, and it became clear that this was a significant and useful model of lupus. However, the genetic background of these mice found in the wild type MRL/MpJ mice included multiple genes involved in autoimmunity so that mice without the *fas* mutation showed autoimmune effects that developed much more slowly. Thus, the *fas* mutation was more of an accelerator than a driver of autoimmunity.

In the course of autoimmune studies in our laboratory about 15 years ago, we noted an unusual wound healing effect in MRL mice. Ear hole punches, which are used to create a lifelong identifier of laboratory mice (the Jackson Laboratory ear punch numbering code), began to close in a scar-free manner and became almost imperceptible within 30 days. The mice employed in this experiment had the *lpr* mutation (MRL/*lpr*), and our initial thought was that the unusual healing was an *lpr* effect, but subsequent studies showed that the standard strain (MRL/MpJ) mouse was fully capable of complete ear hole closure [7] (Fig. 3.1). As noted above, the MRL is a workhorse strain used in many laboratories over decades, with thousands of mice being ear punched, yet there were no reports to our knowledge of this effect in mice. This phenomenon had been previously reported in rabbits and bats [8–11] and had been identified as an epimorphic regenerative response similar to amphibian regeneration [12–14] and developmental processes [15], in which whole tissue forms with normal histological architecture and function. This is very different from the wound repair (scarring response) generally seen in mammals. Ear hole closure occurs without scar formation and with rapid re-epithelialization, formation and growth of a regeneration blastema until all sides come together and closure is accomplished. It is generally complete within 30 days. By 12–16 weeks cartilage regrows and hair follicles are restored [7]. This identified the MRL mouse as a mammalian regenerator. It has also been determined that its parent, the LG/J mouse, is also a regenerator and probably the main contributor of the trait to the MRL mouse [16–18].

MRL regenerative healing has been shown in multiple organ and tissue systems by many different laboratories. This includes the in-vivo restoration of corneal epithelium [19], digit tips and partial healing of lower phalanges [20, 21], cardiac myocardium [22–24], enhanced central nervous system (CNS) and peripheral nervous system (PNS) healing [25–30], articular cartilage growth [31–33), muscle function [34], and myometrial and skin healing [35, 36].

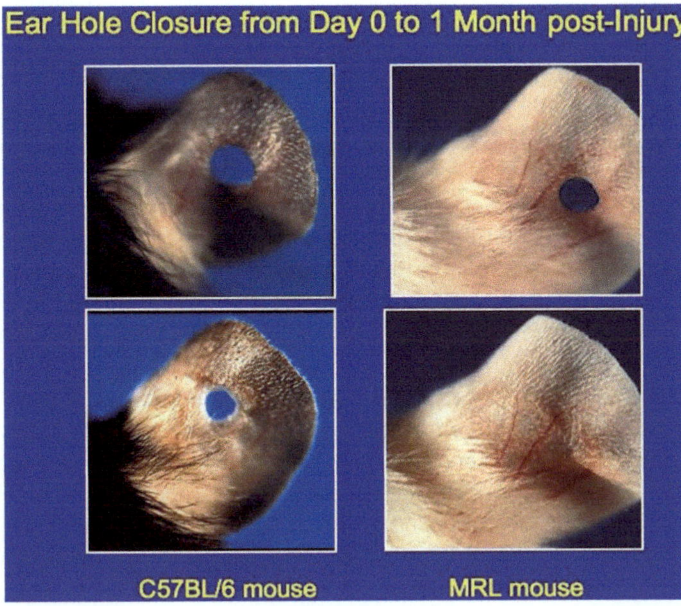

Fig. 3.1 Ear hole closure in C57BL/6 and MRL mice. Holes (*2.1 mm in diameter*) were made in the pinnae of the ears of mice and were seen soon after injury (*upper panels*) and on day 33 (*lower panels*) with complete closure seen in the MRL ear. (These data were part of the original data published in [7])

Besides healing differences, the MRL has shown other unusual phenotypes with many features found associated with the tumor microenvironment even though this mouse has a low incidence of cancer. This chapter will discuss these phenotypes.

Cancer in Regenerating Species

Given this MRL mouse model of regeneration, we were in a position to address issues brought up by observations from other regenerating species. It had been observed that lower species rarely form tumors and that the potential for tumorigenicity increases as one traces the phylogenetic tree to mammals and humans. The inverse was observed for regenerative capacity that decreases through phylogeny. The biological mechanism for this is unknown, though there have been numerous theories [37–43]. Studies examining this issue have focused on two important aspects: (1) the induction of tumorigenicity in regenerating organisms that examines the cellular differences in regenerating tissue and (2) the ability of transplanted tumors to grow in regenerating animals that examines the possibility of a permissive environment for tumor versus regenerating tissue growth.

In the first case, carcinogens were used in an attempt to induce tumors in regeneration-competent tissues [44–47]. In general, carcinogens were not effective in inducing tumors at these sites. Most interesting is the rabbit ear that displayed resistance to carcinogenic induction of tumors. However, after creation of a wound within the regenerating tissue itself, carcinogen-induced tumor growth seemed to be enhanced [47].

In the second case, attempts were made to grow tumors in regenerating species and specifically in regenerating tissue with results suggesting that tumor growth was not generally supported, and a reversion to normal differentiation was sometimes seen [48, 49]. The growth of tumor tissue was also examined in embryonic tissue that showed a similar regulatory process in which tumor cells differentiate into a normal phenotype [50–53].

The Heber–Katz laboratory examined the MRL mouse for tumor growth. In the first case, the induction of epithelial tumors in regenerative wildtype MRL/MpJ and non-regenerative Swiss Webster (SW) C57BL/6 mice involved a two-stage carcinogenesis model[1] (TPA/DMBA) to establish a tumor under the flank skin. TPA causes epithelial proliferation and hyperplasia. Thus, instead of 2–3 layers of keratinocytes, the skin becomes very thick with multiple layers. After a onetime treatment with the carcinogen, the mice were painted daily with TPA/DMBA that produces mutations in skin stem cells, that is, the H-RasV12 oncogenic mutation.

In these experiments, 10 MRL and 11 SW female mice were treated and followed up for 100 days and then pathological analysis of the skin was carried out (Table 3.1). By 70 days, growths that looked like papillomas appeared in the MRL and by 90 days all MRL mice showed growths. By 80 days, growths appeared in the SW and by 100 days, all SW mice showed them. Thus, MRL mice responded to chemical treatment earlier than SW mice. The size of the growths was also different between the two groups of mice. MRL papillomas first increased in diameter and later began to shrink. Growths in SW mice increased in size and never decreased. Furthermore, SW mice had more growths/mouse than MRL mice.

Upon histological analysis, four of the SW mice and none of the MRL mice showed the development of squamous cell carcinomas; the MRL mice showed only the development of papillomas (Table 3.1). These preliminary results suggest that MRL mice show a level of resistance to tumor transformation compared with SW mice and deserve a further look. Moreover, these results are consistent with previous studies in other regenerating animals.

Table 3.1 Tumor induction using TPA/DMBA

Mouse strains	MRL/MpJ	Swiss Webster (SW)
Number of mice with growths	10/10	11/11
Number of growths/mouse	4.5	6.5
Mean size of growth (mm)	2.4	3
Type of growth	Papillomas (10/10)	Papillomas (11/11); sq cell carc (4/11)
Deaths	0	1

[1] *TPA* 12-O-tetradecanoyl Phorbol 13-acetate, *DMBA* 7,12-dimethyl benz anthracene

Table 3.2 Adoptive transfer of human breast cancer cell lines. [113]

Recipient strains	MRL.RAG1 KO		B6.RAG1 KO	
Cell lines injected	MB-321	MCF-7	MB-321	MCF-7
Number of mice with growths	3 mice	2 mice	1 mouse	0
Type of growth	1° tumor (3/3)	1)1° tumor (2/3); 2) metastasis (2/3)	Splenomegaly	
Mean size of growth (mm)	2 mm	1) 2 mm; 2) 3.5 cm		

In the second case, the ability to support the growth of transplanted tumor cells was examined. Immunodeficient MRL.RAG1 KO mice were generated by 10 generations of backcrossing. These mice were still able to show complete ear hole closure, indicating that this regenerative response is not dependent on T or B cells. These mice together with immunodeficient B6.RAG1 KO mice were used as recipients of human breast cancer cell lines that were injected into the rear fat pads of the mice. Two cell lines were examined: one metastatic, MDA-MB231, and the other nonmetastatic, MCF-7. The animals were followed up for 2 months and during this time pathological analysis of tissue was carried out. Out of six B6.RAG1 KO mice, three received MB231 and three received MCF-7, but only one mouse showed a tumor mass (MCF-7). Out of six MRL.RAG1 KO mice, five showed significant tumor growth. The three mice that received the metastatic tumor MB231 had a primary 2-mm mass growing in the skin or muscle. Two of the three mice that received the nonmetastatic tumor MCF-7 showed tumor growth, one metastatic and one local (Table 3.2).

The results of this very preliminary experiment support the notion that MRL mice provide a permissive environment to tumor cell growth and migration. It is surprising that B6.RAG1 KO mice do not support tumor growth since immunodeficient severe combined immunodeficiency (SCID) mice are permissive. However, it has been shown that immunodeficient B6 mice are unusual in that the inflammatory response is poor, and they do not support tumor growth [54]. Thus, it may be that the inflammatory response does play an important role in this, providing a stimulatory tumor microenvironment.

Finally, experiments to analyze the role of *p53* in MRL healing showed that ear hole closure in MRL.p53KO mice is unaffected by the lack of *p53* [55], and these mice do get tumors. However, on comparing the kinetics of tumor generation in the MRL.p53KO with published data using control p53KO mice, the MRL.p53KOs were found to be delayed by at least 1.5 months [56]. Though these two experiments obviously need to be done together, the result is potentially intriguing and adds to the inability to induce tumors in MRL mice using TPA/DMBA.

With these preliminary experiments taken together, it appears that though the MRL mouse presents a permissive environment for the growth of established tumors, induction of intrinsic tumor growth is reduced. This result is consistent with and supports earlier results, showing a direct correlation with the ability of amphibians and reptiles to regenerate but not get tumors. Furthermore, a role for inflammation, our next topic, is implicated in a permissive environment for tumor growth but perhaps not in tumor development.

Role of Inflammation

Early experiments carried out in the Heber–Katz laboratory showed higher levels of neutrophils, mast cells, and macrophages in the histological samples of healing tissue from MRL ear holes versus control strain ear holes, suggesting that inflammatory cells are important in the regenerative process. To test this, several experiments were carried out. First, the nonsteroidal anti-inflammatory drug (NSAID) meloxicam, a cyclooxygenase-2 (COX-2) inhibitor, was given to MRL mice, and it was found to inhibit ear hole closure [57]. Second, gene expression in healing tissue early after injury was examined. When we analyzed functional processes, the majority of those differentially expressed genes were related to innate immune responses and inflammation, tissue remodeling, and metabolic differences (which we will discuss below). In uninjured tissue, gene expression differences between regenerator MRL and non-regenerator C57BL/6 mice showed that out of approximately 19,000 expressed genes, only 158 genes were significantly and differentially expressed and of these 28 were upregulated in the MRL and 130 were downregulated in the MRL relative to C57BL/6.

Upregulated molecules in the MRL included *Tpsab1 (Mcp7)* that encodes a mast cell protease and myeloperoxidase (MPO) produced by neutrophils, macrophages, and mast cells.

Molecules that were downregulated included the mast cell receptor Fcergl (57-fold reduced in MRL), whose elimination enhances healing [58], *TACE* or *Adam 17* that enhances cell invasion [59], and *diamine oxidase (DAO1)* that degrades histamine [60] and may lead to higher levels of histamine. Two molecules related to increased scarring are reduced in the MRL, including connective tissue growth factor (CTGF) [61, 62] and lysyl-oxidase-like 4, involved in collagen crosslinking [63] and bone marrow-derived cell migration.

From these data, we have defined a unique mast cell population in the MRL mouse (Fig. 3.2) that is more immature and less likely to lyse but more likely to be involved in cytokine production, leading to increased growth and bearing some resemblance to a tumor-associated mast cell [64, 65]. One molecule important in tumour-associated mast cells and not mentioned above but which is upregulated in MRL tissue is *Hmox1*. This results in further resistance to degranulation and less sensitivity to stimuli.

Further significant support for the role of inflammation in regenerative healing comes from studies by DeFranco et al. in Brazil who showed that AIRmax mice, bred to be pro-inflammatory, displayed the ability to close ear holes, whereas AIR-min mice, bred to be low-inflammatory, did not close ear holes. Gene-mapping studies identified a macrophage associated molecule NRAMP as being responsible in this case [66, 67]. There is evidence that inflammation enhances axonal regeneration and neural protection [68, 69] with leukotriene B4 and lipoxin A4 having both a positive and negative regulatory effect on stem cell proliferation and differentiation [70]. Eosinophil-derived IL4 enhances liver regeneration [71].

Inflammation not only has an enhancing effect on regenerative responses, but it also has positive effects on cancer development [72, 73]. But noting that the MRL

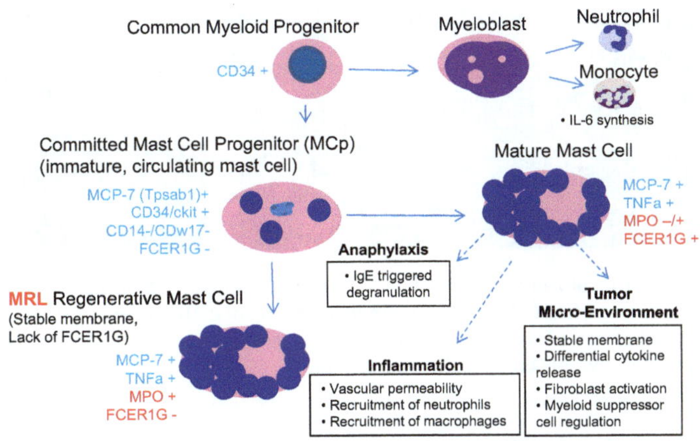

Fig. 3.2 A unique MRL mast cell population. The CD34$^+$ myeloid progenitor matures into a myeloblast and then into circulating neutrophils, monocytes, or mast cell progenitors that now express MCP-7. The mature mast cell which is now tissue-associated loses MCF-7, now expresses TNFa, Fcer1g, and low levels of MPO, can degranulate upon exposure to antigen, is involved with enhanced inflammation, or can be active in the tumor microenvironment by producing multiple cytokines. In the MRL mouse, increased numbers of mast cells at the regenerative site show increased MPO levels, but they are strikingly low in Fcer1g, a marker of mast cell maturity which when crosslinked by immunoglobulin leads to degranulation and maintains high levels in MCP7. (This is from [57])

has a low cancer-induction potential, how might this be explained? Though inflammatory cells increased and are important in the MRL regenerative response, there is another factor that may negatively affect tumorigenesis in the MRL. It has been shown that FcRg on inflammatory cells is required in a mouse model of squamous carcinoma [74]. Given the fact that *Fcerg1* is so severely reduced (57-fold) in the MRL mouse, this may be a major cause of its lowered cancer potential.

Finally, we tested the effect of a high-fat diet (HFD) known to lead to increased inflammation [75–77] on ear hole closure and found a highly significant positive effect on healing. This will be discussed in the section "The Role of Diet in the Regenerative Response: Data from Gene-mapping Studies."

Is the MRL Metabolism Similar to Cancer and a Factor in Healing?

There are several characteristics of the MRL mouse that support a metabolic examination of this strain. Not only does this mouse have a propensity for autoimmune disease and increased inflammation, but it also continues to grow over time and gain weight.

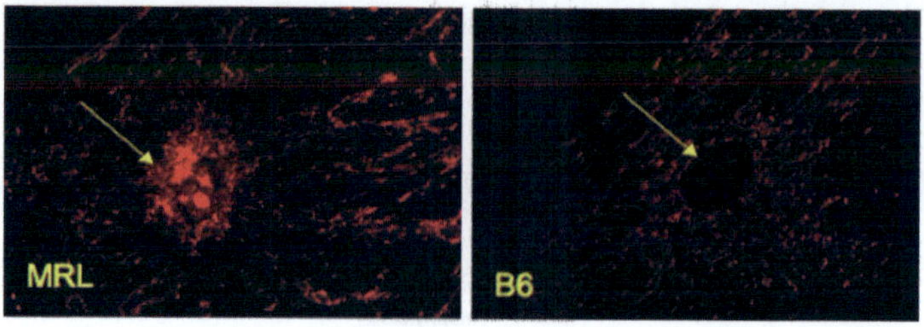

Fig. 3.3 Subcellular localization of mitochondria in MRL and B6 fibroblasts. Cells were stained with Mitotracker Orange CMTMRos. *Arrows* indicate nuclei of cells where in the B6 (*right panel*) no nuclear but rather cytoplasmic localization is seen, and in the MRL (*left panel*) significant perinuclear as well as cytoplasmic localization is seen

As mentioned above, gene expression studies found MRL and B6 differences in metabolism-related genes and showed increases in *Nt5e (converts AMP to adenosine, associated with HIF expression),* and in *Ldha, Pkm2, and Gapdh*, all glycolytic enzymes. It has been suggested that the glycolytic metabolic state may promote inflammation [78]. Additionally, we have shown that the MRL mouse shows many characteristics of glycolysis that seem to play a role in the regenerative response [79]. The evidence will be reviewed below.

Initial studies using ear-derived fibroblasts from regenerating versus non-regenerating mice showed a notable difference in the intracellular localization of mitochondria. Embryonic cells are unique in the localization of their mitochondria to regions surrounding the nucleus. This is thought to be involved in their early developmental stage and pluripotent potential [80]. When we examined fibroblasts from normal non-pre-wounded ear tissue, perinuclear localization of mitochondria was seen in MRL cells but not in B6 cells (Fig. 3.3). This suggests that MRL cells are in a more stem-like state. Increased stem cell markers such as Nanog and Sox2 are found in MRL heart tissue, for example, both pre- and post-injury compared to B6.

Further examination of the mitochondrial state showed decreased membrane potential, but an increase in mitochondrial number. One interpretation of this is that the cell is prepared, pre-activated for a rapid response to future conditions such as proliferation with this "mitochondrial reserve." The lowered mitochondrial membrane potential points to the MRL not using oxidative phosphorylation for its energy source but rather cytoplasmic glycolysis for ATP synthesis. Indeed, lactate production, a measure of glycolysis, was found to be up in MRL; reactive oxygen species (ROS) levels, products of mitochondrial metabolism were down in MRL, as were glutathione pools (Fig. 3.4).

An analysis of oxidative phosphorylation (OXPHOS) in MRL and B6 liver mitochondria showed increased glutamate oxidation and reduced fatty acid oxidation. This was confirmed by a corresponding increase in plasma acyl-carnitine species in MRL (Fig. 3.5). These biochemical features are indicative of increased dependence on glycolysis and a decreased usage of OXPHOS for ATP synthesis. Increased lipid storage was seen in the liver as well. The phenomenon of lipid storage is one that has been documented in many cell types and species responding to environmental stress. For example, during the larval pre-dauer stage in *Caenorhabditis elegans*,

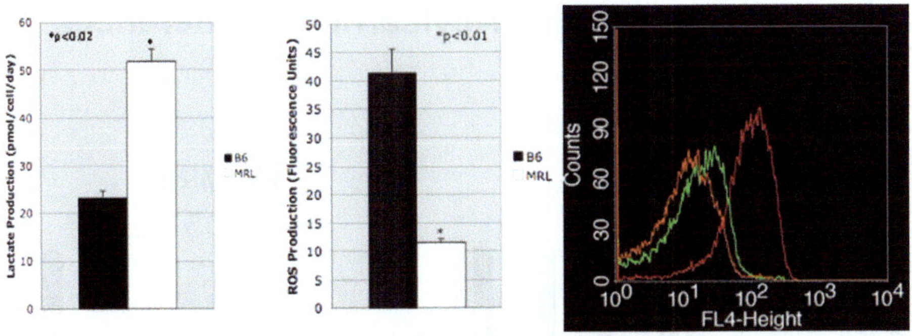

Fig. 3.4 Metabolic differences between MRL and B6 fibroblasts. In the *left-hand panel*, lactate production in primary ear fibroblasts was measured under 21 % oxygen levels in culture using a Lactate Scout portable lactate meter. In the *middle panel*, reactive oxygen species (ROS) production in primary ear fibroblasts was measured by DHR123 Fluorescence by the FACS analysis. In the *right-hand panel*, relative mitochondrial transmembrane potentials were measured using the potentiometric, cationic carbocyanine dye DiIC1 and analyzed by FACS. (This is modified from [79])

fatty acid oxidation is decreased and intracellular lipid is accumulated as a choreographed metabolic response to environmental stress [81]. A similar, physiologically induced state of hypometabolism associated with decreased fatty acid oxidation and increased lipid storage is also seen in the early diapause embryos of many mammalian species. During diapause, the arrested embryo must survive for weeks to months without nutrients from placenta, before fully implanting and reinitiating cell growth, proliferation, and differentiation [82]. The metabolic phenotype of the MRL mouse seems to emulate several of the features of the dauer and diapause stages under baseline conditions, making it preadapted for recovery after stress or injury.

A comparative analysis of mitochondrial DNA sequences of MRL and B6 showed differences in several molecules [83]. Sequence differences were seen in three molecules: the NADH dehydrogenase 3 gene (*mt-Nd3*) and two mitochondrial tRNAs. One was methionyl tRNA, which is known to be exported into the cytoplasm and plays a role in RNA silencing by interacting with Argonaute 2 (*Ago2*), an endonuclease that is part of the RNA-induced silencing complex (RISC) and targets miRNA and siRNA complementary structures [84]. Interestingly, gene expression studies showed that the majority of differences between MRL and B6 involved reduced levels in the MRL. This might be affected by tRNA-met changes. The second molecule, tRNA-arg, has been mapped to hearing loss [85], and this is, in fact, a phenotype associated with the MRL mouse [86, 87]. The biological consequences of these changes are presently unknown.

Early in the past century, Otto Warburg proposed that cancer cells were metabolically different from normal adult cells in that the rate of glycolysis is far greater than the OXPHOS rate in normal tissue even when oxygen is present. The data above support the view that the MRL mouse is a rare instance of a glycolytic metabolizer in the basal adult state. As such, it should provide an environment and system to study the intersection of cancer, regeneration, and metabolism. This has motivated

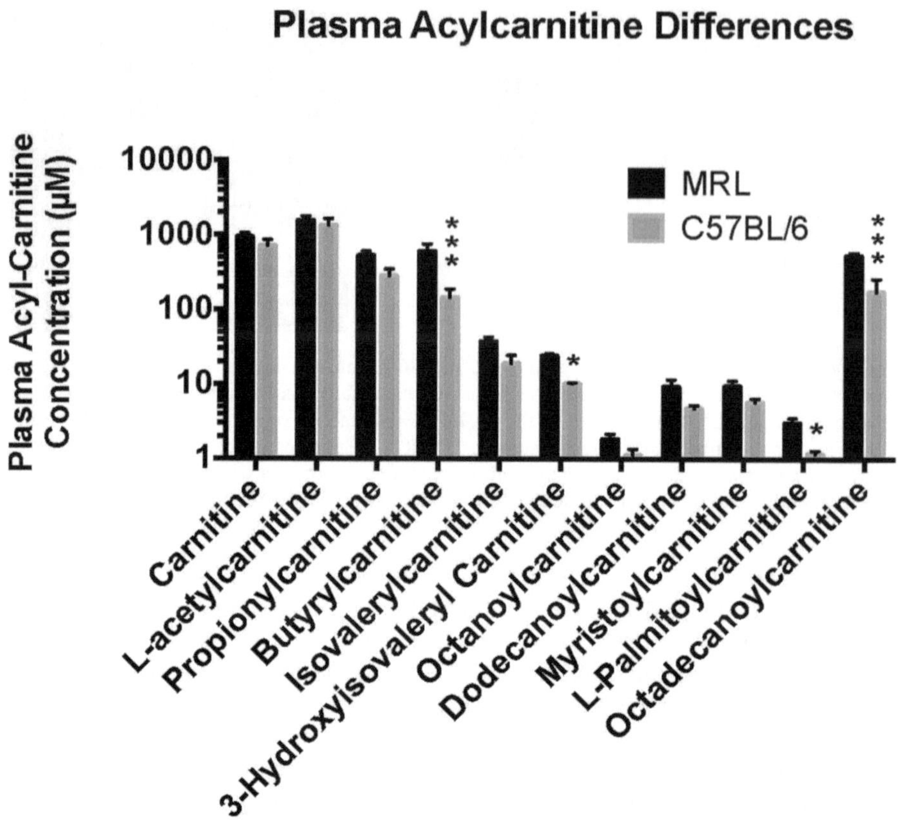

Fig. 3.5 Plasma acyl-carnitine profiles. The C57BL/6J strain is known to have higher plasma acyl-carnitines than most other laboratory mouse strains. Quantitative metabolomic analysis by LC-MS/MS revealed that the MRL mouse had still higher levels. This was particularly evident in C4 (butyryl), C5OH (hydroxyisovaleryl), C16 (palmitoyl), and C18 (octadecanoyl) species. This indicates a generalized decrease in mitochondrial fatty acid oxidation, with more severe defects in short-chain (< C6) and long-chain (C13–C21) fatty acids, with relative preservation of medium-chain (C6–C12) fatty acid oxidation

ongoing studies in our laboratories of the effect of diet on these central processes in the tumor and the tumor microenvironment [88–90].

The Role of Diet in the Regenerative Response: Data from Gene-mapping Studies

As mentioned above, mice were fed a high fat diet (HFD) to determine if its pro-inflammatory effects [75–77] would impact regenerative ear hole closure, and, in fact, it did have a positive effect (data not shown). However, similar results later came from another direction.

Initial studies to determine the genetic complexity of the regenerative ear hole closure response were carried out by using F2 crosses of regenerating MRL and other non-regenerating strains including B6, SJL, and M.m.castaneous mice [91–96]. One attractive aspect of this ear hole closure phenotype is that it is an accessible and quantifiable trait. Thus, one can easily obtain longitudinal data from individual mice. Ear hole closure proved to be a complex genetic trait and was originally mapped to seven loci. Crosses of MRL with multiple non-regenerator mice showed both shared and new quantitative trait loci. Unfortunately, the F2 method generally yields large chromosomal regions of genetic QTL intervals to find candidate genes of interest.

However, the laboratory of James Cheverud was involved in studying the LG/J mouse for genes involved in bone length, obesity, and diabetes and had generated advanced intercross lines (AIL) to greatly narrow the very large mapping intervals. These AIL, through 34 generations of intercrossing, produced smaller chromosomal regions. One of the four parents of the MRL mouse is the LG/J mouse (contributing 75 % of the MRL genome) that displayed many of the features of the MRL mouse, both being large in size, both showing continual growth in adulthood, obesity, and multiple autoimmune syndromes, and both displaying regenerative healing.

The Heber–Katz, Blankenhorn, and Cheverud labs collaborated to map the regenerative loci of the LG/J mouse using these AIL lines. An initial LG × SM F2 cross resulted in loci which, when compared to the MRL × B6 crosses, showed multiple shared loci [97, 98]. Next, ear-punched F34 AIL using approximately 1200 mice were analyzed using 3600 SNPs. The quantitative trait locus (QTL) intervals were 50-fold decreased in size, as predicted, and 19 QTL were identified [99]. The genomic sequencing of the parental LG and SM strains [100] allowed analysis of polymorphic regions, further narrowing the intervals and providing an excellent set of candidate genes.

From the perspective of inflammation and metabolism in the regenerative response, one candidate gene is potentially very important. *RNF7*, found on chromosome 9, is a component of an E3 ligase complex that has been shown to be necessary for the ubiquination and proteosomal breakdown of HIF-1a, similar and in addition to pVHL [101]. HIF-1a is a major factor in promoting glycolysis and acts as a transcription factor for many molecules supporting a glycolytic response, including PGK1, ALD, LDHa, and PDK1. Since HIF-1a is made constitutively and its levels are tightly controlled by its rates of synthesis versus degradation under normoxic conditions [102, 103], RNF7 should play a major role. Since RNF7 showed low to no expression in the MRL or LG/J, this should lead to high levels of HIF-1a expression. This is currently being tested.

A second fortuitous aspect of this experiment is derived from the fact that approximately 1200 AIL mice were being used for multiple phenotype analyses and half of the mice were fed an HFD, whereas the other half were fed a low-fat diet (LFD) for atherosclerosis studies. As mentioned above, an earlier and smaller experiment in the Heber–Katz laboratory suggested that mice fed an HFD display enhanced ear hole closure when compared with mice fed an LFD or normal diet (unpublished data, Heber–Katz). In this larger study analyzing almost 1200 AIL mice,

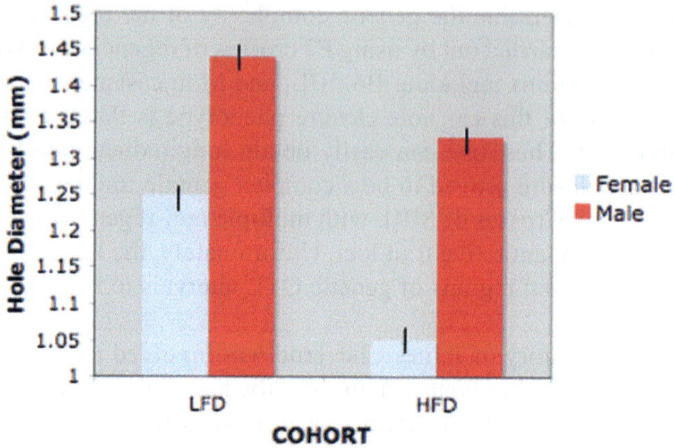

Fig. 3.6 Female and male mice from the F34 AIL line fed a low-fat (L) and high-fat (H) diet. One half of the approximately 1200 mice from advanced intercross F34 line was fed a high-fat diet (HFD) and the other half was fed a low-fat diet (LFD). At 6 weeks of age, mice were ear punched and then at 10 weeks of age, hole diameters were determined. Data is presented as the mean of hole diameter (mm) +/− SE. Females healed better than males, as shown previously. And in both cases, mice healed better on an HFD than an LFD. (These data were generated by James M. Cheverud, Washington University, St. Louis, MO)

it was found that both male and female mice fed an HFD showed highly significant ear hole closure enhancement compared with those mice fed an LFD (Fig. 3.6).

Of the 19 QTL identified for ear hole closure, three of these loci also showed an effect from diet (Cheverud et al. 2014). Interestingly, the candidate genes included molecules that had SNPs in regulatory regions such as HDAC3, which is involved epigenetically in inducing inflammatory genes, especially in macrophages [104]. This molecule is also upregulated in obese individuals who often have a chronic low-level, or "subacute," inflammation associated with increased cancer risk. HDAC3 induces EMT that is involved in metastasis [105]. It is also involved in white adipocyte differentiation [106]. On the other hand, HDAC3 has been reported to repress breast cancer [107].

Another such candidate gene, *FGF1*, is highly active and predictive of some cancer outcomes [108]. *FGF1* mediates differentiation of pre-adipocytes [109] and is regulated by *PPARg* that is induced by an HFD and results in increased levels of insulin and insulin-like growth factor-1 (IGF-1) in the blood, which may promote the development of certain tumors. The *PPARg–FGF1* interaction is important in obesity, *FGF1* is highly upregulated in adipocytes in the presence of an HFD, acts as an angiogenic factor and an endothelial cell factor, and has been reported to be involved in cell migration, inflammation, fat remodeling, and metabolism [110]. Low levels of *FGF1* in the presence of an HFD lead to diabetes [110]. Its expression has been reported to be both reduced and overexpressed in breast cancer [111, 112].

Since there are still a significant number of potential candidate genes in the intervals identified, further focusing on those genes will wait for higher resolution sequencing and further mapping using more advanced AIL.

Acknowledgements
Funding Declaration
The authors declare no competing interests in the studies discussed in this chapter. Funds for this research were supported by the national institutes of health (NIDCR and NIGMS) and the department of defense (DARPA). This research was also supported by a grant from the PA Department of Health Commonwealth Universal Research Enhancement (CURE) Program and from the PA Breast Cancer Coalition. Support for shared resources utilized in these studies was provided by the Cancer Center Support Grant (CCSG) CA010815 to The Wistar Institute EHK is supported by a grant from NCI.

References

1. Andrews BS, Eisenberg RA, Theofilopoulos AN, Izui S, Wilson CB, McConahey PJ, Murphy ED, Roths JB, Dixon FJ. Spontaneous murine lupus-like syndromes. Clinical and immunopathological manifestations in several strains. J Exp Med. 1978;148:1198–215.
2. Cohen, PL, Eisenberg, RA. Lpr and gld: single gene models of systemic autoimmunity and lymphoproliferative disease. Annu Rev Immunol. 1991;9:243–69.
3. Kono DH, Theofilopoulos AN. Genes and genetics of murine lupus. In: Theofilopoulos AN, Bona CA, editors. The molecular pathology of autoimmune diseases. New York: Taylor and Francis; 2002. pp. 353–75.
4. Murphy ED, Roths JB. A single gene model for massive lymphoproliferation with immune complex disease in the new mouse strain MRL. In: Seno S, Takaku F, Irino S, editors. Topics in hematology. Amsterdam: Excerpta Medica; 1977. pp. 69–72.
5. Theofilopoulos AN, Balderas RS, Gozes Y, Aguado MT, Hang L, Morrow PR, Dixon FJ. Association of *lpr* gene with graft-vs-host disease-like syndrome. J Exp Med. 1985;162:1–18.
6. Watson ML, Rao JK, Gilkeson GS, Ruiz P, Eicher EM, Pisetsky DS, Matsuzawa A, Rochelle JM, Seldin MF. Genetic analysis of MRL-*lpr* mice: relationship of *Fas* apoptosis gene to disease manifestations and renal disease-modifying loci. J Exp Med. 1992;176:1645–56.
7. Clark LD, Clark RK, Heber-Katz E. A new murine model for mammalian wound repair and regeneration. Clin Immunol Immunopathol. 1998;88:35–45.
8. Joseph J, Dyson M. Tissue replacement in the rabbit's ear. Brit J Surgery. 1966;53:372–80.
9. Goss RJ, Grimes LN. Tissue interactions in the regeneration of rabbit ear holes. Am Zool. 1975;12:151–7.
10. ten Koppel PGJ, van Osch GJVM, Verwoerd CDA, Verwoerd-Verhoef HL. A new in vivo model for testing cartilage grafts and biomaterials: the `rabbit pinna punch-hole' model. Biomaterials. 2001;22:1407–14.
11. Labandeira-Garcia J, Guerra-Seijas M. Intracellular lipids in rabbit ear cartilage during tissue regeneration. Acta Anat. 1986;127:249–54.
12. Stocum DL. The urodele limb regeneration blastema. Determination and organization of the morphogenetic field. Differentiation. 1984;27:13–28.
13. Brockes JP, Kumar A. Appendage regeneration in adult vertebrates and implications for regenerative medicine. Science. 2005;310:1919–23.
14. Gardiner DM, Bryant SV. Molecular mechanisms in the control of limb regeneration: the role of homeobox genes. Int J Dev Biol. 1996;40:797–805.
15. Edwards RG. From embryonic stem cells to blastema and MRL mice. Reprod Biomed Online. 2008;16:425–61.
16. Kench JA, Russell DM, Fadok VA, Young SK, Worthen GS, Jones-Carson J, Henson JE, Nemazee D. Aberrant wound healing and TGF-beta production in the autoimmune-prone MRL/+ mouse. Clin Immunol. 1999;92:300–10.
17. Blankenhorn EP, Bryan G, Kossenkov AV, Clark LD, Zhang XM, Chang C, Horng W, Pletscher LS, Cheverud JM, Showe LC, Heber-Katz E. Genetic loci that regulate healing and regeneration in LG/J and SM/J mice. Mamm Genome. 2009;20:720–33.

18. Cheverud JM, Lawson HA, Bouckaert K, Kossenkov A, Showe L, Cort L, Blankenhorn EP, Bedelbaeva K, Gourevitch D, Arthur LM, Heber-Katz E. Genetics of murine external ear tissue regeneration is due to differences in cell cycle, dna repair, cell adhesion and migration, and fibrosis. Heredity. 2014;112:508–18.
19. Ueno M, Lyons BL, Burzenski LM, Gott B, Shaffer DJ, Roopenian DC, Shultz LD. Accelerated wound healing of alkali-burned corneas in MRL mice Is associated with a reduced inflammatory signature. Investige Ophthalmol V Sci. 2005;46:4097–106.
20. Chadwick RB, Bu L, Yu H, Hu Y, Wergedal JE, Mohan S, Baylink DJ. Digit tip regrowth and differential gene expression in MRL/MpJ, DBA/2, and C57BL/6 mice. Wound Repair Regen. 2007;15:275–84.
21. Gourevitch DL, Clark L, Bedelbaeva K, Leferovich J, Heber-Katz E. Dynamic changes after murine digit amputation: The MRL mouse digit shows waves of tissue remodeling, growth, and apoptosis. Wound Repair Regen. 2009;17:447–55.
22. Leferovich JM, Bedelbaeva K, Samulewicz S, Zhang XM, Zwas D, Lankford EB, Heber-Katz E. Heart regeneration in adult MRL mice. Proc Natl Acad Sci U S A. 2001;98:9830–5.
23. Haris Naseem R, Meeson AP, Michael DiMaioJ, White MD, Kallhoff J, Humphries C, Goetsch SC, De Windt LJ, Williams MA, Garry MG, Garry DJ. Reparative myocardial mechanisms in adult C57BL/6 and MRL mice following injury. Physiol Genomics. 2007;30:44–52.
24. Alfaro MP, Pagni M, Vincent A, Atkinson J, Hill MF, Cates J, Davidson JM, Rottman J, Lee E, Young PP. The Wnt modulator sFRP2 enhances mesenchymal stem cell engraftment, granulation tissue formation and myocardial repair. Proc Natl Acad Sci U S A. 2008;105:18366–71.
25. Hampton DW, Seitz A, Chen P, Heber-Katz E, Fawcett JW. Altered CNS response to injury in the MRL/MpJ mouse. Neuroscience. 2004;127:821–32.
26. Baker KL, Daniels SB, Lennington JB, Lardaro T, Czap A, Notti RQ, Cooper O, Isacson O, Frasca S, Conover JC. Neuroblast protuberances in the subventricular zone of the regenerative MRL/MpJ mouse. J Comp Neurol. 2006;498:747–61.
27. Balu DT, Hodes GE, Anderson BT, Lucki I. Enhanced sensitivity of the MRL/MpJ mouse to the neuroplastic and behavioral effects of chronic antidepressant treatments. Neuropsychopharmacology. 2009;34:1764–73.
28. Thuret S, Toni N, Aigner S, Yeo GW, Gage FH. Hippocampus-dependent learning is associated with adult neurogenesis in MRL/MpJ mice. Hippocampus. 2009;19:658–69.
29. Thuret S, Thallmair M, Horky LL, Gage FH. Enhanced functional recovery in MRL/MpJ mice after spinal cord dorsal hemisection. PLoS ONE. 2012;7:e30904.
30. Buckley G, Metcalfe AD, Ferguson MWJ. Peripheral nerve regeneration in the MRL/MpJ ear wound model. J Anat. 2011;218:163–72.
31. Fitzgerald J, Rich C, Burkhardt D, Allen J, Herzka AS, Little CB. Evidence for articular cartilage regeneration in MRL/MpJ mice. Osteoarthr Cartil. 2008;16:1319–26.
32. Rai MF, Hashimoto S, Eric E, Johnson EE, Janiszak KL, Fitzgerald J, Ellen Heber- Katz E, Cheverud JM, Sandell LJ. Heritability of articular cartilage regeneration and its association with ear-wound healing. Arthritis Rheum. 2012;64:2300–10.
33. Ward BD, Furman BD, Huebner JL, Kraus VB, Guilak F, Olson SA. Absence of posttraumatic arthritis following intra-articular fracture in the MRL/MpJ mouse. Arthritis Rheum. 2008;58:744–53.
34. Heydemann A, Swaggart KA, Kim GH, Holley-Cuthrell J, Hadhazy M, McNally EM. The superhealing MRL background improves muscular dystrophy. Skelet Muscle. 2012;2:26.
35. Buhimschi CS, Zhao G, Sora N, Madri JA, Buhimschi IA. Myometrial wound healing post-Cesarean delivery in the MRL/MpJ mouse model of uterine scarring. Am J Pathol. 2010;177:197–207.
36. Tolba RH, Schildberg FA, Decker D, Abdullah Z, Buttner R, Minor T, Von Ruecker A. Mechanisms of improved wound healing in Murphy Roths Large (MRL) mice after skin transplantation. Wound Repair Regen. 2010;18:662–70.
37. Prehn RT. Regeneration versus neoplastic growth. Carcinogenesis. 1997;18:1439–44.
38. Prehn RT. Immunosurveillance, regeneration and oncogenesis. Prog Exp Tumor Res. 1971;14:1–24.

39. Donaldson DJ, Mason JM. Cancer related aspects of regeneration research: a review. Growth. 1975;39:475–96.
40. Brockes J. Regeneration and cancer. Biochim Biophys Acta. 1998;1377:M1–11.
41. Del Rio-Tsonis K, Tsonis PA. Amphibian tissue regeneration, a model for cancer regulation. Int J Oncol. 1992;1:1261–4.
42. Tsonis PA, Del Rio-Tsonis K. Spontaneous neoplasms in amphibia. Tumour Biol. 1988;9:221–4.
43. Dinsmore CE, editor. A history of regeneration research: milestones in the evolution of a science. New York: Cambridge University Press; 1991.
44. Breedis C. Induction of accessory limbs and of sarcoma in the newt with carcinogenic substances. Cancer Res. 1965;12:861–6.
45. Zilakos NP, Tsonis PA, Del Rio-Tsonis K, Parchment RE. Newt squamous carcinoma proves phylogenetic conservation of tumors as caricatures of tissue renewal. Cancer Res. 1992;52:4858–65.
46. Tsonis PA, Eguchi G. Carcinogens on regeneration. Effects of N-methyl-N'-nitro-N-nitrosoguanidine and 4-nitroquinoline-1-oxide on limb regeneration in adult newts. Differentiation. 1981;20:52–60.
47. Linell F. On the tumor promoting effect of a single mechanical trauma. Acta Pathol Microbiol Scand. 1947;71:1–100.
48. Ruben LN. The effects of implanting anuran cancer into regenerating adult urodele limbs. J Morphol. 1956;98:389–99.
49. Seilern-Aspang F, Kratochwil K. Induction and differentiation of an epithelial tumor in the newt (*Triturus cristatus*). J Embryol Exp Morphol. 1962;10:337–56.
50. Brinster RL. Effect of cells transferred into the mouse blastocyst on subsequent development. J Exp Med. 1974;140:1049–56.
51. Illmensee K, Mintz B. Totipotency and normal differentiation of single teratocarcinoma cells cloned by injection into blastocytes. Proc Natl Acad Sci U S A. 1976;73:549–53.
52. Gerschenson M, Graves K, Carson SD, Wells RS, Pierce GB. Regulation of melanoma by the embryonic skin. Proc Natl Acad Sci U S A. 1986;83:7307–10.
53. Moroson H, Ioachim HL. Protection by grafts of embryonal rat tissues (teratomas) against induction and transplantation of malignant tumors. Cancer Res. 1995;55:3664–8.
54. Karin de EV, Lidiya VK, Lisa MC. De novo carcinogenesis promoted by chronic inflammation is B lymphocyte dependent. Cancer Cell. 2005;7:41–23.
55. Arthur LM, Demarest RM, Clark L, Gourevitch D, Bedelbaeva K, Anderson R, Snyder A, Capobianco AJ, Lieberman P, Feigenbaum L, Heber-Katz E. Epimorphic Regeneration in Mice is p53-independent. Cell Cycle. 2010;9:3667–73.
56. De laCE1, García-Cao I, Herranz M, López P, García-Palencia P, Flores JM, Serrano M, Fernández-Piqueras J, Martín-Caballero J. Tumorigenic activity of p21Wafl/Cip1 in thymic lymphoma. Oncogene. 2006;25:4128–32.
57. Gourevitch D, Kossenkov AV, Zhang Y, Clark, Chang C, Showe LC, Heber-Katz E. Inflammation and its correlates in regenerative wound healing: an alternate perspective. Adv Wound Care. 2014;3:592–603.
58. Takaya N, Katoh Y, Iwabuchi K, Hayashi I, Konishi H, Itoh S, Okumura K, Ra C, Nagaoka I, Daida H. Platelets activated by collagen through the immunoreceptor tyrosine-based activation motif in the Fc receptor gamma-chain play a pivotal role in the development of myocardial ischemia-reperfusion injury. J Mol Cell Cardiol. 2005;39:856–64.
59. Zheng X, Jiang F, Katakowski M, Kalkanis SN, Hong X, Zhang X, Zhang ZG, Yang H, Chopp M. Inhibition of ADAM17 reduces hypoxia-induced brain tumor cell invasiveness. Cancer Sci. 2007;98:674–84.
60. Furukawa F, Yoshimasu T, Yamamoto Y, Kanazawa N, Tachibana T. Mast cells and histamine metabolism in skin lesions from MRL/MP-lpr/lpr mice. Autoimmun Rev. 2009;8:495–9.
61. Babic AM, Chen CC, Lau LF. Fisp12/mouse connective tissue growth factor mediates endothelial cell adhesion and migration through integrin alphavbeta3, promotes endothelial cell survival, and induces angiogenesis in vivo. Mol Cell Biol. 1999;19:2958–66.

62. Sonnylal S, Shi-Wen X, Leoni P, Naff K, Van Pelt CS, Nakamura H, Leask A, Abraham D, Bou-Gharios G, de Crombrugghe B. Selective expression of connective tissue growth factor in fibroblasts in vivo promotes systemic tissue fibrosis. Arthritis Rheum. 2010;62:1523–32.
63. Wong CC, Gilkes DM, Zhang H, Chen J, Wei H, Chaturvedi P, Fraley SI, Wong CM, Khoo US, Ng IO, Wirtz D, Semenza GL. Hypoxia-inducible factor 1 is a master regulator of breast cancer metastatic niche formation. Proc Natl Acad Sci U S A. 2011;108:16369–74.
64. Xia Z-W, Xu L-Q, Zhong W-W, JWei J-J, Li N-L, Shao J, Li Y-Z, Yu S-C, Zhang Z-L. Heme oxygenase-1 attenuates ovalbumin-induced airway inflammation by up-regulation of Foxp3 T-regulatory cells, interleukin-10, and membrane-bound transforming growth factor-β1. Am J Pathol. 2007;171:1904–14.
65. Theoharis CT, Pio Conti. The JEKYLL and HYDE of tumor growth. Trends Immunol. 2004;25:235–41.
66. De Franco M, Carneiro P, Peters L, Vorraro F, Borrego A, Ribeiro O, Starobinas N, Cabrera W, Ibanez O. Slc11a1 (Nramp1) alleles interact with acute inflammation loci to modulate wound-healing traits in mice. Mamm Genome. 2007;18:263–9.
67. Canhamero T, Valino Garcia L, De Franco M. Acute inflammation loci influence wound healing in mice. Adv Wound Care. 2014;3:582–91.
68. Yin Y, Henzl MT, Lorber B, Nakazawa T, Thomas TT, Jiang F, Langer R, Benowitz LI. Oncomodulin is a macrophage-derived signal for axon regeneration in retinal ganglion cells. Nat Neurosci. 2006;9:843–52.
69. Kyritsis N, Kizil C, Zocher S, Kroehne V, Kaslin J, Freudenreich D, Iltzsche A, Brand M. Acute inflammation initiates the regenerative response in the adult zebrafish brain. Science. 2012;338(6112):1353–6.
70. Wada K, Arita M, Nakajima A, Katayama K, Kudo C, Kamisaki Y, Serhan CN. Leukotriene B4 and lipoxin A4 are regulatory signals for neural stem cell proliferation and differentiation. FASEB J. 2006;20:1785–92.
71. Goh YP, Henderson NC, Heredia JE, Red Eagle A, Odegaard JI, Lehwald N, Nguyen KD, Sheppard D, Mukundan L, Locksley RM, Chawla A. Eosinophils secrete IL-4 to facilitate liver regeneration. Proc Natl Acad Sci U S A. 2013;110:9914–9.
72. Coussens LM, Werb ZZ. Inflammation and cancer. Nature. 2002;420:860–7.
73. Balkwill F, Charles KA, Mantovani A. Smoldering and polarized inflammation in the initiation and promotion of malignant disease. Cancer Cell. 2005;7:211–7.
74. Andreu P, Johansson M, Affara NI, Pucci F, Tan T, Junankar S, Korets L, Lam J, Tawfik D, DeNardo DG, Naldini L, de Visser KE, De Palma M, Coussens LM. FcRgamma activation regulates inflammation-associated squamous carcinogenesis. Cancer Cell. 2010;17:121–34.
75. Monteiro R, Azevedo I. Chronic inflammation in obesity and the metabolic syndrome. Mediators Inflamm. 2010;2010:289645.
76. Kim KA1, Gu W, Lee IA, Joh EH, Kim DH. High fat diet-induced gut microbiota exacerbates inflammation and obesity in mice via the TLR4 signaling pathway. PLoS ONE. 2012;7:e47713.
77. Weisberg SP, McCann D, Desai M, Rosenbaum M, Leibel RL, et al. Obesity is associated with macrophage accumulation in adipose tissue. J Clin Invest. 2003;112:1796–808.
78. O'Neill LAJ, Hardie DG. Metabolism of inflammation limited by AMPK and pseudo-starvation. Nature. 2013;493:346.
79. Naviaux RK, Le TP, Bedelbaeva K, Leferovich J, Gourevitch D, Sachadyn P, Zhang XM, Clark L, Heber-Katz E. Retained features of embryonic metabolism in the adult MRL mouse. Mol Genet Metab. 2009;96:133–44.
80. Katayama M, Zhong Z, Lai L, Sutovsky P, Prather RS, Schatten H. Mitochondrial distribution and microtubule organization in fertilized and cloned porcine embryos: implications for developmental potential. Dev Biol. 2006;299:206–20.
81. Lant B, Storey KB. An overview of stress response and hypometabolic strategies in Caenorhabditis elegans: conserved and contrasting signals with the mammalian system. Int J Biol Sci. 2010;6:9–50.
82. Storey KB, Storey JM. Metabolic rate depression in animals: transcriptional and translational controls. Biol Rev Camb Philos Soc. 2004;79:207–33.

83. Sachadyn P, Zhang XM, Clark LD, Naviaux RK, Heber-Katz E. Naturally occurring mitochondrial DNA heteroplasmy in the MRL mouse. Mitochondrion. 2008;8:358–66.
84. Maniataki E, Mourelatos Z. Human mitochondrial tRNAMet is exported to the cytoplasm and associates with the Argonaute 2 protein. RNA. 2005;11:849–52.
85. Johnson KR, Zheng QY, Bykhovskaya Y, Spirina O, Fischel-Ghodsian N. A nuclear-mitochondrial DNA interaction affecting hearing impairment in mice. Nat Genet. 2001;27:191–4.
86. Jones SM, Jones TA, Johnson KR, Yu H, Erway LC, Zheng QY. A comparison of vestibular and auditory phenotypes in inbred mouse strains. Brain Res. 2006;1091:40–6.
87. Zhou X, Jen PH, Seburn KL, Frankel WN, Zheng QY. Auditory brainstem responses in 10 inbred strains of mice. Brain Res. 2006;1091:16–26.
88. Gatenby RA, Gillies RJ. Why do cancers have high aerobic glycolysis? Nat Rev Cancer. 2004;4:891–9.
89. Lopez-Lazaro M. The Warburg effect: why and how do cancer cells activate glycolysis in the presence of oxygen? Anticancer Agents Med Chem. 2008;8:305–12.
90. Pavlides S, Whitaker-Menezes D, Castello-Cros R, Flomenberg N, Witkiewicz AK, Frank PG, Casimiro MC, Wang C, Fortina P, Addya S, Pestell RG, Martinez-Outschoorn UE, Sotgia F, Lisanti MP. The reverse Warburg effect: aerobic glycolysis in cancer associated fibroblasts and the tumor stroma. Cell Cycle. 2009;8:3984–4001.
91. McBrearty BA, Clark LD, Zhang XM, Blankenhorn EP, Heber-Katz E. Genetic analysis of a mammalian wound-healing trait. Proc Natl Acad Sci U S A. 1998;95:11792–7.
92. Masinde GL, Li X, Gu W, Davidson H, Mohan S, Baylink DJ. Identification of wound healing/regeneration quantitative trait loci (QTL) at multiple time points that explain seventy percent of variance in (MRL/MpJ and SJL/J) mice F2 population. Genome Res. 2001;11:2027–33.
93. Blankenhorn EP, Troutman S, Clark LD, Zhang XM, Chen P, Heber-Katz E. Sexually dimorphic genes regulate healing and regeneration in MRL mice. Mamm Genome. 2003;14:250–60.
94. Heber-Katz E, Chen P, Dvm LC, Zhang X-M, Troutman S, Blankenhorn EP. Regeneration in MRL mice: further genetic loci controlling the ear hole closure trait using MRL and M. m. castaneus mice. Wound Repair Regen. 2004;12:384–92.
95. Yu H, Mohan S, Masinde G, Baylink D. Mapping the dominant wound healing and soft tissue regeneration QTL in MRL x CAST. Mamm Genome. 2005;16:918–24.
96. Li XM, Gu WK, Masinde G, Hamilton-Ulland M, Xu SZ, Mohan S, Baylink D. Genetic control of the rate of wound healing in mice. Heredity. 2010;86:668–74.
97. Blankenhorn E, Bryan G, Kossenkov A, Clark L, Zhang XM, Chang C, Horng W, Pletscher L, Cheverud J, Showe L, Heber-Katz E. Genetic loci that regulate healing and regeneration in LG/J and SM/J mice. Mamm Genome. 2009;20:720–33.
98. Cheverud JM, Lawson HA, Funk R, Zhou J, Blankenhorn EP, Heber-Katz E. Healing quantitative trait loci in a combined cross analysis using related mouse strain crosses. Heredity. 2012;108:441–6.
99. Cheverud JM, Lawson HA, Bouckaert K, Kossenkov A, Showe L, Cort L, Blankenhorn EP, Bedelbaeva K, Gourevitch D, Arthur LM, Heber-Katz E. Genetics of murine external ear tissue regeneration is due to differences in cell cycle, dna repair, cell adhesion and migration, and fibrosis. Heredity. 2014;112:508–18.
100. Lawson HA, Cady JE, Partridge C, Wolf JB, Semenkovich CF, Cheverud JM. Genetic effects at pleiotropic loci are context-dependent with consequences for the maintenance of genetic variation in populations. PLoS Genet. 2011;7:e1002256.
101. Tan M, Gu Q, He H, Pamarthy D, Semenza GL, Sun Y. SAG/ROC2/RBX2 is a HIF-1 target gene that promotes HIF-1[alpha] ubiquitination and degradation. Oncogene. 2007;27:1404–11.
102. Semenza GL. HIF-1 and mechanisms of hypoxia sensing. Curr Opin.Cell Biol. 2001;13:167–71.
103. Weidemann A, Johnson RS. Biology of HIF-1alpha. Cell Death Differ. 2008;15:621–7.

104. Chen X, Barozzi I, Termanini A, Prosperini E, Recchiuti A, Dalli J, Mietton F, Matteoli G, Hiebert S, Natoli G. Requirement for the histone deacetylase Hdac3 for the inflammatory gene expression program in macrophages. Proc Natl Acad Sci U S A. 2012;109:E2865–74.
105. Wu MZ, Tsai YP, Yang MH, Huang CH, Chang SY, Chang CC, Teng SC, Wu KJ. Interplay between HDAC3 and WDR5 is essential for hypoxia-induced epithelial-mesenchymal transition. Mol Cell. 2011;43:811–22.
106. Razidlo DF, Whitney TJ, Casper ME, McGee-Lawrence ME, Stensgard BA, Li X, Secreto FJ, Knutson SK, Hiebert SW, Westendorf JJ. Histone deacetylase 3 depletion in osteo/chondroprogenitor cells decreases bone density and increases marrow fat. PLoS ONE. 2010;J5:e11492
107. Kim HC, Choi KC, Choi HK, Kang HB, Kim MJ, Lee YH, Lee OH, Lee J, Kim YJ, Jun W, Jeong JW, Yoon HG. HDAC3 selectively represses CREB3-mediated transcription and migration of metastatic breast cancer cells. Cell Mol Life Sci. 2010;67:3499–510.
108. Smith G, Ng MT, Shepherd L, Herrington CS, Gourley C, Ferguson MJ, Wolf CR. Individuality in FGF1 expression significantly influences platinum resistance and progression free survival in ovarian cancer. Br J Cancer. 2012;107:1327–36.
109. Hutley L, Shurety W, Newell F, McGeary R, Pelton N, Grant J, Herington A, Cameron D, Whitehead J, Prins J. Fibroblast growth factor 1: a key regulator of human adipogenesis. Diabetes. 2004;53:3097–106.
110. Jonker JW, Suh JM, Atkins AR, Ahmadian M, Li P, Whyte J, He M, Juguilon H, Yin YQ, Phillips CT, Yu RT, Olefsky JM, Henry RR, Downes M, Evans RM. A PPARγ-FGF1 axis is required for adaptive adipose remodeling and metabolic homeostasis. Nature. 2012;485:391–411.
111. Bansal GS, Yiangou C, Coope RC, Gomm JJ, Luqmani YA, Coombes RC, Johnston CL. Expression of fibroblast growth factor 1 is lower in breast cancer than in the normal human breast. Br J Cancer. 1995;72:1420–6.
112. Yoshimura N, Sano H, Hashiramoto A, Yamada R, Nakajima H, Kondo M, Oka T. The expression and localization of fibroblast growth factor-1 (FGF-1) and FGF receptor-1 (FGFR-1) in human breast cancer. Clin Immunol Immunopathol. 1998;89:28–34.
113. Abel EL, Angel JM, Kiguchi K, DiGiovanni J. Multi-stage chemical carcinogenesis in mouse skin: Fundamentals and applications. Nat Protoc. 2000;4:1350–62.

Chapter 4
Living Large: What Mouse Models Reveal about Growth Hormone and Obesity

Darlene E. Berryman, Lara Householder, Vivian Lesende, Edward O. List and John J. Kopchick

Abstract Growth hormone (GH) regulates a broad spectrum of biological processes in addition to promoting longitudinal growth. As such, GH influences most organ and cellular systems in the body with adipose tissue being one of its well-established targets. This chapter will describe mouse lines with specific alteration in GH action. Mice with increased, decreased, and absence of GH action have a unique phenotype for which clinical equivalents exist (acromegaly, GH deficiency, and Laron syndrome, respectively). Interestingly, these mouse lines demonstrate adiposity profiles that are counterintuitive to health and longevity. That is, mice with excess GH action are lean but insulin resistant, prone to cancer, and short-lived (or "unhealthy lean"). On the other hand, mice with no GH action are obese but insulin sensitive, resistant to cancer, and long-lived (or "healthy obesity"). These extremes in GH action provide fascinating mouse strains with which to study the features of fat that are responsible for metabolic dysfunction and to explore traits that are obligatory for cancer or lifespan.

Keywords Growth hormone · Obesity · White adipose tissue · Depot-specific effects

D. E. Berryman (✉)
Edison Biotechnology Institute, Ohio University, Athens, OH, USA

School of Applied Health Sciences and Wellness, College of Health Sciences and Professions, Ohio University, Athens, OH, USA

Department of Biomedical Sciences, Heritage College of Osteopathic Medicine, Ohio University, Athens, OH, 45701, USA

Konneker Research Laboratories, Ohio University,
The Ridges 172 Water Tower Drive, Athens, OH,45701 USA
e-mail: berrymad@ohio.edu

L. Householder
Edison Biotechnology Institute, Ohio University, Athens, OH, USA

N. A. Berger (ed.), *Murine Models, Energy Balance, and Cancer,*
Energy Balance and Cancer, DOI 10.1007/978-3-319-16733-6_4

School of Applied Health Sciences and Wellness, College of Health Sciences and Professions, Ohio University, Athens, OH, USA

Konneker Research Laboratories, Ohio University, The Ridges 172 Water Tower Drive, Athens, OH 45701 USA
e-mail: Lb155602@ohio.edu

V. Lesende
Edison Biotechnology Institute, Ohio University, Athens, OH, USA

School of Applied Health Sciences and Wellness, College of Health Sciences and Professions, Ohio University, Athens, OH, USA
e-mail: Vl295212@ohio.edu

E. O. List
Edison Biotechnology Institute, Ohio University, Athens, OH, USA

Konneker Research Laboratories, Ohio University, The Ridges 172 Water Tower Drive, Athens, OH 45701 USA
e-mail: list@ohio.edu

J. J. Kopchick
Edison Biotechnology Institute, Ohio University, Athens, OH, USA

Department of Biomedical Sciences, Heritage College of Osteopathic Medicine, Ohio University, Athens, OH, 45701, USA

Konneker Research Laboratories, Ohio University, The Ridges 172 Water Tower Drive, Athens, OH 45701 USA
e-mail: kopchick@ohio.edu

Introduction

Growth hormone (GH) is a pleiotropic hormone that promotes growth, as well as alters a broad range of metabolic actions. Due to its multiple actions, the use of in vitro or ex vivo techniques to understand its role in energy balance, obesity, and cancer is limited. In vivo analysis allows for tissue "crosstalk", which may well be critical to understand how endocrine factors affect overall metabolism. To that end, this review describes several decades of research in mutant, transgenic, and gene-disrupted strains of mice with specific alterations in GH-induced intracellular signaling. Analysis of these mice has led to a better understanding of the varied functions of GH. Relevant to this review, these mice provide evidence that GH, which reduces obesity, can simultaneously cause metabolic dysregulation, increase the incidence and severity of certain types of cancers, promote insulin resistance, and shorten lifespan. In this overview, we will first introduce the newer complexity of adipose tissue (AT) to provide the foundational knowledge to understand not only the importance of the quantity of AT but also the quality of that tissue for whole body homeostasis. We will then introduce GH, the general phenotypes of the selected mouse lines that have extreme alterations in GH action, and highlight data describing their unique obese phenotype and cancer incidence.

Adipose Tissue Introduction

AT is an intricate, multi-depot organ with important functions in energy storage, metabolism, and mechanical support [1]. Once regarded as a simple storage space for extra energy in the form of triglycerides, AT is now recognized as an active endocrine organ, secreting hormones and cytokines that initiate signaling in many tissues throughout the body [2]. The discovery of AT's endocrine function as well as the obesity epidemic has instigated a surge of AT interest and research. The rise in world-wide incidence of obesity has been accompanied by a concomitant rise in comorbidities, or associated diseases, of obesity, including some cancers. Thus, the need to understand both the normal physiology of AT and its pathophysiology in obesity is paramount. This section will focus on the basic morphology and physiology of AT, its dysfunctions in the obese state, and its relationship with the development and progression of cancers.

Cell Types in Adipose Tissue

AT has many layers of complexity, the first of which is its cell types. Though the characteristic cell type is the adipocyte, there are actually three distinct types of adipocytes: white, brown, and beige. Each of these adipocyte types has a unique structure and function. White adipocytes, the primary cell type of white adipose tissue (WAT), have a large, unilocular lipid droplet in which they store energy in the form of triglycerides. In periods of excess energy, such as obesity, the lipid droplet expands, resulting in hypertrophy of white adipocytes [3] (Fig. 4.1). When energy is needed, white adipocytes release energy through a process called lipolysis [4]. Conversely, brown adipocytes, the primary cell type of brown adipose tissue (BAT), are multilocular, featuring multiple, small lipid droplets. They have many mitochondria and consume energy to produce heat (non-shivering thermogenesis). Brown adipocytes accomplish this through a unique protein, uncoupling protein-1 (UCP-1), that works in the mitochondria to uncouple fuel oxidation from ATP by creating a proton leak across the inner mitochondrial membrane. The third type of adipocytes, beige cells, are UCP-1-positive cells found within WAT. Once thought to be brown adipocytes residing in WAT, they are now understood to be inducible cells that can have brown-like activity. In essence, when dormant, they behave like white adipocytes by storing energy in a large lipid droplet. However, under certain stimuli, they can take on a brown adipocyte phenotype. While in their active state they are similar to a brown adipocyte, they differ in both gene expression and lineage and are thus a unique cell type apart from both white and brown adipocytes.

Though adipocytes are the main cell type, AT also hosts a variety of other cells types as well as other tissue components (Fig. 4.1). Non-adipocyte cells are collectively referred to as the stromal vascular fraction (SVF) and are important contributors to the function and health of AT. Cells of the SVF include preadipocytes, immune cells (macrophages, lymphocytes, and mast cells), vascular tissue cells, neural tissue cells, and fibroblasts [5, 7]. These cells contribute to the secretory

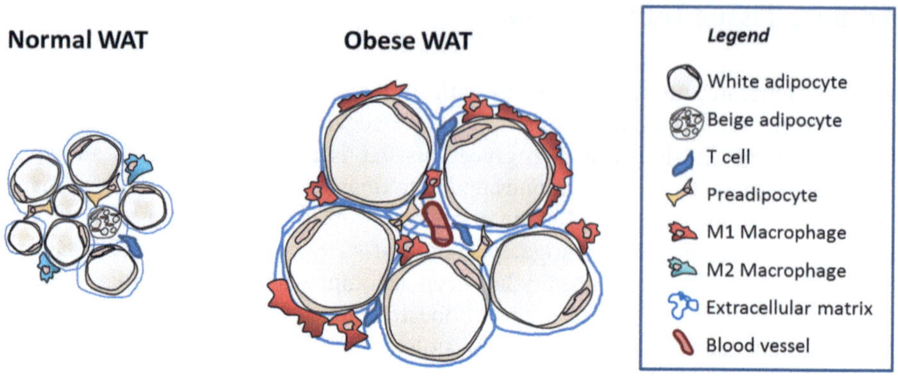

Fig. 4.1 Structure and composition of normal and obese WAT. As obesity develops, WAT undergoes several functional and structural characteristic changes. First, adipocytes expand and become hypertrophic. The immune cell population is also altered. Normal WAT contains anti-inflammatory M2 macrophages whereas obese WAT sees a large influx of inflammatory M1 macrophages and T cells. Often, M1 macrophages in obese WAT will surround a dead or dying adipocyte to form what is called a crown-like structure. In addition to changes in the cells of obese WAT, the extracellular matrix (ECM) also accumulates and becomes fibrotic. These structural changes lead to the development of metabolic dysfunction in obese WAT. Adapted with permission [91]

profile of AT that affects AT as well as other organs. Importantly, the cellular population in AT is altered in obesity, which contributes to the overall dysfunction of the tissue in the obese state [5, 7].

Adipose Tissue Depots

The adipose organ is distributed throughout the body in well-defined depots that can be distinguished not only by location but also by composition and function [8, 9]. WAT depots are categorized as either subcutaneous (SubQ) or intra-abdominal, although there are many different depots within those generic categories. In rodents, there are two SubQ depots found just under the skin, and several intra-abdominal depots, which are found within the peritoneum. In mice, the SubQ depots are located both anteriorly and posteriorly. The posterior depots are considered less complex than the anterior SubQ depots [10]. There are several intra-abdominal depots, which are further categorized as visceral and non-visceral although there is some controversy as to how these are defined. Some researchers define visceral depots more strictly and consider only depots that drain into the portal vein to be visceral. Other researchers refer to all depots in the intra-abdominal and visceral regions as visceral [10]. There are three frequently studied intra-abdominal depots: perigonadal (surrounding the sex organs), retroperitoneal (behind the kidneys), and mesenteric (surrounding the intestines). Although the perigonadal depot (also called epididymal in males and paraovarian in females) is the most commonly studied as it is often large and easy to collect, the mesenteric depot is the only one of these three depots that drains into the portal vein and, as such, is a true visceral depot. In rodents, the main adult BAT depot is found in the interscapular region (between the shoulder blades) although there are other smaller BAT depots (cervical, mediastinal, pericardial, and perirenal depots) [11] that have not been well studied.

Not only is AT found in distinct depots, but these depots have divergent characteristics, including differences in glucose metabolism, endocrine function, and adipocyte size. Depot differences are an important factor for the health of obese individuals as the pattern of obesity has been shown to have an effect on the overall health of the individual. Specifically, those with increased visceral adiposity, sometimes called central pattern obesity, generally have more severely depressed metabolic health than those whose adiposity is found subcutaneously [12]. Transplantation studies in mice have demonstrated this effect as well and have shown that the results are due to intrinsic differences in the depot itself, not merely due to anatomical location [13]. Thus, due to the distinct characteristics of individual depots, it is important to study multiple depots when examining the adipose organ.

Endocrine Function of Adipose Tissue

Although it was once considered a passive storage organ, AT is now understood to carry out many other functions. Importantly, adipocytes and other cell types in AT can secrete and respond to many bioactive peptides in both autocrine/paracrine and endocrine manners [6, 14]; thus, it is considered to be an endocrine organ. The secreted proteins of AT are collectively referred to as adipokines and can have drastic effects on whole body metabolism and energy homeostasis. A classic example is the hormone leptin, which is secreted by adipocytes and interacts with the central nervous system (CNS) to regulate energy intake, metabolism, and satiety [15]. Mice with mutations in the leptin or leptin receptor gene clearly demonstrate the importance of this hormone, as they are strikingly obese [16, 18]. In addition to the CNS, leptin signaling also affects other tissues and cell types such as bone and immune cells [19]. Adiponectin is another classic adipokine with whole body effects. Secreted by adipocytes, adiponectin circulates at high levels in the blood and acts primarily on receptors in the muscles and liver [20]. Although adiponectin can be found in circulation in several multimeric complexes, the high-molecular-weight form has been suggested to have the predominant bioactivity [21, 22]. A defect in the adiponectin gene results in metabolic dysregulation, whereas overexpression improves insulin sensitivity and other markers of metabolic health [23, 25]. Thus, both leptin and adiponectin demonstrate the importance of AT's endocrine function and serve to deepen its complexity.

Pathophysiology of Obesity

Obesity is defined as excess WAT, which occurs through both hypertrophy and hyperplasia of adipocytes. However, this definition belies the complexity of the pathophysiological changes that occur in WAT in the obese state. It is now understood that characteristic changes accompany WAT expansion, including insulin resistance

[26], immune cell infiltration [27, 28], pathologically accelerated remodeling [29], and fibrosis [30]. These changes promote the development of obesity-related diseases such as diabetes, cardiovascular disease (CVD), and some cancers. A classic example of these changes is the formation of crown-like structures (CLS), which can be observed histologically. These structures are seen when M1 macrophages surround a dead or dying adipocyte [28, 31, 32]. A syncytium of macrophages is formed around the adipocyte as the macrophages consume the lipid droplet. Another well-established alteration in AT function during obesity is the alteration of adipokine secretion. Both adiponectin and leptin expression and secretion are altered with increased fat mass. Adiponectin levels usually have an inverse relationship with adiposity, being higher in lean states [33]. Since adiponectin promotes insulin sensitivity, low levels of adiponectin may further contribute to insulin resistance and metabolic dysfunction during obesity [34]. Leptin concentration, on the other hand, increases with increasing fat mass [5, 12]. Although this could be seen as beneficial due to leptin's effect on satiety and metabolic rate, obese individuals actually become leptin insensitive [35, 36]. As both CLS and altered adipokines demonstrate, obesity not only involves increased WAT but also includes drastic changes to WAT morphology and physiology. These changes radiate beyond local tissue dysfunction and initiate whole body consequences. Some of these dysfunctions will be discussed below, along with their potential role in the initiation and progression of cancers.

Dysfunctional Adipose Tissue and Cancer

Epidemiological studies have established a clear relationship between obesity and many cancers, including breast, liver, kidney, thyroid, colon, and esophageal as well as some leukemias and lymphomas [37]. In fact, the association is so great that obesity is considered a major risk factor for cancer and is known to drastically increase the mortality rate for all cancers [38]. However, molecular mechanisms that would serve to explain this relationship are still not well understood. A number of potential processes have been proposed including alterations to adipokines and inflammatory cytokines, increased immune cell infiltration, altered aromatization of sex hormones, and increased insulin and IGF-1-induced signaling.

Altered Adipokines and Cytokines

Secreted proteins from AT can act on both normal and cancer cells in both paracrine and endocrine manners. Many organs have proximal fat pads or stromal adipocytes that can affect the microenvironment of a tumor. Specifically, secretion of extracellular matrix (ECM)-related proteins, such as matrix metalloproteinase-11, are upregulated in obesity, which can promote ECM remodeling and create a favorable environment for tumor invasion [39, 40]. In addition, increased BMI is associated with higher levels of inflammatory cytokines secreted from immune cells within AT, including tumor necrosis factor-α (TNF-α), interleukin-6 (IL-6), and plasminogen activator inhibitor 1 (PAI1). Inflammatory cytokines have been associated with the promotion of obesity-related tumorigenesis and metastasis in both cell culture and in obese mice though the specific mechanisms are still not well understood. Other adipokines are also altered in the obese state as previously mentioned. Importantly, both leptin and adiponectin have been shown to play a role in cancer risk although the mechanisms are still unknown and may be tissue dependent. Similar to their role in metabolic disease, leptin and adiponectin have opposing actions; that is, increased leptin promotes tumors [41, 44], whereas increased adiponectin decreases tumor growth [45, 46].

Immune Cell Infiltration

Immune cells can make up a substantial portion of the SVF in WAT. The resident and infiltrating immune cells vary in type, number, activation status, and production of various cytokines based on the level of adiposity as well as the anatomical location of the WAT depot [47, 49]. In lean WAT, anti-inflammatory (M2-like) macrophages, myeloid-derived suppressor cells, regulatory T cells, and eosinophils predominate [50, 52] whereas with increasing adiposity, pro-inflammatory (M1-like) macrophages, neutrophils, T cells (CD4+and CD8+cells), B cells, and mast cells have been found to infiltrate the tissue, particularly in visceral depots [28, 48, 53]. Moreover, recent evidence indicates that different depots (e.g., mesenteric, retroperitoneal, epididymal) have distinct immune profiles [47], suggesting that intra-abdominal fat pads can be considered distinct mini-organs at least with respect to their leukocyte populations. Taking all of this into account, it is important to consider WAT-associated immune cells as a contributing factor to the chronic inflammation and insulin resistance present in the obese state.

Insulin and IGF-1 Signaling

The obese state is associated with many metabolic changes, including insulin resistance and resultant hyperinsulinemia. Alterations to IGF-binding proteins increase the bioavailability of insulin-like growth factor 1 (IGF-1), further increasing the activity of the IGF axis in obesity. Both insulin and IGF-1 are potent growth factors; thus, it is no surprise that their increased activity is associated with increased risk and progression of obesity-associated cancers [54]. Tumors also take advantage of this environment and are known to overexpress the IGF-1 receptor (IGF1R) [55]. The relationship between IGF-1 and cancer will be discussed more thoroughly later in the chapter.

Growth Hormone (Structure/General Function)

GH is a peptide secreted from somatotrophs of the anterior pituitary in two primary forms: the 22 kDa form, which is the most abundant and active form, and the 20 kDa form, which is less active than the 22 kDa form [56, 57]. GH exerts its many effects by binding to the growth hormone receptor (GHR) present on the cell surface of most tissues, including liver, AT, heart, kidneys, intestine, lung, pancreas, cartilage and skeletal muscle (see Fig. 4.2). GH binding initiates an intracellular signaling cascade (discussed in more detail in the subsequent section) that, among other effects, induces the production and secretion of IGF-1, another potent growth

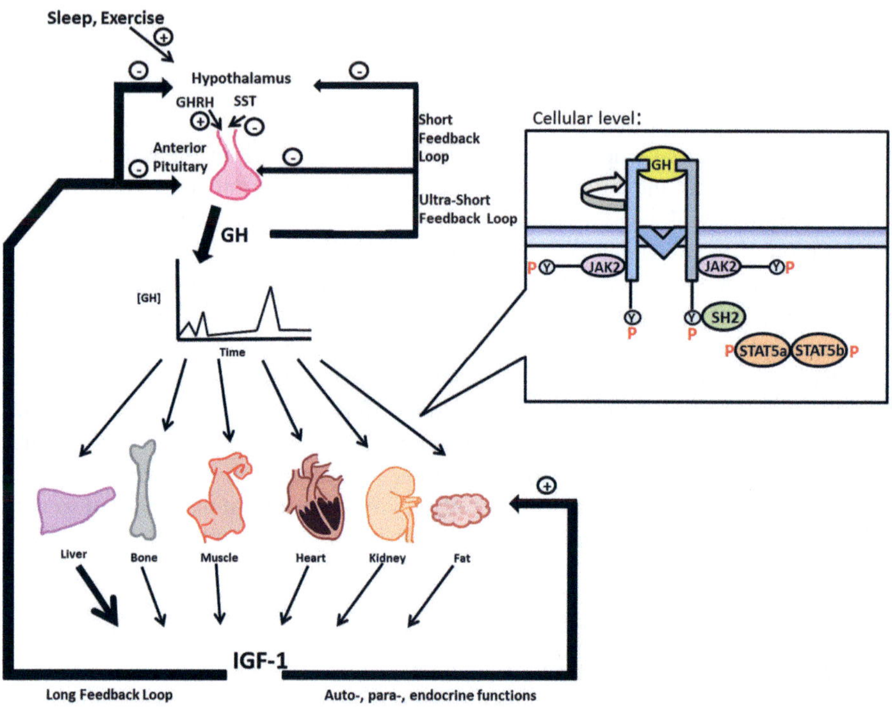

Fig. 4.2 Growth hormone (GH)-induced signaling in target tissues. Growth hormone releasing hormone (GHRH) from the hypothalamus stimulates the pulsatile release of GH from the anterior pituitary. Somatostatin (SST) has an inhibitory effect on the release of GH. GH exerts its effects on cells of target tissues by interacting with the dimerized growth hormone receptor (GHR). Binding of GH induces rotation of the GHRs, stimulating the associated Janus kinase 2 (JAK2) molecules to promote cross-phosphorylation of the tyrosine residues (Y) of JAK2 as well as of the cytoplasmic domains of the dimerized GHR. These tyrosine phosphorylated residues recruit signaling molecules with Src homology 2 (SH2) domains. This allows STAT5 to bind, become phosphorylated, dimerize, and travel to the nucleus where they bind to STAT responsive elements of enhancer regions to increase transcription of genes related to metabolism, body growth, and production of GH's effector molecule, IGF-1. GH-induced signaling may also occur through JAK2-independent signaling (not pictured).

factor. GH has many functions, but perhaps the most well-studied is its stimulation of childhood bone growth [58]. Both GH and IGF-1 from the bone (acting in a paracrine/autocrine manner) act on the growth plate to stimulate longitudinal bone growth [59]. In adults, GH plays roles in determining body composition and metabolism, having an anabolic effect on many tissues but a catabolic effect on AT. That is, GH not only decreases body fat by increasing lipolysis and fat oxidation, but also reduces glucose uptake into cells [58].

In humans, GH is secreted in a pulsatile manner, with the largest pulses occurring during the night during rapid eye movement (REM) sleep [60, 61]. The amplitude of these pulses is under the control of several hormones. The secretion of GH is stimulated by growth hormone-releasing hormone (GHRH) and inhibited by somatostatin, both released from the hypothalamus through the portal system [60]. Ghrelin, secreted by the stomach, also acts as a stimulator of GH secretion [60]. Interactions between these hormones result in the increase in amplitude of GH pulses during adolescence and the decrease thereafter [60]. Physiological factors such as stress and hypoglycemia inhibit GH secretion [60, 62] whereas exercise and sleep stimulate GH secretion [60, 62]. Finally, IGF-1 acts as a negative feedback regulator of GH at the level of the hypothalamus and the anterior pituitary [62].

Signaling Pathways of GH

GH contains two binding sites for the GHR, one on each side of the molecule, allowing it to bind to two different sites on a membrane bound predimerized GHR molecule (Herington & Lobie 2011) (Fig. 4.2, inset). The dimerized GHRs are each associated with the cytoplasmic kinase, Janus kinase 2 (JAK2) and/or kinases of the Src family. The binding of GH to its binding sites on the GHR results in the rotation of the intracellular domains of a GHR; this activates the associated kinases by initiating their cross-phosphorylation at tyrosine residues, as well as the phosphorylation of the GHR molecules at tyrosine residues of their cytoplasmic tails [63].

Two types of signaling are initiated by binding of GH to the GHR: JAK2-dependent signaling and JAK2-independent signaling. In JAK2-dependent signaling, the tyrosine phosphorylation of the GHR cytoplasmic domain results in the recruitment of additional signaling molecules containing Src homology 2 (SH2)-domains [63]. As a result, Signal Transducer and Activator of Transcription 5 (STAT5) a and b are phosphorylated and dimerize, allowing them to bind to STAT-responsive elements of promoter/enhancer regions of genes related to metabolism, body growth, and IGF-1 production [63]. JAK2-independent signaling utilizes the Src family kinases rather than JAK2 to stimulate p44/42 mitogen-activated protein kinase (MAPK) activity and activity of the transcription activator Elk-1 [63, 64]. This pathway involves Src activation of GTP-bound RalA and RalB. Zhu et al. explain that this increased RalA activity results in the activation of phospholipase D and the production of its metabolite, phosphatidic acid, which is needed for the activation of p44/42 MAPK and Elk-1 transcription.

Growth Hormone's Effects on Metabolism and Energy Balance

GH acts predominantly when the body is in a postabsorptive or fasting state to switch primary metabolism from the utilization of glucose and proteins to the oxidation of lipids [65]. When in the postabsorptive state, GH stimulates lipolysis and the oxidation of lipids, as well as ketogenesis, resulting in increased levels of circulating free fatty acid (FFA) and ketone bodies [65]. The lipolytic effects of GH are thought to be mediated through the stimulation of hormone-sensitive lipase (HSL [65, 66]. Furthermore, there is evidence to suggest that GH, through IGF-1, can inhibit the conversion of cortisone to cortisol in human abdominal WAT, by inhibiting the activity of the enzyme 11β-hydroxysteroid dehydrogenase 1 [65]. The resulting low cortisol levels prevent the accumulation of WAT in this region [65]. In the postabsorptive state, GH inhibits glucose oxidation and prevents the uptake of glucose by the muscle though IGF-1's suppression of insulin secretion [67] and/or by its direct effect at the tissue level in terms of glucose uptake [68, 70]. In general, GH increases protein synthesis and prevents breakdown at the whole body level, as well as in muscle tissue specifically, while decreasing the degradation of amino acids [65]. Thus, in a fasting state, GH functions to preserve positive nitrogen balance and prevent muscle breakdown, maintaining lean body mass [65]. Finally, high GH levels are known to increase resting energy expenditure (REE). Although the primary mechanism is unknown, GH increases resting cardiac output and blood flow in the skeletal muscle and kidneys that contributes to the elevation in REE [65]. A study by Mulligan et al. demonstrated the effects of chronic recombinant human growth hormone (rhGH) on the energy balance of patients suffering from the wasting of lean body mass due to human immunodeficiency virus (HIV). After 3 months of rGH treatment, patients have increased REE, increased lipid oxidation, and decreased protein oxidation [71]. However, the anabolic effects of GH are sufficient to increase body weight and lean mass despite increased REE [71]. While starvation increases GH secretion, obesity decreases its secretion [72]. In obese humans, the GH/IGF-1 axis is altered, but when corrected with GH treatment, REE and body fat loss is increased [72].

Growth Hormone's Implications on Cancer

There is significant evidence to suggest that the GH/IGF-1 axis is capable of initiating and promoting the growth of selected types of cancer. The GH/IGF-1 axis is thought to affect cancer development in several ways: by decreasing the rate of apoptosis of cells sustaining DNA damage, increasing the rate of proliferation of many cell types, and by increasing the rates of angiogenesis and metastasis [73]. In fact, mutations in Ras and Akt, signaling proteins downstream of the IGF-1 receptor, are two of the most frequently detected mutations in human cancers [74]. Epidemiological studies suggest a link between levels of circulating GH and/or IGF-1 and the development of many types of cancers, including colon, breast, prostate, and lung; however, causality has not been determined, as many tumors possess the ability to secrete IGF-1 [73, 75]. For this chapter, we will focus on the known effects of the GH/IGF-1 axis on breast and prostate cancer, as well as the incidence of cancer in acromegalic patients.

Although GH is thought to be necessary for normal mammary development in humans, there is clinical evidence suggesting that circulating levels of GH/IGF-1 are associated with increased risk of breast cancer [54, 76, 77]. Studies in dogs have shown that GH is produced locally in the mammary glands, where its secretion is stimulated by progestins rather than the usual hormones, GHRH and ghrelin [77]. Normally, mammary gland GH plays a role in the proliferation and differentiation of the mammary epithelium, but it is also highly expressed in mammary neoplasia and carcinoma cells, as well as in neoplastic human breast tissue, suggesting a role in the development of cancer [77]. GHR expression has been detected in several breast cancer cell lines and in human tumor biopsies [78]. Furthermore, GHR expression has been shown to be two times higher in breast carcinoma tissue than in adjacent mammary tissue, suggesting that cancerous breast tissue has a heightened sensitivity to GH [79].

As with breast development, GH is one of the many hormones involved in the normal development of the prostate, but is also thought to play a role in tumorogenesis [80]. However, studies regarding the presence of GH expression in human prostate cancer tissues and cell cultures are somewhat contradictory. A study by Untergasser et al. found no GH expression in the tissues and cell cultures examined, but did, however, detect expression of active full-length GHR as well as the exon 3 deletion variant [81]. Both of these receptor types were shown to respond to low levels of GH, increasing cell proliferation. In contrast, a study by Chopin et al. tested expression of GH and GHR in four prostate cancer cell lines, revealing that prostate cancer cell lines expressed 22 kDa and 20 kDa GH isoforms, while normal prostate cancer cells expressed the 22 kDa isoform only [82]. This study also detected expression of the full-length GHR and the exon 3 deletion variant, suggesting an autocrine/paracrine manner of GH-induced signaling. This differential expression of isoforms in normal and prostate cancer cells suggests that the GH and GHR variants expressed may be involved in the development of prostate cancer. Furthermore, the increased expression of GHR seen in prostate adenocarcinoma tissues compared to those in benign prostate hyperplasia further suggests a role of GH in tumor development and progression [83]. In this tissue, levels of the nonfunctional truncated GHR, GHR_{tr}, were also reduced, preventing the formation of nonfunctional heterodimers, and thus further increasing GH signaling [83]. Levels of circulating IGF-1, GH's effector molecule, are also thought to be associated with prostate cancer risk [84].

Patients with acromegaly, a disease characterized by chronically elevated GH and IGF-1, are at an elevated risk of developing several types of cancers [85]. The most common type of cancer in acromegalic patients is thyroid cancer, with one study suggesting a 60-fold increase in incidence compared to the general population [85, 86]. Acromegalic patients are also considered to be at an increased risk of colorectal cancer compared to the general population, but the extent of this risk is debated [73].

Whereas high IGF-1 levels are considered to be a risk factor for prostate cancer in the general population, acromegalic patients do not exhibit an elevated risk of prostate cancer, possibly due to their elevated IGF binding-protein 3 (IGFBP3) levels, which is thought to prevent prostate cancer development [87]. Acromegalic women do not show increased rates of developing breast cancer, but do have higher mortality rates with breast cancer; this trend is thought to be caused by GH inducing resistance to chemotherapeutic drugs [75, 77]. The risk of acromegalic patients developing other forms of cancer has been studied, but results are not yet conclusive due to the rarity of the condition [73]. In contrast, a study involving patients with GH deficiency or Laron syndrome showed no cases of malignant tumors, as compared with their first- and second-degree relatives [75]. Thus, although the increased cancer risk due to high GH levels is yet to be fully understood, it is clear that lowered or absent GH signaling is associated with an extreme reduction in cancer risk.

Mouse Lines with Altered GH Action

Although many animal models have been used to study GH, mice are particularly useful because of their genetic and physiological similarities to humans, as well as the ease with which one can manipulate their genome. To that end, mouse strains (aka lines) have been genetically engineered to have extremes in GH action. That is, there are several mouse lines with excess GH-induced intracellular signaling due to introduction of a GH transgene (rat GH, bovine GH, ovine GH, or human GH mice), a line with a reduction in GH action due to the expression of a GH antagonist transgene (GHA mice), and mice with no GH action due to gene disruption of the GH receptor (GHR−/− mice) (See Fig. 4.3; Table 4.1 for a comparison) (as recently reviewed [88]). Bovine GH transgenic (bGH), GH antagonist transgenic (GHA), and GHR gene-disrupted (GHR−/−) mice provide a unique opportunity to evaluate the in vivo effects of excess, deficiency, or absence of GH signaling, respectively. Also available are spontaneously mutant mice that lack proper pituitary development (Ames and Snell dwarf strains), resulting in defects not only in GH production but also in thyroid-stimulating hormone (TSH) and prolactin (PRL) [89, 90]. In addition, there are mouse lines with alterations in molecules upstream or downstream of GH signaling. For example, Lit/Lit and GHRH−/− mice have defects in GHRH receptor or GHRH genes, resulting in a relative GH deficiency [91]. Further, many genetically engineered mice with alterations in IGF-1, its receptor or its associated signaling molecules have been studied. Because these mutants are less specific to GH action, they will not be discussed further in this review. However, the commonalities with the aforementioned mice are invaluable to discern the specific contribution of GH to a specific phenotype.

Mice with Excess GH Action

Multiple independent laboratories have generated GH transgenic mice using varying species of the GH transgene. Importantly, while GH from some species (such

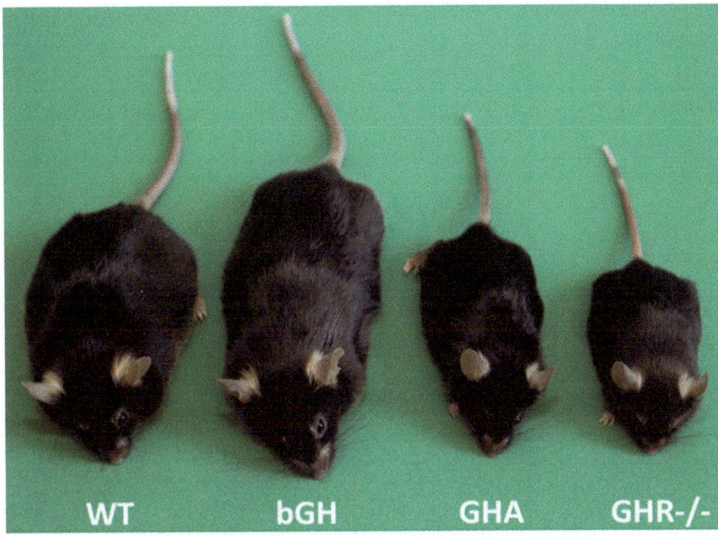

Fig. 4.3 Photo comparing the size of mice with altered GH action in the same genetic background (C57BL/6 J). Shown from left to right are wild-type (*WT*), a giant bGH transgenic (*bGH*), a dwarf GHA transgenic (*GHA*), and a dwarf GHR gene disrupted (GHR−/−) mouse. These mice represent normal, elevated, decreased, and absent levels of GH action, respectively.

as human) can bind the PRL receptor as well as the GH receptor, bovine GH binds exclusively to the GH receptor. Therefore, studies using bGH transgenic mice will be emphasized in this chapter. bGH mice have chronically elevated levels of GH and IGF-1 [92], comparable to the clinical condition of acromegaly. As might be expected, the excess GH signaling results in accelerated growth starting at 3 weeks of age and in greater adult body weight as compared to control mice [93, 96]. Thus, these mice are notably "giant" as can be seen in Fig. 4.3 Other key attributes of these mice include impaired glucose metabolism and drastically reduced lifespan (~50 % of littermate controls) [97, 98]. Interestingly, the hyperinsulinemia and impaired glucose tolerance that has been reported in young adult life, presumably due to the well-documented insulin-antagonizing effect of GH, appears to be somewhat attenuated in later life [99, 100]. This may be an adaptation to the chronic high levels of GH and/or a consequence of disease progression and the end organ damage reported in these mice [101, 106]. The cause of the premature death is likely multifactorial although increased neoplastic lesions are at least, in part, responsible as will be discussed in the later sections.

Mice with Absence or Reduction in GH action

In the mid-1990s, our laboratory generated GHR gene deleted mice (GHR−/−) [107], which have no GH action. Because of their complete lack of GH signaling from birth, these mice are a valuable tool to study the clinical condition of Laron Syndrome (LS), a disease resulting from mutations in the GHR gene or downstream

Table 4.1 Phenotypic comparison of genetically modified mice with altered activity of GH. (Adapted with permission [88])

Mouse Model	GH transgenic	GHA transgenic	GHR−/−
Level of GH signaling	↑↑	↓	none
Growth/body weight	↑↑	↓	↓↓
Body composition			
lean mass	↑↑	↓	↓↓
fat mass	↑ (younger age) ↓↓ (older ages)	↑ (younger ages) ↑↑ (older ages)	↑ (all ages)
depot differences	↓↓All depots	↑↑SubQ ↑ all depots older ages	↑↑(SubQ
Plasma levels			
GH	↑↑	↑	↑
IGF-1	↑↑	↓	↓↓
glucose	↑/↔	↓ (younger ages) ↔ (older ages)	↓
insulin	↑	↔ (younger ages) ↑ (older ages)	↓↓
leptin	↓	↑↑	↑
adiponectin	↓	↑	↑↑
resistin	↓	↑	↑
ghrelin	↔	↔	↔
Reproduction time to reach sexual maturity	↓↓	n.d.	↑
fertility	↓	n.d.	↓
Energy balance			
energy intake (total)	↑	↑	↓
energy expenditure	↑/↔	n.d.	↑
body temperature	↑	n.d.	↓
respiratory quotient	↑	n.d.	↓
Morbidity			
tumor incidence	↑	n.d.	↓↓
cardiac/vascular deficits	↑	n.d.	↔
kidney impairments	↑	↓	↓
Lifespan	↓	↔	↑

n.d. = not determined

intracellular signaling intermediates. In fact, a recent book was dedicated to comparing LS with GHR−/− mice [108] and researchers world-wide have used the mice for a variety of studies (for a comprehensive review, see [109]). GHR−/− mice are dwarf (~ 50 % the size of littermate controls) [110]. As expected, they also have very low levels of IGF-1 (~20 % of the levels of control mice) [110] and concomitant increases in levels of GH [107, 111]. These mice have a number of striking features, including extreme insulin sensitivity throughout life [111, 120]. However, one of the most unique features of GHR−/− mice is their increased longevity with mean lifespans being consistently 21–40% longer than control mice, depending on the laboratory conducting the study, genetic background, and the sex of the mice [121]. In fact, one GHR−/− mouse still holds the record for the longest-lived laboratory mouse (http://www.methuselahfoundation.org/), living 1 week shy of 5 years.

In 1991, another useful mouse line for studying the physiological effects of reduced GH action was generated in our laboratory, the GHA mice [122]. These mice express a bovine GHR antagonist (GHA) transgene in which the glycine at position 119 in the bovine GH (bGH) gene is replaced by lysine (G119K) [123, 125]. Thus, they exhibit a reduction in GH action but not GH resistance as seen in GHR−/− mice. Expression of the mutated GHA transgene results in a classic receptor antagonist [126]. That is, the glycine for lysine amino acid substitution in the third helix of GH allows GHA via Site 1 to bind properly to one of the two monomers in the GHR dimer on the surface of target tissues, but prevents GHA from binding properly to the second GHR found in the homodimer [123, 125]. Consequently, GHA fails to induce intracellular GHR signaling [122, 123, 125], and thus acts as a competitive inhibitor of endogenous mouse GH. GHA mice provided a means to understand the physiological impact of GH deficiency, but more importantly, laid the foundation for the development of a pharmaceutical agent, Somavert (pegvisomant for injection) for the treatment of acromegaly [127]. Although this discovery is not the focus of this review, it has been the focus of many other reviews [88, 127], and it nicely illustrates that the findings in GH-modified mice have been translated into clinical therapies.

The overall phenotype of GHA mice is generally intermediate to that of control and GHR−/− mice. That is, GHA mice have an overall dwarf phenotype [122, 123, 125], but not as severe as is seen with the GHR−/− mice [111, 128]. Male GHA mice have a ~20–42% reduction in serum IGF-1 [95, 111, 129, 131], again, less dramatic than in GHR−/− mice. Glucose and insulin levels appear to be somewhat dependent on age. That is, young male GHA mice tend to have lower levels of glucose and insulin [111, 131] but higher levels of insulin with advancing age [111, 128]. Unlike GHR−/− mice and other models with a reduction in the GH/IGF-1 axis, GHA mice are not long-lived although there is a tendency, albeit not significant, for GHA mice to live longer than littermate controls [111]. Thus, GHA mice provide a unique opportunity to determine key traits necessary for lifespan extension and may be considered a more clinically relevant model than the long-lived GHR−/− mouse since repression of GH action is attainable therapeutically with the use of a pegvisomant.

Obesity and Depot-specific Alterations in GH Mouse Lines

Excess, reduction, and absence of GH signaling have profound effects on adiposity, which is age- and sex-dependent. For bGH mice, fat mass is increased at younger ages in most studies, but dramatically reduced compared to controls with advancing age [94, 96, 132, 133]. In mass, all bGH WAT depots appear similarly decreased although specific properties of the WAT depots vary and become more dramatic with

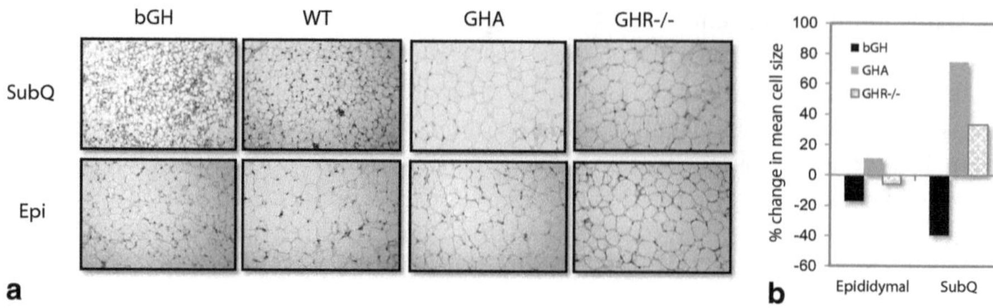

Fig. 4.4 Depot Differences in WAT from bGH, GHA, and GHR−/− mice. **a** Hematoxylin and eosin staining of inguinal SubQ and epididymal adipose tissue. Adipose tissue samples were obtained from 6-month-old GHR−/−, GHA, bGH, and control mice. Adipose tissue was fixed in 10 % buffered formalin, paraffin-embedded, and then 5 μm-sections were stained. **b** Shown is mean adipocyte size relative to the size of WT adipocytes within a given depot. Adapted with permission [91]

advancing age as we will discuss in more detail below [96]. In contrast, GHR−/− mice are relatively obese throughout life [134, 135]. An interesting feature of the increased adiposity in the GHR−/− mice is the disproportionate accumulation of WAT in the SubQ depots [112, 115, 134, 136]; thus, GH resistance influences WAT in a depot-dependent manner. GHA mice are also obese with specific enlargement of the SubQ fat pad at least in early life [128, 131, 134]. Interestingly, by approximately 11 months of age, the body weight of male GHA mice gradually catches up with controls despite deficits in body length [111, 128]. This eventual weight gain is caused by dramatic increases in adiposity without major gains in lean mass [128]. Whereas SubQ WAT depots are also disproportionately enlarged in GHA mice at all ages, the marked increase in obesity with advancing age is a result of all WAT depots being relatively enlarged. Notably, female bGH, GHR−/−, and GHA mice show a similar trend as their male counterparts; however, the difference is less significant as seen in males, and the age at which significant differences are seen is somewhat delayed [96, 128, 135, 137]. Depot differences in AT are apparent in these mice. Overall, the most striking differences are seen with SubQ fat pads whereas minimal changes are visible in the intra-abdominal fat pads, such as epididymal (Fig. 4.4). For example, major fluctuations in adipocyte sizes are obvious in the SubQ fat pads among bGH, GHA, and GHR−/− mice. Because these mice have preferential enlargement of SubQ WAT and, as previously mentioned, SubQ WAT is thought to be "healthier" than other adipose depots, it is tempting to speculate that the SubQ enlargement is in part responsible for the favorable metabolic phenotype of the dwarf mice.

Multiple laboratories have explored various aspects of AT in these mice with recent attention given also to the significance of the depot differences in these mouse lines. For example, intrinsic depot differences have been shown for differentiation and proliferation of isolated adipocytes from female mice. That is, Flint et al. [137] demonstrate that isolated preadipocytes from GHR−/− SubQ depots are able to proliferate, differentiate, and respond to hormones in an identical fashion as the preadipocytes isolated from control animals. In contrast, cells isolated from the perigonadal fat pads in GHR−/− are not able to proliferate and differentiate in vitro. The fundamental requirement for GH action in one depot and the independence of GH action in the SubQ depot for proliferation/differentiation demonstrate an inherent difference in AT depots of GHR−/− mice. Further differences in adipokine secretions and expression of genes related to lipid metabolism among various depots (most notably, epididymal and SubQ depots) have been reported [138, 140, 141]. Cellular senescence, although apparently depot-dependent, is altered in the AT of these mice. In a study that directly compared senescence in bGH and GHR−/− mice (Stout reference), bGH and GHR−/− mice were reported to have greater and decreased cellular senescence, respectively, in most depots. This finding is significant in that cellular senescence triggers a stress response accompanied by a senescence-associated secretory phenotype (SASP) that likely influences aging and preempts a variety of pathologic processes [142].

Although not yet published, our laboratory and others have identified many other unique features of AT in these mice that influence the health of the animals. For example, we have evidence of significant alterations in ECM deposition with bGH mice having marked increases, specifically in the SubQ depot [143]. We also have preliminary data suggesting genotype and depot differences in gene or protein expression related to macrophage and T cell infiltration, angiogenic capacity, beige adipocyte composition, and glucose/lipid metabolism [179]. In fact, most of the unpublished data generated to explore the AT transcriptome in these mice reveal greater differences among depots, especially between SubQ and perigonadal, than among genotypes themselves [179]. Clearly, the use of these lines may help unravel the underlying mechanism of how depot differences are established and maintained and provide insight into the novel means to modulate AT mass and obesity.

Adipokine Levels in GH Mouse Lines

GH has clear and dramatic effects on adipokine levels in GHA-altered mouse lines, as might be expected based on their altered fat mass. However, adipokines that are normally oppositely regulated, such as leptin and adiponectin, follow a unique trend in GH-modified mice. For example, bGH mice have reduced leptin, as might be expected due to their reduced adiposity, but they also have reduced adiponectin levels [134, 138, 144]. At the opposite end of the spectrum, leptin levels are elevated in GHR−/− mice [114, 120, 134] and in GHA mice [131, 134]; this is consistent with their relative obesity. However, leptin levels surge to a much higher level with advancing age in GHA mice, a trend not observed with GHR−/− mice [138]. Adiponectin levels (both total and high molecular weight forms) also are increased in GHR−/− and GHA mice, which is counterintuitive since adiponectin levels are usually negatively correlated with increased adiposity [120, 134, 138, 145]. However, since adiponectin is also positively associated with improved insulin sensitivity, this adipokine appears to be more highly correlated with insulin sensitivity at least in GHR−/− mice [131, 134]. Other adipokines have also been assessed, but not in all mouse strains, and results are less consistent. For example, resistin has been reported to be either increased [144] or decreased [146] in bGH mice whereas it has been shown to be marginally elevated [146] or decreased in GHR−/− mice [139]. Some of the differences reported relate to whether adipokines were measured in circulation or in the tissue itself and the age of mice studied. In general, inflammatory cytokines, such as TNF-α and IL-6, are reported to be increased in AT of bGH WAT [144] and decreased in WAT of GHR−/− mice [139].

Impact of High Fat Feeding and Calorie Restriction on GH Mouse Lines

Providing a high fat (HF) diet in mice can cause obesity and associated conditions, such as increased rates of cancer, type 2 diabetes, and liver steatosis [147, 151], whereas calorie restriction (CR) is an effective intervention to delay aging, extend life span, and improve insulin action in peripheral tissues in many species [150]. All of the aforementioned mouse lines have been challenged with HF feeding and some with CR although the parameters monitored and the dietary manipulations vary. Two separate groups have challenged male bGH mice with HF diets, one providing the diet from 10 weeks of age till 22 weeks of age [112] and the other study with older mice from 5 to 6 months of age followed by 4–8 weeks of HF feeding [153]. The HF diet is comparable between the studies, with 40 % of kcals coming from fat, whereas the LF diet had only 8–9 % of kcals from fat. In both studies, bGH mice are resistant to diet-induced obesity despite significant weight gain in lean mass due to the extra calorie intake and have greater impairment in glucose homeostasis than wild-type mice-fed HF diets [112]. Although the studies agree that bGH mice increase energy intake to a greater extent than controls when HF fed, one study showed no difference in energy expenditure or respiratory exchange ratio (RER) in bGH mice on either diet [112], whereas the other revealed a higher energy expenditure and RER (indicative of greater glucose oxidation) on the LF diet, a difference that was further exaggerated by HF feeding [153]. This difference may be due to the age differences between the studies. Only one study has evaluated CR in bGH mice. Although calorie restricted bGH mice have a 26 % reduction in body weight compared to *ad libitum* fed bGH mice, few details were provided regarding the relative proportion of fat versus lean tissue loss. Also, CR did not improve glucose or insulin levels in bGH mice [144].

Two separate groups have reported the effect of HF feeding on GHR−/− mice [112, 154]. There were several differences in experimental design (background strain, control diet, length of dietary manipulation), but the results are surprisingly similar. Overall, both studies report that GHR−/− mice appear more susceptible to diet-induced obesity with greater percentage of body weight gains than control mice and most fat pads increased with calorically dense diets. Importantly, despite a greater susceptibility to obesity with HF feeding, GHR−/− mice remain relatively protected from the hyperglycemia that often accompanies increased adiposity [112]. Several studies have also calorie restricted GHR−/− mice [119, 155]. Interestingly, although CR does marginally reduce body weight in GHR−/− mice, it does not further increase the lifespan or promote additional improvements in signaling as seen in control mice or even in other dwarf lines such as Ames dwarf mice [154]. This suggests that the GH/IGF-1 axis and CR might share similar or overlapping mechanisms for lifespan extension.

One study has compared GHA mice on HF and LF diets; this is the only study that also included female mice [179]. On an HF diet, GHA gain weight more dramatically than WT, with males gaining more than females. Most of this weight gain is due to an increase in fat mass with WT mice gaining primarily in the perigonadal depots, whereas GHA mice gain in both the SubQ and perigonadal regions. Like GHR−/− mice, GHA mice are also somewhat protected from detrimental glucose metabolism changes on an HF diet. Sex differences are observed in many measures with males reacting more dramatically to both GH-induced signaling reduction and diet stimulation. GHA mice have never been challenged with CR.

GH Mouse Lines, Body Temperature and Energy Balance

Alterations in the quantity of lean and fat mass are usually a result of fluctuations in energy balance (i.e., energy intake and energy expenditure). Several groups have directly measured these parameters or other factors that influence them such as body temperature and BAT content (summarized in Table 4.1). Regarding energy intake, almost all studies show that GHR−/− mice consume less [111, 112, 114, 134, 157], GHA mice consume levels comparable to controls, and bGH mice consume more total energy [112, 153], which would be expected based on their relative size. However, several studies, but not all, show that GHR−/− and GHA mice consume signficantly more energy when values are normalized to body weight [111, 112, 134, 157] whereas bGH mice consume energy levels proportional to body weight [112, 153]. One variable related to energy intake that has been assessed in these mice is ghrelin. Ghrelin is an endogenous ligand for the GH secretagogue receptor [158] and has been shown repeatedly to increase food intake and stimulate appetite [159]. Ghrelin mRNA expression and serum concentrations are not significantly different in bGH, GHR−/−, and GHA mice [114, 160] although one of these studies does show that GHR−/− mice have a blunted feeding response to ghrelin when centrally injected.

In terms of the other half of the energy balance equation, several groups have assessed energy expenditure in at least bGH and GHR−/− mice. As for bGH mice, higher energy expenditure, higher respiratory quotient (RQ) indicative of greater glucose utilization, and higher core temperature have been reported [112, 153]. However, when the data are normalized to lean mass instead of total body weight, the difference in energy expenditure and RQ is nonsignificant, suggesting that the measures are proportional to their increased lean mass. Five studies have been published measuring energy expenditure of GHR−/− mice, three of which used male adult mice [112, 139, 161] and two used female GHR−/− mice [157, 162]. Surprisingly, GHR−/− mice were reported to have increased oxygen consumption. Most studies also report that the GHR−/− mice have lower RQ values [157, 161], indicative of increased reliance on fat as a metabolic fuel. Further, GHR−/− mice have been suggested to be more energy efficient [157], to have greater metabolic complexity, and to have lower energy costs of locomotor activity [162]. Although they experience an increase in energy expenditure, GHR−/− mice also have reductions in core body temperature [117]. Of note, dwarf mice are much smaller than normal mice and, consequently, have higher body surface to body mass ratio; thus, the loss of heat by radiation can be assumed to be greater (in spite of increased or normal insulation by SubQ fat), which may account for their increased energy expenditure for thermogenesis.

The differences in energy expenditure in these mice may also relate to differences in the content of brown AT. That is, GHR−/− and GHA mice have been reported to have enlarged interscapular BAT depots [114, 136]. The enlargement of BAT tissue is accompanied by an increase in UCP-1 expression [136]. These authors further suggest that UCP1 gene expression may be negatively regulated by GH, as GH transgenic mice with excess GH have lowered UCP1 expression [136].

GH Mouse Lines and Cancer

There is mounting evidence from animal studies that implicate GH in oncogenesis, as recently reviewed by Chhabra, Waters and Brooks [162]. Indeed, studies in the aforementioned mice and in clinical conditions that are comparable to the mouse strains, such as LS and acromegaly as has previously been described, would suggest a strong positive correlation between the activity of the GH/IGF-1 axis and cancer.

Mouse lines with excess GH action appear more susceptible to at least some forms of cancer. bGH mice of both sexes exhibit hepatomegaly, hepatocyte hypertrophy with enlarged nuclear size, and exacerbated hepatocellular proliferation compared with normal controls [164]. Enhanced expression of proto-oncogenes (c-myc, c-jun, and c-fos) as well as dysregulation of several other relevant oncogenic pathways were also reported in the livers of bGH mice [164]. However, whether the hepatic abnormalities are due to a direct effect of GH or secondary to other physiologic or endocrine alterations in these mice is debated. Regardless, chronic expression of IGF-1 results in different pathologies in liver, suggesting that at least some of the liver alterations are a direct effect of GH [165]. The incidence of other cancers is also elevated in GH transgenic mice, as human GH transgenic mice have been shown to have a greater incidence and earlier onset of mammary tumors [166]. However, it is important to note that the role of PRL versus GH signaling in this process has not been resolved.

Whereas increased cancer incidence is common with excess GH expression, GH resistance or a reduction in GH action as seen in GHR−/− mice as well as GHA mice is well documented to offer protection from cancer. Ikeno and colleagues [167] conducted a detailed end-of-life pathological analysis on GHR−/− mice and littermate controls. They report that GHR–/– mice have an overall reduced occurrence of tumors with neoplastic lesions found in more than 90 % of WT mice at death but in only 68 % of the GHR–/– mice. In addition, the tumor burden (number of different types of tumors) in the GHR–/– mice is reduced by 47 % compared to WT mice. As for the cause of death, 83 % of WT mice die from neoplastic disease whereas only 42 % (~50 % reduction) of the GHR–/– mice die from such lesions. GHR–/– mice have also been used to examine the role of the GH/IGF-1 axis in mammary carcinogenesis. GHR–/– mice have been crossed with C3(1)/TAg mice, a mouse line that develops prostate cancer in male mice and estrogen receptor alpha-negative mammary tumors in female mice [168]. Whereas control mice develop an average of nearly 10 tumors per animal, the crossed line (TAg/GHR–/– mice) only shows ~3 tumors per animal. Even more striking is that the tumor size in control animals grows to 10 times the volume of those found in the crossed Tag/GHR–/– mice. These experiments demonstrate that the disruption of the GH/IGF-1 axis significantly retards the progression of estrogen-independent breast cancer. A similar study in males was used to examine the role of the GH/IGF-1 in prostate carcinogenesis [169]. Seven out of eight control mice developed prostate neoplasia, whereas only one out of eight TAg/GHR–/– mice developed such lesions, suggesting that disruption of GH signaling also confers resistance to prostate carcinogenesis. Like GHR–/– mice, GHA mice also show protection from breast cancer in vivo [170]. In this study, 8-week-old GHA and control mice were given a weekly gavage of a carcinogen and monitored for tumor incidence. At the end of the 39-week-study period, 68 % of GHA animals remained without tumors, whereas only 32 % of the control animals were tumor-free. Collectively, these data provide strong evidence in vivo that suppressing GH/IGF-1 axis could be an effective therapeutic target to inhibit the growth and/or progression of cancer. Understanding the unique adiposity profiles and energy balance in these animals, as will be discussed in the subsequent sections, will likely provide important clues as to the relevant traits of cancer progression.

Future Directions

New technologies for genetic manipulation of mouse lines have recently been developed, which will allow us to further elucidate the role of GH in obesity, glucose homeostasis, energy metabolism, and cancer. For example, Luque et al. describe an adult GH-deficient mouse in which they ablate the somatotrophs with a Cre system that utilizes diphtheria toxin [171], thus attempting to eliminate some of the confounding effects such as the dwarf size of the GHR–/– mice. Circulating levels of GH and IGF-1 are decreased but are still detectable in these mice. Interestingly, even partial GH deficiency, as shown in these mice, has a dramatic impact on metabolic function resulting in improved insulin sensitivity. Tissue-specific disruption of GHR has also been reported by several groups using Cre/lox technology. Mice with liver- [172, 173], muscle- [174, 175], pancreatic β-cell- [176], adipose- [177], and macrophage-[178] specific disruptions of GHR have been initially characterized and all have unique phenotypes with respect to obesity, adipokines, and glucose homeostasis. For example, fat-specific GHR gene-disrupted mice are obese with all depots enlarged, but do not see improvement in insulin sensitivity or increases in adiponectin [175]. Further, these tissue specific mice provide evidence for significant tissue "crosstalk" in mice with altered GH action. The field anxiously awaits the additional data from these mice, including a more careful assessment of adipose quality, cancer susceptibility, and lifespan.

References

1. Trayhurn P, Beattie JH. Physiological role of adipose tissue: white adipose tissue as an endocrine and secretory organ. Proc Nutr Soc. 2001;60(3):329–39.
2. Galic S, Oakhill JS, Steinberg GR. Adipose tissue as an endocrine organ. Mol Cell Endocrinol. 2010;316(2):129–39.
3. Bahceci M, Gokalp D, Bahceci S, Tuzcu A, Atmaca S, Arikan S. The correlation between adiposity and adiponectin, tumor necrosis factor alpha, interleukin-6 and high sensitivity C-reactive protein levels. Is adipocyte size associated with inflammation in adults? J Endocrinol Invest. 2007;30(3):210–14.
4. Lass A, Zimmermann R, Oberer M, Zechner R. Lipolysis—a highly regulated multi-enzyme complex mediates the catabolism of cellular fat stores. Prog Lipid Res. 2011;50(1):14–27.
5. Fain JN, Madan AK, Hiler ML, Cheema P, Bahouth SW. Comparison of the release of adipokines by adipose tissue, adipose tissue matrix, and adipocytes from visceral and subcutaneous abdominal adipose tissues of obese humans. Endocrinology. 2004;145(5):2273–82.
6. Kershaw EE, Flier JS. Adipose tissue as an endocrine organ. J Clin Endocrinol Metab. 2004; 89(6):2548–56.
7. Wajchenberg BL, Giannella-Neto D, da Silva ME, Santos RF. Depot-specific hormonal characteristics of subcutaneous and visceral adipose tissue and their relation to the metabolic syndrome. Horm Metab Res. 2002;34(11–12):616–21.
8. Smorlesi A, Frontini A, Giordano A, Cinti S. The adipose organ: white-brown adipocyte plasticity and metabolic inflammation. Obes Rev. 2012;13(Suppl 2):83–96.
9. Vanderburgh PM. Fat distribution: its physiological significance, health implications, and its adaptation to exercise training. Mil Med. 1992;157(4):189–92.
10. Cinti S. The adipose organ. Prostaglandins Leukot Essent Fatty Acids. 2005;73(1):9–15.
11. Giordano A, Frontini A, Castellucci M, Cinti S. Presence and distribution of cholinergic nerves in rat mediastinal brown adipose tissue. J Histochem Cytochem. 2004;52(7):923–30.
12. Wajchenberg BL. Subcutaneous and visceral adipose tissue: their relation to the metabolic syndrome. Endocr Rev. 2000;21(6):697–738.
13. Tran TT, Yamamoto Y, Gesta S, Kahn CR. Beneficial effects of subcutaneous fat transplantation on metabolism. Cell Metab. 2008;7(5):410–20.
14. Trayhurn P, Wood IS. Adipokines: inflammation and the pleiotropic role of white adipose tissue. Br J Nutr. 2004;92(3):347–55.
15. Klok MD, Jakobsdottir S, Drent ML. The role of leptin and ghrelin in the regulation of food intake and body weight in humans: a review. Obes Rev. 2007;8(1):21–34.
16. Lindstrom P. The physiology of obese-hyperglycemic mice [ob/ob mice]. ScientificWorldJournal. 2007;7:666–85.

17. Drel VR, Mashtalir N, Ilnytska O, Shin J, Li F, Lyzogubov VV, Obrosova IG. The leptin-deficient (ob/ob) mouse: a new animal model of peripheral neuropathy of type 2 diabetes and obesity. Diabetes. 2006;55(12):3335–43.
18. Kobayashi K, Forte TM, Taniguchi S, Ishida BY, Oka K, Chan L. The db/db mouse, a model for diabetic dyslipidemia: molecular characterization and effects of Western diet feeding. Metabolism. 2000;49(1):22–31.
19. Houseknecht KL, Baile CA, Matteri RL, Spurlock ME. The biology of leptin: a review. J Anim Sc.i 1998;76(5):1405–20.
20. Yamauchi T, Kamon J, Ito Y, Tsuchida A, Yokomizo T, Kita S, Sugiyama T, Miyagishi M, Hara K, Tsunoda M, et al. Cloning of adiponectin receptors that mediate antidiabetic metabolic effects. Nature. 2003;423(6941):762–9.
21. Hara K, Horikoshi M, Yamauchi T, Yago H, Miyazaki O, Ebinuma H, Imai Y, Nagai R, Kadowaki T. Measurement of the high-molecular weight form of adiponectin in plasma is useful for the prediction of insulin resistance and metabolic syndrome. Diabetes Care. 2006;29(6):1357–62.
22. Wang Y, Lam KS, Yau MH, Xu A. Post-translational modifications of adiponectin: mechanisms and functional implications. Biochem J. 2008;409(3):623–33.
23. Kubota N, Terauchi Y, Yamauchi T, Kubota T, Moroi M, Matsui J, Eto K, Yamashita T, Kamon J, Satoh H, et al. Disruption of adiponectin causes insulin resistance and neointimal formation. J Biol Chem. 2002;277(29):25863–6.
24. Maeda N, Shimomura I, Kishida K, Nishizawa H, Matsuda M, Nagaretani H, Furuyama N, Kondo H, Takahashi M, Arita Y, et al. Diet-induced insulin resistance in mice lacking adiponectin/ACRP30. Nat Med. 2002;8(7):731–7.
25. Combs TP, Pajvani UB, Berg AH, Lin Y, Jelicks LA, Laplante M, Nawrocki AR, Rajala MW, Parlow AF, Cheeseboro L, et al. A transgenic mouse with a deletion in the collagenous domain of adiponectin displays elevated circulating adiponectin and improved insulin sensitivity. Endocrinology. 2004;145(1):367–83.
26. Martyn JA, Kaneki M, Yasuhara S. Obesity-induced insulin resistance and hyperglycemia: etiologic factors and molecular mechanisms. Anesthesiology. 2008;109(1):137–48.
27. Strissel KJ, Stancheva Z, Miyoshi H, Perfield JW, 2nd, DeFuria J, Jick Z, Greenberg AS, Obin MS. Adipocyte death, adipose tissue remodeling, and obesity complications. Diabetes. 2007;56(12):2910–8.
28. Weisberg SP, McCann D, Desai M, Rosenbaum M, Leibel RL, Ferrante AW, Jr. Obesity is associated with macrophage accumulation in adipose tissue. J Clin Invest. 2003;112(12):1796–808.
29. Strawford A, Antelo F, Christiansen M, Hellerstein MK. Adipose tissue triglyceride turnover, de novo lipogenesis, and cell proliferation in humans measured with 2 H2O. Am J Physiol Endocrinol Metab. 2004;286(4):E577–88.
30. Khan T, Muise ES, Iyengar P, Wang ZV, Chandalia M, Abate N, Zhang BB, Bonaldo P, Chua S, Scherer PE. Metabolic dysregulation and adipose tissue fibrosis: role of collagen VI. Mol Cell Biol. 2009;29(6):1575–91.
31. Morris DL, Singer K, Lumeng CN. Adipose tissue macrophages: phenotypic plasticity and diversity in lean and obese states. Curr Opin Clin Nutr Metab Care. 2011;14(4):341–6.
32. Cinti S, Mitchell G, Barbatelli G, Murano I, Ceresi E, Faloia E, Wang S, Fortier M, Greenberg AS, Obin MS. Adipocyte death defines macrophage localization and function in adipose tissue of obese mice and humans. J Lipid Res. 2005;46(11):2347–55.
33. Turer AT, Khera A, Ayers CR, Turer CB, Grundy SM, Vega GL, Scherer PE. Adipose tissue mass and location affect circulating adiponectin levels. Diabetologia. 2011;54(10):2515–24.
34. Lim S, Quon MJ, Koh KK. Modulation of adiponectin as a potential therapeutic strategy. Atherosclerosis. 2014;233(2):721–8.
35. Levin BE, Dunn-Meynell AA. Reduced central leptin sensitivity in rats with diet-induced obesity. Am J Physiol Regul Integr Comp Physiol. 2002;283(4):R941–8.

36. Heymsfield SB, Greenberg AS, Fujioka K, Dixon RM, Kushner R, Hunt T, Lubina JA, Patane J, Self B, Hunt P, et al. Recombinant leptin for weight loss in obese and lean adults: a randomized, controlled, dose-escalation trial. JAMA. 1999;282(16):1568–75.
37. Renehan AG, Tyson M, Egger M, Heller RF, Zwahlen M. Body-mass index and incidence of cancer: a systematic review and meta-analysis of prospective observational studies. Lancet. 2008;371(9612):569–78.
38. Calle EE, Rodriguez C, Walker-Thurmond K, Thun MJ. Overweight, obesity, and mortality from cancer in a prospectively studied cohort of U.S. adults. N Engl J Med. 2003;348(17):1625–38.
39. Maquoi E, Munaut C, Colige A, Collen D, Lijnen HR. Modulation of adipose tissue expression of murine matrix metalloproteinases and their tissue inhibitors with obesity. Diabetes. 2002;51(4):1093–01.
40. Motrescu ER, Rio MC. Cancer cells, adipocytes and matrix metalloproteinase 11: a vicious tumor progression cycle. Biol Chem. 2008;389(8):1037–41.
41. Howard JM, Pidgeon GP, Reynolds JV. Leptin and gastro-intestinal malignancies. Obes Rev. 2010;11(12):863–74.
42. Hardwick JC, Van Den Brink GR, Offerhaus GJ, Van Deventer SJ, Peppelenbosch MP. Leptin is a growth factor for colonic epithelial cells. Gastroenterology. 2001;121(1):79–90.
43. Dieudonne MN, Machinal-Quelin F, Serazin-Leroy V, Leneveu MC, Pecquery R, Giudicelli Y. Leptin mediates a proliferative response in human MCF7 breast cancer cells. Biochem Biophys Res Commun. 2002;293(1):622–8.
44. Onuma M, Bub JD, Rummel TL, Iwamoto Y. Prostate cancer cell-adipocyte interaction: leptin mediates androgen-independent prostate cancer cell proliferation through c-Jun NH2-terminal kinase. J Biol Chem. 2003;278(43):42660–7.
45. Bub JD, Miyazaki T, Iwamoto Y. Adiponectin as a growth inhibitor in prostate cancer cells. Biochem Biophys Res Commun. 2006;340(4):1158–66.
46. Kim AY, Lee YS, Kim KH, Lee JH, Lee HK, Jang SH, Kim SE, Lee GY, Lee JW, Jung SA, et al. Adiponectin represses colon cancer cell proliferation via AdipoR1- and -R2-mediated AMPK activation. Mol Endocrinol. 2010;24(7):1441–52.
47. Cohen CA, Shea AA, Heffron CL, Schmelz EM, Roberts PC. Intra-abdominal fat depots represent distinct immunomodulatory microenvironments: a murine model. PLoS One. 2013;8(6):e66477.
48. Tateya S, Kim F, Tamori Y. Recent advances in obesity-induced inflammation and insulin resistance. Front Endocrinol (Lausanne). 2013;4:93.
49. Patel PS, Buras ED, Balasubramanyam A. The role of the immune system in obesity and insulin resistance. J Obes. 2013;2013:616193.
50. Wu D, Molofsky AB, Liang HE, Ricardo-Gonzalez RR, Jouihan HA, Bando JK, Chawla A, Locksley RM: Eosinophils sustain adipose alternatively activated macrophages associated with glucose homeostasis. Science. 2011;332(6026):243–247.
51. Rothenberg ME, Hogan SP. The eosinophil. Annu Rev Immunol. 2006;24:147–74.
52. Sun S, Ji Y, Kersten S, Qi L. Mechanisms of inflammatory responses in obese adipose tissue. Annu Rev Nutr. 2012;32:261–86.
53. Altintas MM, Azad A, Nayer B, Contreras G, Zaias J, Faul C, Reiser J, Nayer A. Mast cells, macrophages, and crown-like structures distinguish subcutaneous from visceral fat in mice. J Lipid Res. 2011;52(3):480–88.
54. Pollak M. Insulin and insulin-like growth factor signalling in neoplasia. Nat Rev Cancer. 2008;8(12):915–28.
55. Renehan AG, Frystyk J, Flyvbjerg A. Obesity and cancer risk: the role of the insulin-IGF axis. Trends Endocrinol Metab. 2006;17(8):328–36.
56. Baumann G. Growth hormone heterogeneity: genes, isohormones, variants, and binding proteins. Endocr Rev. 1991;12(4):424–49.
57. Kopchick JJ. Growth hormone receptor antagonists and methods of reducing growth hormone activity in mammals. In: United States Patent. U. S. A.: Ohio University; 1999.
58. Ho KK, O'Sullivan AJ, Hoffman DM. Metabolic actions of growth hormone in man. Endocr J 1996;43 Suppl:S57–63.

59. Ohlsson C, Sjogren K, Jansson JO, Isaksson OG. The relative importance of endocrine versus autocrine/paracrine insulin-like growth factor-I in the regulation of body growth. Pediatr Nephrol. 2000;14(7):541–3.
60. Krysiak R, Gdula-Dymek A, Bednarska-Czerwinska A, Okopien B. Growth hormone therapy in children and adults. Pharmacol Rep. 2007;59(5):500–16.
61. Khan AS, Sane DC, Wannenburg T, Sonntag WE. Growth hormone, insulin-like growth factor-1 and the aging cardiovascular system. Cardiovasc Res. 2002;54(1):25–35.
62. Romero CJ, Pine-Twaddell E, Sima DI, Miller RS, He L, Wondisford F, Radovick S. Insulin-like growth factor 1 mediates negative feedback to somatotroph GH expression via POU1F1/CREB binding protein interactions. Mol Cell Biol. 2012;32(21):4258–69.
63. Brooks AJ, Dai W, O'Mara ML, Abankwa D, Chhabra Y, Pelekanos RA, Gardon O, Tunny KA, Blucher KM, Morton CJ, et al. Mechanism of activation of protein kinase JAK2 by the growth hormone receptor. Science. 2014;344(6185):1249783.
64. Zhu T, Ling L, Lobie PE. Identification of a JAK2-independent pathway regulating growth hormone (GH)-stimulated p44/42 mitogen-activated protein kinase activity. GH activation of Ral and phospholipase D is Src-dependent. J Biol Chem. 2002;277(47):45592–603.
65. Moller N, Jorgensen JO. Effects of growth hormone on glucose, lipid, and protein metabolism in human subjects. Endocr Rev. 2009;30(2):152–77.
66. Kraemer FB, Shen WJ. Hormone-sensitive lipase: control of intracellular tri-(di-)acylglycerol and cholesteryl ester hydrolysis. J Lipid Res. 2002;43(10):1585–94.
67. Moller N, Jorgensen JO, Abildgard N, Orskov L, Schmitz O, Christiansen JS: Effects of growth hormone on glucose metabolism. Horm Res. 1991;36(Suppl 1):32–35.
68. Rabinowitz D, Klassen GA, Zierler KL. Effect of human growth hormone on muscle and adipose tissue metabolism in the forearm of man. J Clin Invest. 1965;44:51–61.
69. Jorgensen JO, Moller L, Krag M, Billestrup N, Christiansen JS. Effects of growth hormone on glucose and fat metabolism in human subjects. Endocrinol Metab Clin North Am. 2007;36(1):75–87.
70. Jorgensen JO, Jessen N, Pedersen SB, Vestergaard E, Gormsen L, Lund SA, Billestrup N. GH receptor signaling in skeletal muscle and adipose tissue in human subjects following exposure to an intravenous GH bolus. Am J Physiol Endocrinol Metab. 2006;291(5):E899–905.
71. Mulligan K, Tai VW, Schambelan M. Effects of chronic growth hormone treatment on energy intake and resting energy metabolism in patients with human immunodeficiency virus-associated wasting–a clinical research center study. J Clin Endocrinol Metab. 1998;83(5):1542–47.
72. Karlsson C, Stenlof K, Johannsson G, Marin P, Bjorntorp P, Bengtsson BA, Carlsson B, Carlsson LM, Sjostrom L. Effects of growth hormone treatment on the leptin system and on energy expenditure in abdominally obese men. Eur J Endocrinol. 1998;138(4):408–14.
73. Jenkins PJ, Mukherjee A, Shalet SM. Does growth hormone cause cancer? Clin Endocrinol (Oxf). 2006;64(2):115–21.
74. Guevara-Aguirre J, Balasubramanian P, Guevara-Aguirre M, Wei M, Madia F, Cheng CW, Hwang D, Martin-Montalvo A, Saavedra J, Ingles S, et al. Growth hormone receptor deficiency is associated with a major reduction in pro-aging signaling, cancer, and diabetes in humans. Sci Transl Med. 2011;3(70):70ra13.
75. Zatelli MC, Minoia M, Mole D, Cason V, Tagliati F, Margutti A, Bondanelli M, Ambrosio MR, degli Uberti E. Growth hormone excess promotes breast cancer chemoresistance. The J Clin Endocrinol Metab. 2009;94(10):3931–8.
76. Rinaldi S, Peeters PH, Berrino F, Dossus L, Biessy C, Olsen A, Tjonneland A, Overvad K, Clavel-Chapelon F, Boutron-Ruault MC, et al. IGF-1, IGFBP-3 and breast cancer risk in women: The European Prospective Investigation into Cancer and Nutrition (EPIC). Endocr Relat Cancer. 2006;13(2):593–605.
77. Thijssen JH. On the possible role of mammary-derived growth hormone in human breast cancer. Maturitas. 2009;65 Suppl 1:S13–6.
78. Decouvelaere C, Peyrat JP, Bonneterre J, Djiane J, Jammes H. Presence of the two growth hormone receptor messenger RNA isoforms in human breast cancer. Cell Growth Differ. 1995;6(4):477–83.

79. Gebre-Medhin M, Kindblom LG, Wennbo H, Tornell J, Meis-Kindblom JM. Growth hormone receptor is expressed in human breast cancer. Am J Pathol. 2001;158(4):1217–1222.
80. Nakonechnaya AO, Jefferson HS, Chen X, Shewchuk BM. Differential effects of exogenous and autocrine growth hormone on LNCaP prostate cancer cell proliferation and survival. J Cell Biochem. 2013;114(6):1322–35.
81. Untergasser G, Rumpold H, Hermann M, Dirnhofer S, Jilg G, Berger P. Proliferative disorders of the aging human prostate: involvement of protein hormones and their receptors. Exp Gerontol. 1999;34(2):275–87.
82. Chopin LK, Veveris-Lowe TL, Philipps AF, Herington AC. Co-expression of GH and GHR isoforms in prostate cancer cell lines. Growth Horm IGF Res. 2002;12(2):126–36.
83. Weiss-Messer E, Merom O, Adi A, Karry R, Bidosee M, Ber R, Kaploun A, Stein A, Barkey RJ. Growth hormone (GH) receptors in prostate cancer: gene expression in human tissues and cell lines and characterization, GH signaling and androgen receptor regulation in LNCaP cells. Mol Cell Endocrinol. 2004;220(1–2):109–23.
84. Chokkalingam AP, Pollak M, Fillmore CM, Gao YT, Stanczyk FZ, Deng J, Sesterhenn IA, Mostofi FK, Fears TR, Madigan MP, et al. Insulin-like growth factors and prostate cancer: a population-based case-control study in China. Cancer Epidemiol Biomarkers Prev. 2001;10(5):421–7.
85. Dagdelen S, Cinar N, Erbas T. Increased thyroid cancer risk in acromegaly. Pituitary 2013.
86. Renehan AG, Brennan BM: Acromegaly, growth hormone and cancer risk. Best Pract Res Clin Endocrinol Metab. 2008;22(4):639–57.
87. Svensson J, Kindblom J, Shao R, Moverare-Skrtic S, Lagerquist MK, Andersson N, Sjogren K, Venken K, Vanderschueren D, Jansson JO, et al. Liver-derived IGF1 enhances the androgenic response in prostate. J Endocrinol. 2008;199(3):489–97.
88. Kopchick JJ, List EO, Kelder B, Gosney ES, Berryman DE. Evaluation of growth hormone (GH) action in mice: discovery of GH receptor antagonists and clinical indications. Mol Cell Endocrinol. 2013.
89. Li S, Crenshaw EB, 3rd, Rawson EJ, Simmons DM, Swanson LW, Rosenfeld MG. Dwarf locus mutants lacking three pituitary cell types result from mutations in the POU-domain gene pit-1. Nature. 1990;347(6293):528–33.
90. Buckwalter MS, Katz RW, Camper SA: Localization of the panhypopituitary dwarf mutation (df) on mouse chromosome 11 in an intersubspecific backcross. Genomics. 1991;10(3):515–26.
91. Berryman DE, List EO, Sackmann-Sala L, Lubbers E, Munn R, Kopchick JJ. Growth hormone and adipose tissue: beyond the adipocyte. Growth Horm IGF Res. 2011;21(3):113–23.
92. Kopchick JJ, Bellush LL, Coschigano KT. Transgenic models of growth hormone action. Annu Rev Nutr. 1999;19:437–61.
93. Palmiter RD, Brinster RL, Hammer RE, Trumbauer ME, Rosenfeld MG, Birnberg NC, Evans RM. Dramatic growth of mice that develop from eggs microinjected with metallothionein-growth hormone fusion genes. Nature. 1982;300(5893):611–5.
94. Kaps M, Moura AS, Safranski TJ, Lamberson WR. Components of growth in mice hemizygous for a MT/bGH transgene. J Anim Sci. 1999;77(5):1148–54.
95. Knapp JR, Chen WY, Turner ND, Byers FM, Kopchick JJ. Growth patterns and body composition of transgenic mice expressing mutated bovine somatotropin genes. J Anim Sci. 1994;72(11):2812–9.
96. Palmer AJ, Chung MY, List EO, Walker J, Okada S, Kopchick JJ, Berryman DE. Age-related changes in body composition of bovine growth hormone transgenic mice. Endocrinology. 2009;150(3):1353–60.
97. Bartke A. Can growth hormone (GH) accelerate aging? Evidence from GH-transgenic mice. Neuroendocrinology. 2003;78(4):210–6.
98. Wolf E, Kahnt E, Ehrlein J, Hermanns W, Brem G, Wanke R. Effects of long-term elevated serum levels of growth hormone on life expectancy of mice: lessons from transgenic animal models. Mech Ageing Dev. 1993;68(1–3):71–87.

99. Jara A, Benner CM, Sim D, Liu X, List EO, Householder LA, Berryman DE, Kopchick JJ. Elevated systolic blood pressure in male GH transgenic mice is age dependent. Endocrinology. 2014;155(3):975–86.
100. Ding J, Berryman DE, Kopchick JJ. Plasma proteomic profiles of bovine growth hormone transgenic mice as they age. Transgenic Res. 2011;20(6):1305–20.
101. Izzard AS, Emerson M, Prehar S, Neyses L, Trainer P, List EO, Kopchick JJ, Heagerty AM. The cardiovascular phenotype of a mouse model of acromegaly. Growth Horm IGF Res. 2009.
102. Bogazzi F, Russo D, Raggi F, Ultimieri F, Urbani C, Gasperi M, Bartalena L, Martino E. Transgenic mice overexpressing growth hormone (GH) have reduced or increased cardiac apoptosis through activation of multiple GH-dependent or -independent cell death pathways. Endocrinology. 2008;149(11):5758–69.
103. Bollano E, Omerovic E, Bohlooly-y M, Kujacic V, Madhu B, Tornell J, Isaksson O, Soussi B, Schulze W, Fu ML, et al. Impairment of cardiac function and bioenergetics in adult transgenic mice overexpressing the bovine growth hormone gene. Endocrinology. 2000;141(6):2229–35.
104. Doi T, Striker LJ, Kimata K, Peten EP, Yamada Y, Striker GE. Glomerulosclerosis in mice transgenic for growth hormone. Increased mesangial extracellular matrix is correlated with kidney mRNA levels. J Exp Med. 1991;173(5):1287–90.
105. Kumar PA, Brosius FC, 3rd, Menon RK. The glomerular podocyte as a target of growth hormone action: implications for the pathogenesis of diabetic nephropathy. Curr Diabetes Rev. 2011;7(1):50–5.
106. Bohlooly YM, Olsson B, Bruder CE, Linden D, Sjogren K, Bjursell M, Egecioglu E, Svensson L, Brodin P, Waterton JC, et al. Growth hormone overexpression in the central nervous system results in hyperphagia-induced obesity associated with insulin resistance and dyslipidemia. Diabetes. 2005;54(1):51–62.
107. Zhou Y, Xu BC, Maheshwari HG, He L, Reed M, Lozykowski M, Okada S, Cataldo L, Coschigamo K, Wagner TE, et al. A mammalian model for Laron syndrome produced by targeted disruption of the mouse growth hormone receptor/binding protein gene (the Laron mouse). Proc Natl Acad Sci U S A. 1997;94(24):13215–20.
108. Laron Z, Kopchick J. Laron syndrome—from man to mouse lessons from clinical and experimental experience. Berlin; New York: Springer; 2011: 1 online resource (xiv, 531 p.).
109. List EO, Sackmann-Sala L, Berryman DE, Funk K, Kelder B, Gosney ES, Okada S, Ding J, Cruz-Topete D, Kopchick JJ. Endocrine parameters and phenotypes of the growth hormone receptor gene disrupted (GHR-/-) mouse. Endocr Rev. 2011;32(3):356–86.
110. Coschigano KT, Clemmons D, Bellush LL, Kopchick JJ: Assessment of growth parameters and life span of GHR/BP gene-disrupted mice. Endocrinology. 2000;141(7):2608–13.
111. Coschigano KT, Holland AN, Riders ME, List EO, Flyvbjerg A, Kopchick JJ. Deletion, but not antagonism, of the mouse growth hormone receptor results in severely decreased body weights, insulin and IGF-1 levels and increased lifespan. Endocrinology. 2003;144(9):3799–810.
112. Berryman DE, List EO, Kohn DT, Coschigano KT, Seeley RJ, Kopchick JJ. Effect of growth hormone on susceptibility to diet-induced obesity. Endocrinology. 2006;147(6):2801–8.
113. Panici JA, Wang F, Bonkowski MS, Spong A, Bartke A, Pawlikowska L, Kwok PY, Masternak MM. Is altered expression of hepatic insulin-related genes in growth hormone receptor knockout mice due to GH resistance or a difference in biological life spans? J Gerontol A Biol Sci Med Sci. 2009;64(11):1126–33.
114. Egecioglu E, Bjursell M, Ljungberg A, Dickson SL, Kopchick JJ, Bergstrom G, Svensson L, Oscarsson J, Tornell J, Bohlooly YM. Growth hormone receptor deficiency results in blunted ghrelin feeding response, obesity, and hypolipidemia in mice. Am J Physiol Endocrinol Metab. 2006;290(2):E317–25.
115. Liu JL, Coschigano KT, Robertson K, Lipsett M, Guo Y, Kopchick JJ, Kumar U, Liu YL. Disruption of growth hormone receptor gene causes diminished pancreatic islet size and increased insulin sensitivity in mice. Am J Physiol Endocrinol Metab. 2004;287(3):E405–13.

116. Bartke A, Peluso MR, Moretz N, Wright C, Bonkowski M, Winters TA, Shanahan MF, Kopchick JJ, Banz WJ. Effects of Soy-derived diets on plasma and liver lipids, glucose tolerance, and longevity in normal, long-lived and short-lived mice. Horm Metab Res. 2004;36(8):550–8.
117. Hauck SJ, Hunter WS, Danilovich N, Kopchick JJ, Bartke A. Reduced levels of thyroid hormones, insulin, and glucose, and lower body core temperature in the growth hormone receptor/binding protein knockout mouse. Exp Biol Med (Maywood). 2001;226(6):552–8.
118. Dominici FP, Arostegui Diaz G, Bartke A, Kopchick JJ, Turyn D. Compensatory alterations of insulin signal transduction in liver of growth hormone receptor knockout mice. J Endocrinol. 2000;166(3):579–90.
119. Bonkowski MS, Dominici FP, Arum O, Rocha JS, Al Regaiey KA, Westbrook R, Spong A, Panici J, Masternak MM, Kopchick JJ, et al. Disruption of growth hormone receptor prevents calorie restriction from improving insulin action and longevity. PLoS One 2009;4(2):e4567.
120. Al-Regaiey KA, Masternak MM, Bonkowski M, Sun L, Bartke A. Long-lived growth hormone receptor knockout mice: interaction of reduced insulin-like growth factor i/insulin signaling and caloric restriction. Endocrinology. 2005;146(2):851–60.
121. Junnila RK, List EO, Berryman DE, Murrey JW, Kopchick JJ. The GH/IGF-1 axis in ageing and longevity. Nat Rev Endocrinol. 2013;9(6):366–76.
122. Chen WY, Wight DC, Wagner TE, Kopchick JJ. Expression of a mutated bovine growth hormone gene suppresses growth of transgenic mice. Proc Natl Acad Sci U S A. 1990;87(13):5061–5.
123. Chen WY, White ME, Wagner TE, Kopchick JJ. Functional antagonism between endogenous mouse growth hormone (GH) and a GH analog results in dwarf transgenic mice. Endocrinology. 1991;129(3):1402–8.
124. Chen WY, Wight DC, Chen NY, Coleman TA, Wagner TE, Kopchick JJ. Mutations in the third alpha-helix of bovine growth hormone dramatically affect its intracellular distribution in vitro and growth enhancement in transgenic mice. J Biol Chem. 1991;266(4):2252–8.
125. Chen WY, Wight DC, Mehta BV, Wagner TE, Kopchick JJ. Glycine 119 of bovine growth hormone is critical for growth-promoting activity. Mol Endocrinol. 1991;5(12):1845–52.
126. Okada S, Chen WY, Wiehl P, Kelder B, Goodman HM, Guller S, Sonenberg M, Kopchick JJ. A growth hormone (GH) analog can antagonize the ability of native GH to promote differentiation of 3T3-F442A preadipocytes and stimulate insulin-like and lipolytic activities in primary rat adipocytes. Endocrinology. 1992;130(4):2284–90.
127. Kopchick JJ. Discovery and mechanism of action of pegvisomant. Eur J Endocrinol. 2003;148 Suppl 2:S21–5.
128. Berryman DE, Lubbers ER, Magon V, List EO, Kopchick JJ. A dwarf mouse model with decreased GH/IGF-1 activity that does not experience life-span extension: potential impact of increased adiposity, leptin, and insulin with advancing age. J Gerontol A Biol Sci Med Sci. 2014;69(2):131–41.
129. Flyvbjerg A, Bennet WF, Rasch R, Kopchick JJ, Scarlett JA. Inhibitory effect of a growth hormone receptor antagonist (G120K-PEG) on renal enlargement, glomerular hypertrophy and urinary albumin excretion in experimental diabetes in mice. Diabetes. 1999;48:377–82.
130. Chen NY, Chen WY, Striker LJ, Striker GE, Kopchick JJ. Co-expression of bovine growth hormone (GH) and human GH antagonist genes in transgenic mice. Endocrinology. 1997;138(2):851–4.
131. Yakar S, Setser J, Zhao H, Stannard B, Haluzik M, Glatt V, Bouxsein ML, Kopchick JJ, LeRoith D. Inhibition of growth hormone action improves insulin sensitivity in liver IGF-1-deficient mice. J Clin Invest. 2004;113(1):96–105.
132. Pomp D, Oberbauer AM, Murray JD. Development of obesity following inactivation of a growth hormone transgene in mice. Transgenic Res. 1996;5(1):13–23.
133. Eckstein F, Lochmuller EM, Koller B, Wehr U, Weusten A, Rambeck W, Hoeflich A, Wolf E. Body composition, bone mass and microstructural analysis in GH-transgenic mice reveals that skeletal changes are specific to bone compartment and gender. Growth Horm IGF Res. 2002;12(2):116–25.

134. Berryman DE, List EO, Coschigano KT, Behar K, Kim JK, Kopchick JJ. Comparing adiposity profiles in three mouse models with altered GH signaling. Growth Horm IGF Res. 2004;14(4):309–18.
135. Berryman DE, List EO, Palmer AJ, Chung MY, Wright-Piekarski J, Lubbers E, O'Connor P, Okada S, Kopchick JJ. Two-year body composition analyses of long-lived GHR null mice. J Gerontol A Biol Sci. Med Sci. 2010;65(1):31–40.
136. Li Y, Knapp JR, Kopchick JJ. Enlargement of interscapular brown adipose tissue in growth hormone antagonist transgenic and in growth hormone receptor gene-disrupted dwarf mice. Exp Biol Med (Maywood). 2003;228(2):207–15.
137. Flint DJ, Binart N, Boumard S, Kopchick JJ, Kelly P. Developmental aspects of adipose tissue in GH receptor and prolactin receptor gene disrupted mice: site-specific effects upon proliferation, differentiation and hormone sensitivity. J Endocrinol. 2006;191(1):101–11.
138. Lubbers ER, List EO, Jara A, Sackman-Sala L, Cordoba-Chacon J, Gahete MD, Kineman RD, Boparai R, Bartke A, Kopchick JJ, et al. Adiponectin in mice with altered GH action: links to insulin sensitivity and longevity? J Endocrinol. 2013;216(3):363–74.
139. Masternak MM, Bartke A, Wang F, Spong A, Gesing A, Fang Y, Salmon AB, Hughes LF, Liberati T, Boparai R. et al. Metabolic effects of intra-abdominal fat in GHRKO mice. Aging Cell. 2012;11(1):73–81.
140. Kelder B, Berryman DE, Clark R, Li A, List EO, Kopchick JJ: CIDE-A gene expression is decreased in white adipose tissue of growth hormone receptor/binding protein gene disrupted mice and with high-fat feeding of normal mice. Growth Horm IGF Res 2007;17(4):346–351.
141. Stout MB, Tchkonia T, Pirtskhalava T, Palmer AK, List EO, Berryman DE, Lubbers ER, Escande C, Spong A, Masternak MM, et al. Growth hormone action predicts age-related white adipose tissue dysfunction and senescent cell burden in mice. Aging. 2014;6(7):575–86.
142. Kirkland J, Tchkonia T, Pirtskhalava T, Giorgadze N, Bartke A, Masternak M, Miller R, Kopchick J, Berryman D, List E, et al. Age-related fat redistribution and cellular senescence are delayed in growth hormone/IGF-1 deficient mice. Obesity. 2010;18:S53–53.
143. Householder LA, Troike K, Lubbers ER, Duran-Ortiz S, Jara A, List EO, Kopchick JJ, Berryman DE. Excess growth hormone promotes white adipose tissue fibrosis in a depot-dependent manner in GH transgenic mice. Endocrine Rev. 2014;35:3, MON-0884
144. Wang Z, Masternak MM, Al-Regaiey KA, Bartke A. Adipocytokines and the regulation of lipid metabolism in growth hormone transgenic and calorie-restricted mice. Endocrinology. 2007;148(6):2845–53.
145. Nilsson L, Binart N, Bohlooly YM, Bramnert M, Egecioglu E, Kindblom J, Kelly PA, Kopchick JJ, Ormandy CJ, Ling C, et al. Prolactin and growth hormone regulate adiponectin secretion and receptor expression in adipose tissue. Biochem Biophys Res Commun. 2005;331(4):1120–6.
146. Vijeyta F. Effects of growth hormone on circulating resistin levels in mice. Ohio University Masters Thesis 2012:1–141.
147. Ip BC, Liu C, Smith DE, Ausman LM, Wang XD. High-refined-carbohydrate and high-fat diets induce comparable hepatic tumorigenesis in male mice. J Nutr. 2014;144(5):647–53.
148. Surwit RS, Feinglos MN, Rodin J, Sutherland A, Petro AE, Opara EC, Kuhn CM, Rebuffe-Scrive M. Differential effects of fat and sucrose on the development of obesity and diabetes in C57BL/6 J and A/J mice. Metabolism. 1995;44(5):645–51.
149. Surwit RS, Kuhn CM, Cochrane C, McCubbin JA, Feinglos MN. Diet-induced type II diabetes in C57BL/6 J mice. Diabetes. 1988;37(9):1163–7.
150. Surwit RS, Wang S, Petro AE, Sanchis D, Raimbault S, Ricquier D, Collins S. Diet-induced changes in uncoupling proteins in obesity-prone and obesity-resistant strains of mice. Proc Natl Acad Sci U S A. 1998;95(7):4061–5.
151. Zhao Y, Tan YS, Aupperlee MD, Langohr IM, Kirk EL, Troester MA, Schwartz RC, Haslam SZ. Pubertal high fat diet: effects on mammary cancer development. Breast Cancer Res. 2013;15(5):R100.
152. Smith DL, Jr., Nagy TR, Allison DB. Calorie restriction: what recent results suggest for the future of ageing research. Eur J Clin Invest. 2010;40(5):440–50.

153. Olsson B, Bohlooly YM, Fitzgerald SM, Frick F, Ljungberg A, Ahren B, Tornell J, Bergstrom G, Oscarsson J. Bovine growth hormone transgenic mice are resistant to diet-induced obesity but develop hyperphagia, dyslipidemia, and diabetes on a high-fat diet. Endocrinology. 2005;146(2):920–30.
154. Robertson K, Kopchick JJ, Liu JL. Growth hormone receptor gene deficiency causes delayed insulin responsiveness in skeletal muscles without affecting compensatory islet cell overgrowth in obese mice. Am J Physiol Endocrinol Metab. 2006;291(3):E491–8.
155. Bonkowski MS, Rocha JS, Masternak MM, Al Regaiey KA, Bartke A. Targeted disruption of growth hormone receptor interferes with the beneficial actions of calorie restriction. Proc Natl Acad Sci U S A. 2006;103(20):7901–5.
156. Bartke A, Wright JC, Mattison JA, Ingram DK, Miller RA, Roth GS. Extending the lifespan of long-lived mice. Nature. 2001;414(6862):412.
157. Longo KA, Berryman DE, Kelder B, Charoenthongtrakul S, Distefano PS, Geddes BJ, Kopchick JJ. Daily energy balance in growth hormone receptor/binding protein (GHR -/-) gene-disrupted mice is achieved through an increase in dark-phase energy efficiency. Growth Horm IGF Res. 2010;20(1):73–9.
158. Howard AD, Feighner SD, Cully DF, Arena JP, Liberator PA, Rosenblum CI, Hamelin M, Hreniuk DL, Palyha OC, Anderson J, et al. A receptor in pituitary and hypothalamus that functions in growth hormone release. Science. 1996;273(5277):974–7.
159. Chen CY, Fujimiya M, Asakawa A, Chang FY, Cheng JT, Lee SD, Inui A. At the cutting edge: ghrelin gene products in food intake and gut motility. Neuroendocrinology. 2009;89(1):9–17.
160. Nass R, Liu J, Hellmann P, Coschigano KT, Gaylinn B, Berryman DE, Kopchick JJ, Thorner MO. Chronic changes in peripheral growth hormone levels do not affect ghrelin stomach mRNA expression and serum ghrelin levels in three transgenic mouse models. J Neuroendocrinol. 2004;16(8):669–75.
161. Westbrook R, Bonkowski MS, Strader AD, Bartke A. Alterations in oxygen consumption, respiratory quotient, and heat production in long-lived GHRKO and Ames dwarf mice, and short-lived bGH transgenic mice. J Gerontol A Biol Sci Med Sci. 2009;64(4):443–51.
162. Hong SL, Longo KA, Gosney E, Kopchick JJ. Increased metabolic flexibility and complexity in a long-lived growth hormone insensitive mouse model. J Gerontol A Biol Sci Med Sci. 2014;69(3):274–81.
163. Chhabra Y, Waters MJ, Brooks AJ. Role of the growth hormone-IGF-1 axis in cancer. Expert Rev Endocrinol Metab. 2011;6(1):71–84.
164. Miquet JG, Freund T, Martinez CS, Gonzalez L, Diaz ME, Micucci GP, Zotta E, Boparai RK, Bartke A, Turyn D, et al. Hepatocellular alterations and dysregulation of oncogenic pathways in the liver of transgenic mice overexpressing growth hormone. Cell Cycle. 2013;12(7):1042–57.
165. Quaife CJ, Mathews LS, Pinkert CA, Hammer RE, Brinster RL, Palmiter RD. Histopathology associated with elevated levels of growth hormone and insulin-like growth factor I in transgenic mice. Endocrinology. 1989;124(1):40–8.
166. Cecim M, Bartke A, Yun JS, Wagner TE. Expression of human, but not bovine, growth hormone genes promotes development of mammary tumors in transgenic mice. Transgenics. 1994;1:431–7.
167. Ikeno Y, Hubbard GB, Lee S, Cortez LA, Lew CM, Webb CR, Berryman DE, List EO, Kopchick JJ, Bartke A. Reduced incidence and delayed occurrence of fatal neoplastic diseases in growth hormone receptor/binding protein knockout mice. J Gerontol A Biol Sci Med Sci. 2009;64(5):522–9.
168. Zhang X, Mehta RG, Lantvit DD, Coschigano KT, Kopchick JJ, Green JE, Hedayat S, Christov KT, Ray VH, Unterman TG, et al. Inhibition of estrogen-independent mammary carcinogenesis by disruption of growth hormone signaling. Carcinogenesis. 2007;28(1):143–50.
169. Wang Z, Prins GS, Coschigano KT, Kopchick JJ, Green JE, Ray VH, Hedayat S, Christov KT, Unterman TG, Swanson SM. Disruption of growth hormone signaling retards early stages of prostate carcinogenesis in the C3(1)/T antigen mouse. Endocrinology. 2005;146(12):5188–96.

170. Pollak M, Blouin MJ, Zhang JC, Kopchick JJ. Reduced mammary gland carcinogenesis in transgenic mice expressing a growth hormone antagonist. Br J Cancer. 2001;85(3):428–30.
171. Luque RM, Lin Q, Cordoba-Chacon J, Subbaiah PV, Buch T, Waisman A, Vankelecom H, Kineman RD. Metabolic impact of Adult-Onset, Isolated, Growth Hormone Deficiency (AOiGHD) due to destruction of pituitary somatotropes. PLoS One 2011;6(1):e15767.
172. List EO, Berryman DE, Funk K, Jara A, Kelder B, Wang F, Stout MB, Zhi X, Sun L, White TA, et al. Liver-specific GH receptor gene disrupted (LiGHRKO) mice have decreased endocrine IGF-1, increased local IGF-1 as well as altered body size, body composition and adipokine profiles. Endocrinology. 2014:en20132086.
173. Fan Y, Menon RK, Cohen P, Hwang D, Clemens T, Digirolamo DJ, Kopchick JJ, Leroith D, Trucco M, Sperling MA. Liver-specific Deletion of the Growth Hormone Receptor Reveals Essential Role of GH Signaling in Hepatic Lipid Metabolism. J Biol Chem. 2009.
174. Mavalli MD, DiGirolamo DJ, Fan Y, Riddle RC, Campbell KS, van Groen T, Frank SJ, Sperling MA, Esser KA, Bamman MM, et al. Distinct growth hormone receptor signaling modes regulate skeletal muscle development and insulin sensitivity in mice. J Clin Invest. 2010;120(11):4007–20.
175. Vijayakumar A, Wu Y, Sun H, Li X, Jeddy Z, Liu C, Schwartz GJ, Yakar S, LeRoith D. Targeted loss of GHR signaling in mouse skeletal muscle protects against high-fat diet-induced metabolic deterioration. Diabetes. 2012;61(1):94–103.
176. Wu Y, Liu C, Sun H, Vijayakumar A, Giglou PR, Qiao R, Oppenheimer J, Yakar S, LeRoith D. Growth hormone receptor regulates beta cell hyperplasia and glucose-stimulated insulin secretion in obese mice. J Clin Invest. 2011;121(6):2422–6.
177. List EO, Berryman DE, Funk K, Gosney ES, Jara A, Kelder B, Wang X, Kutz L, Troike K, Lozier N, et al. The role of GH in adipose tissue: lessons from adipose-specific GH receptor gene-disrupted mice. Mol Endocrinol. 2013;27(3):524–35.
178. Lu C, Kumar PA, Sun J, Aggarwal A, Fan Y, Sperling MA, Lumeng CN, Menon RK. Targeted deletion of Growth Hormone (GH) receptor in macrophage reveals novel osteopontin-mediated effects of GH on glucose homeostasis and insulin sensitivity in diet-induced obesity. J Biol Chem. 2013;288(22):15725–35.
179. Benencia F, Harshman S, Duran-Ortiz S, Lubbers ER, List EO, Householder L, Al-Naeeli M, Liang X, Welch L, Kopchick JJ, Berryman DE. Endocrinol. 2015;156(5):1794–803.

Chapter 5
Mouse Models to Study Obesity Effects on Hematologic Malignancies

Jonathan Tucci and Steven D. Mittelman

Abstract In the present chapter, we discuss how mouse models have been used to investigate the interactions between obesity and hematologic malignancies. We begin with a brief overview of mouse models of obesity, followed by a description of how hematologic malignancies have been studied in these models. Unfortunately, the study of obesity and hematologic malignancies in mice is not as advanced as in other cancers such as breast and colorectal carcinoma. Therefore, where no studies exist, we describe how current models of hematologic malignancies could be made obese so that the effects of obesity on these malignancies could be studied.

Keywords Leukemia · Lymphoma · Obesity · Mouse · Diet

Introduction

Hematologic malignancies represent a substantial cancer burden in the USA. Approximately 55,000 people die per year from leukemia, lymphoma, and myeloma [1]. Leukemia is by far the most common cancer in children, while overall leukemia and non-Hodgkin lymphoma (NHL) are the 6th and 7th leading causes of cancer mortality, respectively [2].

S. D. Mittelman (✉)
Center for Endocrinology, Diabetes and Metabolism, Children's Hospital Los Angeles
Departments of Pediatrics and Physiology & Biophysics, Keck School of Medicine, University of Southern California, 4650 Sunset Blvd., Mailstop #61, Los Angeles, CA 90027, USA
e-mail: smittelman@chla.usc.edu

J. Tucci
Department of Pediatrics, Keck School of Medicine, University of Southern California, 4650 Sunset Blvd., Mailstop #93, Los Angeles, CA 90027, USA
e-mail: jtucci@chla.usc.edu

N. A. Berger (ed.), *Murine Models, Energy Balance, and Cancer,*
Energy Balance and Cancer, DOI 10.1007/978-3-319-16733-6_5

As explored in a previous volume [3], there are many ways in which obesity and hematologic malignancies interact. Eugenia Calle's landmark paper demonstrating that obesity increases cancer mortality identified associations between obesity and mortality from multiple myeloma (MM) and NHL. Subsequently, studies have shown that overweight and obesity are associated with an increased risk of developing all four common subtypes of leukemia [4]: acute lymphoblastic leukemia (ALL), chronic lymphocytic leukemia (CLL), acute myelogenous leukemia (AML), and chronic myelogenous leukemia (CML), as well as NHL [5] and MM [6]. Beyond incidence, obesity also impacts survival in hematologic malignancies. Obese children diagnosed with ALL have an increased rate of minimal residual disease [7] and a substantially increased relapse rate [8, 9]. Obesity also increases risk of treatment-related toxicity in children with ALL [10] and AML [11].

With its high prevalence in our population, obesity could significantly impair survival rates in these malignancies. A recent study found that 36 % of children with ALL were overweight or obese at diagnosis, and this increased to 49 % by the end of therapy [12]. Therefore, it is important to explore mechanisms whereby obesity affects hematologic malignancies, so that treatment strategies can be developed.

Mouse Models of Obesity

Diet-Induced Obesity

Perhaps the most popular mouse model of obesity is the diet-induced obese (DIO) mouse model. This term generally refers to C57BL/6 mice raised on a high-fat diet. While many high-fat diets have been used, the most common ones are based on diets originally developed by Surwit [13], and contain either 45 or 60 % of calories from fat. Excess fat in these defined ingredient diets comes from lard, and there are several control diets from which to choose, with variable amounts of sucrose based on the investigator's needs. It should be noted that these high-fat diets also have higher calorie density than the control diets or standard mouse chow; so technically speaking, the mice are exposed to a high-fat/high-calorie diet. C57BL/6 mice on these diets consume more calories per day and gain weight rapidly [14, 15]. They become glucose intolerant and develop diabetes with age, as well as some of the cardiovascular complications associated with human obesity [16, 17].

There are several commercially available variations of the high-fat diet. The Western diet contains high carbohydrate, primarily from sucrose, and high fat, primarily from anhydrous milk fat. This diet also results in obesity and can be more atherogenic than high-fat diet. Other diets adjust the form of fat used (e.g., cocoa butter) to increase atherogenicity.

It is possible to categorize most mouse strains into diet sensitive or diet resistant based on their response to a high-fat diet. C3H/HeJ, A/J, C57 L/J, and Balb/C mice are relatively resistant to obesity, while AKR/J, DBA/2J, and C57BL/6 mice are diet sensitive [14]. It is important to keep in mind that a portion (10–15 %) of even sensitive strains of mice will be diet resistant when raised on a high-fat diet.

Diets high in fructose, sucrose, or both have been used to induce insulin resistance and hypertriglyceridemia in rodent models; however, they are less effective in producing obesity in mice than in larger rodents [18–21].

Genetic Models of Obesity

There are several genetic mutations in mice that can cause varying degrees of obesity. Most of these involve perturbations in the function of leptin, an anorexigenic adipokine released in proportion to adipose tissue mass. The ob/ob leptin-deficient mouse and the db/db mouse with a nonsignaling leptin receptor were discovered as spontaneous mutations. These mice exhibit hyperphagia and develop severe obesity and diabetes, which is exacerbated when they are put on a high-fat diet [22–24]. Similar to the db/db mouse, the s/s mouse carries an STAT3 mutation that disrupts downstream leptin receptor signaling [25]. Leptin acts primarily on proopiomelanocortin (POMC)-expressing neurons in the hypothalamus, inducing signaling through α-melanocyte-stimulating hormone (α-MSH) on melanocortin (MC) receptors, predominantly MC4 and to a lesser extent MC3 [26]. Mice with knockout of POMC, MC3, or MC4 all develop obesity to varying degrees. Similarly, mice which overexpress the MC receptor antagonists agouti or agouti-related protein (AgRP) are characterized by hyperphagia and obesity [27, 28].

The fat mass and obesity-associated gene, *FTO*, was the first obesity susceptibility gene identified by genome wide association studies [29]. Single nucleotide polymorphisms which are associated with human obesity have been shown to increase the expression of the FTO product [30]. Likewise, increased copy number of FTO in mice leads to increased food intake and obesity [31].

Other lesser utilized genetically altered murine models have been created with varying effects on obesity and/or diabetes. Transgenic mice overexpressing corticotrophin releasing factor stimulate adrenocorticotrophic hormone (ACTH) release and truncal obesity in a Cushing's syndrome-like phenotype [32]. Similar overexpression of the GLUT4 glucose transporter in adipose tissue leads to early onset obesity in mice [33]. Neuropeptide Y (NPY) stimulates feeding through its action on its receptors, NPY1R and NPY2R. Surprisingly, mice lacking NPY1R or NPY2R develop obesity from decreased energy expenditure in the absence of hyperphagia [34, 35]. Tissue-specific overexpression of the glucocorticosteroid-converting enzyme, 11beta-hydroxysteroid dehydrogenase type 1 (11beta HSD-1), has varying effects on the development of obesity and diabetes. Selective overexpression in adipose tissue induces hyperphagia, obesity, and insulin resistance, especially after high-fat feeding. Yet, liver-specific overexpression only stimulates mild obesity and diabetes [36, 37]. Homozygous mutations in the Tubby gene lead to maturity-onset obesity without concomitant diabetes, although the precise mechanism behind this phenotype is unknown [38].

Some genetic models seem to uncouple the effects of obesity on adiposity and metabolism. The adiponectin overexpressing mouse, made on an ob/ob background, exhibits severe adiposity, but is relatively protected from dyslipidemia, insulin resistance, and deposition of ectopic fat [39]. Likewise, the adipose triglyceride lipase (aTGL) null mouse stores excess lipid in adipose tissue, leading to increased adiposity yet a lean metabolic phenotype [40]. These models could be used for the study of hematologic malignancy, and other pathologies, to test for the effects of adipose tissue per se without the metabolic derangements generally observed in obesity.

Early Life Programming

It has long been known that the intrauterine environment can have important effects on later obesity and glucose metabolism. Infants of diabetic mother are at increased risk for future obesity and metabolic syndrome [41, 42]. Similarly, the prenatal environment of a newborn pup can be a determinative factor of its adult metabolic status. Maternal diet-induced obesity has been linked with offspring obesity, hyperphagia, and insulin resistance [43]. These findings have also been illustrated in the offspring of normal weight mothers which were fed a high-fat diet during pregnancy [44]. Conversely, switching obese C57BL/6J mothers onto a low-fat diet can attenuate offspring metabolic alterations [45].

Perinatal nutrition also appears to have long-term effects on body weight regulation. For example, infants who are overfed early in life are also at increased risk of lifelong obesity [46]. Interestingly, children who are born small for gestational age and who exhibit catch-up growth early in life are also at increased risk for later obesity and metabolic syndrome [47]. Both of these phenomena have been simulated in rodents. Approximately 5 days after delivery, milk production of a mother mouse or a rat reaches a steady state based on milk consumption by the litter, and does not significantly change thereafter. Therefore, when a litter is reduced in size on day of life #5 ("selective culling"), the remaining pups will have access to increased milk. This model has been used extensively in rats [48], and more recently in mice [49], to accentuate obesity.

As in humans, undernourishment in early life followed by calorie surfeit can lead to later obesity. Mice exposed to perinatal undernourishment are small as pups, but later in life exhibit obesity and other features of the metabolic syndrome when put on a high-fat diet [50]. Mice exposed to perinatal malnutrition display changes in the expression of adipose tissue lipid metabolism genes that induce a greater responsiveness to a high-fat diet [51]. Hypotheses into this phenomenon have centered on the development of hypothalamic signaling in the offspring. In particular, undernourished offspring experience an earlier surge in leptin that is associated with later overfeeding and obesity. Normally nourished pups provided with exogenous leptin to simulate this early surge display a similar obese phenotype in adulthood [50].

Immunodeficient Mouse Models for Xenotransplantation

To improve the relevance of mouse models to human cancer, much recent work has been done developing xenografts, wherein human cancer cells are engrafted onto an immunodeficient mouse. This allows the study of human cancer cells in an in vivo environment, albeit one lacking an adaptive immune system. Most of this work has utilized varieties of Nude, Rag1 null, and severe combined immune deficiency (SCID) mice.

Nude mice are characterized by their abnormal hair growth and defective thymic development. This athymia results in a lack of T cells, as well as a partial B cell developmental defect [52]. In an attempt to make obese nude mice, Moiola et al. raised 4-week-old Swiss nu/nu mice on a high-fat diet for 16 weeks. This diet resulted in a modest elevation in cholesterol level, though only a tendency for increased body weight [53]. A high-fat diet for 6 months in BALB/c nu/nu mice led to a modest increase in body fat, but no difference in overall body weight compared to control fed mice [54]. Nkhata et al. used gold thioglucose injections to induce obesity in CD-1 ovariectomized female nude mice [55]. These injections induce obesity by causing hyperphagia secondary to ventromedial hypothalamus damage [56]. Injections of 0.3 and 0.5 mg/kg body weight of gold thioglucose induced substantial obesity, though the higher dose was associated with a significant mortality rate [55].

Rag1 null mice lack both B and T cells, due to a deficiency of the recombinant-activating gene necessary for B and T cell maturation. Rag1 mice are considered a "non-leaky" alternative to "leaky" severe combined immunodeficiency (SCID) mice that still develop small levels of B cells and immunoglobulin M (IgM) [57]. These mice, which are available on a C57BL/6 background, develop substantial obesity when put on a high-fat diet [58]. In fact, they become even more obese and insulin resistant than C57BL/6 controls, possibly due to absence of specific CD4+ T cells [58, 59].

SCID mice put onto a high-fat diet develop significant obesity, and weighed in one study on average 45 % more than control fed mice [60]. However, Lucas et al. found that C.B-17 SCID mice on a high-fat diet develop increased adiposity, but no increase in overall body weight. Nod/SCID IL2Rγc −/− (NSG) mice lack lymphocytes and natural killer (NK) cells, and have been shown to yield better engraftment of human hematopoietic cells. However, we found that NSG mice were relatively resistant to diet-induced obesity when raised on a 60 % fat diet, weighing only about 15 % more than control fed mice [61]. To accentuate this obesity, we reduced litters to 2 mice on day of life #5 as described above leading to fat-fed mice that were 33 % heavier than control-fed (non-litter reduced) mice.

These immunodeficient mouse models have all been used to xenograft human hematopoietic malignancies [62–71]. However, to our knowledge, no studies have been reported in which obese phenotypes of these mice have been used to evaluate the effects of obesity on human hematopoietic cancer in vivo.

Mouse Models to Study Obesity and Hematologic Malignancies

Acute Lymphoblastic Leukemia (ALL)

ALL is the most common subtype of leukemia diagnosed in children, and represents malignant outgrowth of subclones of B or T cell precursors. While survival from ALL in children has improved tremendously over the preceding decades, patients with certain risk factors, including obesity, have a worse prognosis. In addition, survival in adults with ALL is much lower than children.

In 1990, Nora Heisterkamp and John Groffen introduced the human *BCR/ABL* transgene into C57BL/CBA mice, demonstrating for the first time that this product of the Philadelphia chromosome caused leukemia [72]. This "P190 mouse" develops leukemia by about 2 months of age, characterized by BCR/ABL expressing pre-B ALL cells. Malignant cells accumulate in the bone marrow and spleen, as in the clinical disease; however, they also form lymphoma-like tumors, which are not commonly seen in the human disease. Mice become rapidly ill if not treated, and need to be sacrificed within days of disease becoming clinically apparent (to avoid death as an endpoint). This model has since been bred onto a C57BL/6 background.

In 2010, we weaned male C57BL/6 P190 mice onto either high-fat diet (60% fat from research diets, see above) or chow diet. Like the background strain, these P190 mice developed diet-induced obesity, with significantly heavier weight within 1 week of weaning. While median lifespan was not different between obese and control P190 mice (107 vs. 113 days, $p = 0.2$), there was a time-dependent effect of obesity to increase ALL hazard ratio ($p < 0.05$). Thus, obesity accelerated the risk of ALL at older ages [73].

There are several other mouse models of ALL that have not to our knowledge been used in the setting of obesity. Most ALL models are developed through the transgenic expression of gene fusions responsible for human leukemogenesis. Fortunately, most of these models utilize the C57BL/6 mouse, which is susceptible to diet-induced obesity. The *TEL-AML1* (*ETV6-RUNX1*) fusion gene, marked by the t(12;21) translocation, is the most common genetic alteration identified in childhood cancer and is associated with pediatric B-ALL [74, 75]. p16 and p19 are tumor suppressors, mutated in human pre-B ALL and cooperate with *TEL-AML1* to initiate leukemogenesis. One quarter of transgenic C57BL/6J mice expressing *TEL-AML1* developed B-ALL within a year and 75% of C57BL/6J $p16^{INK4a}/p19^{ARF}$-deficient mice expressing *TEL-AML1* developed B-ALL over the same period [76]. In pediatric T-ALL, the *LMO2* locus is the most common site for genetic alteration and *LMO2* overexpression induces T-ALL in C57BL/6J mice [77, 78]. Fifty percent of T-ALL cases contain activating mutations in *NOTCH-1* [79]. In two $KRAS^{G12D}$-induced T-ALL models, about half of the mice developed activating mutations in *NOTCH-1* signaling accelerating leukemia progression [80, 81].

Acute Myelogenous Leukemia (AML)

AML is less common than ALL, and more commonly seen in adults, with a slight bias toward men. Both AML incidence and mortality increase with age. However, AML can develop secondary to treatment, especially radiation, in children with cancer. AML is characterized by the uncontrolled proliferation of myeloid progenitors and is classified by the type and maturity of the myeloblast. About half of the patients with AML enter remission with rates varying by myeloblast classification and genetic alteration [82, 83].

Clinical studies in children and adults have demonstrated obesity as a risk factor in the development and prognosis of AML. A retrospective review of 768 children with AML by the Children's Cancer Group showed that those children who were overweight at diagnosis had a poorer survival (hazard ratio [HR] 1.88), and were more likely to exhibit treatment-related mortality (HR 3.49) than normal weight patients [84]. Similarly, a meta-analysis of prospective cohort studies uncovered a significant association between obesity and AML incidence and mortality [4, 85]. Obesity may also be associated with incidence and severity of the specific AML subtype, acute promyelocytic leukemia (APL). One clinical study demonstrated an association between increasing BMI and APL diagnosis [86]. In a second study, the majority of subjects diagnosed with APL were overweight/obese, and increasing BMI was associated with increased risk of differentiation syndrome and relapse [87].

One recent study has translated these clinical findings into a murine model. Transgenic C57BL/6J x C3H mice expressing the *PML-RAR* alpha fusion gene, found in 90% of APL patients, were fed standard or high-fat chow [88]. The resulting diet-induced obesity in high-fat fed mice led to increased APL penetrance (100% vs. 61% in control mice) and decreased survival (282 days vs. 332 days in control mice). APL cells were then harvested and transplanted into mice fed a normal or 30% calorie-restricted diet to test the effect of dieting on APL growth. Calorie restriction reduced circulating insulin-like growth factor 1 (IGF-1) and significantly increased survival in transplanted mice (23.5 days vs. 16 days in control mice). Addition of IGF-1 to calorie-restricted mice reversed this benefit.

There are several other murine models which exhibit transgenic expression of gene fusions that drive AML leukemogenesis. While none of these have been used to evaluate the effects of obesity, these models have been created on backgrounds susceptible to diet-induced obesity, such as C57BL/6 and SCID mice. The *Mll-AF4* [t(4;11)(q21;q23)] and *Mll-AF9* [t(9;11)(q21;q23)] translocations comprise a majority of mixed lineage leukemia (MLL) gene fusions which are found in two third of infantile ALL cases [89]. Yet, transgenic expression of these fusions primarily induces AML in murine models. Cre-driven conditional expression of *Mll-AF4* in C57BL/6 × 129 mice induces ALL, AML, or MLL within an average latency of 3–4 months [90]. Transgenic expression of *Mll-AF9* in embryonic stem cells or hematopoietic progenitors transplanted into C57BL/6 mice initiates AML leukemogenesis in as little as 4 months [91, 92]. Other *Mll* transgenes, such as *Mll-ENL* and *Mll-AF10*, have also proved successful in obesity-sensitive C57BL/6J and SCID models [93–97].

Chronic Lymphocytic Leukemia (CLL)

CLL is primarily a disease of older adults, generally diagnosed after 70 years of age. CLL differs from ALL and AML as it is characterized by the expansion of more mature lymphocytes, which proliferate at a slower rate. Patients with the most indolent form of CLL are diagnosed with monoclonal lymphocytosis and generally do not require treatment.

In a retrospective cohort analysis of over 4 million USA veterans, researchers identified an increased risk of CLL in obese white and African-American men [98], and this effect of obesity to increase CLL risk was confirmed in meta-analysis [4]. However, studies exploring the mechanism behind this association are limited. Plasma adiponectin concentrations are decreased in obesity, and in a small study of 19 patients with CLL, adiponectin levels were lower in CLL patients than matched controls [99]. Another study has demonstrated that CLL cells can utilize adipocyte-derived lipids, lipoproteins, and short-chain free-fatty acids to mediate resistance to dexamethasone, a glucocorticoid used in leukemia treatment. Adding peroxisome proliferator-activator receptor (PPAR)-alpha or fatty acid oxidation inhibitors attenuated dexamethasone resistance in vitro and in an NSG CLL xenograft model [100].

Early in vivo CLL studies have employed New Zealand black (NZB) mice that harbor a mutation in the microRNA-16 locus akin to the 13q14 mutation in human CLL [101]. These CLL cells can recapitulate the disease following transplantation into recipient mice [102]. While no studies have assessed CLL in NZB mice in the context of obesity, NZB were shown to be susceptible to diet-induced obesity following an 8-week high-fat diet [103]. The New Zealand obese (NZO) mouse strain, derived from the NZB strain, develop obesity, hyperinsulinemia, and insulin resistance on a normal diet. The NZO strain also displays a similar proliferation of B cells of identical marker expression as the NZB sister strain [104]. Thus, this strain could likely be utilized to assess the effect of obesity on CLL initiation and pathogenesis.

Multiple transgenic models have also been created to study CLL in mice. Overexpression of the *TCL1* gene, which has been observed in human CLL [105], results in the proliferation of CD5+ B cells in mice, similar to the proliferation in NZB and NZO mice [106]. CLL cells also often overexpress *Bcl-2* and tumor necrosis factor receptor associated factor 1/2 (TRAF1/2) [107, 108]. In a study by Zapata et al., transgenic BALB/c mice with *Bcl-2* overexpression were crossed with transgenic FVB/N mice overexpressing *TRAF2*, resulting in progeny which developed CLL-like disease [109]. The BALB/c strain is diet-induced obesity resistant, whereas FVB/N mice develop obesity and metabolic dysfunction on a high-fat diet [110]; thus it is not clear whether this cross would be amenable to testing the effects of obesity on CLL.

Chronic Myelogenous Leukemia (CML)

CML affects about 6000 people each year and comprises 10% of all new leukemia diagnoses, the least of the four main types of leukemia. Like CLL, CML primarily affects older individuals with half of new cases occurring at age 65 or older [111]. As with the other leukemias, obesity is a risk factor for CML development. In a recent case-control study, patients with CML were 4.29 times more like to have been obese at age 25 and 5.12 times at age 40. The strongest association between CML and obesity was observed in patients who gained more than 1 kg per year between 25 and 40 years of age [112]. Despite the similarities with other leukemias, CML is driven by only one genetic alteration, the BCR/ABL translocation.

Studies with human CML cells suggest that obesity may play a role in CML development and progression via increased leptin levels. CML cells express leptin receptor, with increased expression in the blast phase [113]. Leptin stimulates CML proliferation and prevents apoptosis, suggesting a role in obesity's effect on leukemia pathogenesis [114].

As discussed above, the P190 BCR/ABL isoform induces ALL in a transgenic mouse model. The P190 isoform also has the ability to induce CML, though it is rarely found in human CML. The two other BCR/ABL isoforms, P210 and P230, share the ability to induce myeloid, but not lymphoid, leukemogenesis [115]. Transduction of the P210 BCR/ABL gene fusion in BALB/c mice bone marrow cells induced CML within 3 weeks [116].

Another receptor tyrosine kinase, Axl, was found by Neubauer et al. to be aberrantly expressed in CML cells in about half of the patients they studied [117]. In an attempt to generate a CML mouse model, transgenic mice were developed which expressed *axl* in myeloid cells, under the control of the GCSF receptor [118]. Interestingly, this mouse model did not exhibit disordered myeloid proliferation, but they did develop obesity, hyperinsulinemia, and hyperglycemia, possibly due to elevated tumor necrosis factor alpha (TNFα) levels. Further work needs to be done to determine if *axl* plays a mechanistic role in linking obesity and CML.

Non-Hodgkin's Lymphoma (NHL)

NHL is the most common hematologic cancer in the USA, totaling about 4% of all new cancer diagnoses [119]. NHL is also the most diverse of the hematologic malignancies with a multitude of unique classification types. At least four out of five NHL diagnoses are of B cell origin with diffuse large B cell lymphoma (DLBCL) being the greatest of these [119]. Follicular, mantle cell, marginal zone, and Burkitt lymphoma comprise the other major B cell lymphomas. Various T cell lymphoma types also exist but are generally less common.

Numerous clinical studies exist supporting an association between obesity and NHL. Obesity and excess calorie intake have been associated with increase NHL risk in case-controlled studies [120, 121], while regular physical exercise was associated with a decreased risk [122]. A meta-analysis of 16 case-control and cohort studies concluded that overweight (RR=1.07, CI=1.01–1.14) and obese (RR=1.20, CI=1.07–1.34) subjects had increased the risk of NHL, specifically DLBCL (RR=1.40, CI=1.18–1.66) but not follicular lymphoma (RR=1.10, CI=0.82–1.47) [5].

Further clinical studies have explored the effect of obesity on NHL incidence. Several have illustrated associations between total fat, saturated fat, and monounsaturated fat intake with increased NHL risk [123–126]. One recent study has also shown that an increase in trans- and omega-6 fatty acids was positively associated with NHL risk while omega-3 fatty acid and seafood intake was inversely associated with NHL risk [127]. Two separate studies have also demonstrated a relationship between NHL risk and polymorphisms in obesity-related genes including neuropeptide Y, ghrelin and leptin hormones, and the leptin receptor [128, 129].

Some murine strains develop spontaneous lymphoma and can be used to study the influence of obesity on this cancer. Up to 30 % of C57BL/6 mice develop lymphoma over their lifespan [130, 131]. Exposure to radiation and the mutagen, N-ethyl-N-nitrosourea can induce T cell lymphomas in these mice [132]. Crossing C57BL/6 mice with 129 mice resulted in an increased incidence of lymphomas, with 42 % of male and 67 % of female developing lymphoma over a 2-year period. This B cell lymphoma often arose in mesenteric lymph nodes and disseminated to the gut associated lymphoid tissue and spleen [133]. While there are no studies in which B6;129 mice were put on a high-fat diet, both parent strains are susceptible to diet-induced obesity [110], and so there is a good chance that combined strain will be sensitive as well. Finally, AKR mice develop spontaneous T cell leukemia/lymphoma (thymomas) due to a recombinant retrovirus which targets thymocytes [134].

Other murine models of lymphoma that could be used to study obesity include transgenic mice which overexpress either the *myc* oncogene or the anti-apoptotic signal, *Bcl-2*. The Eμ-Myc model inserts the *myc* gene in the IgH region and initiates B cell lymphomas of either DLBCL or Burkitt type in 100 % of mice with variable onset [135]. *Bcl-2* overexpression in mice stimulates follicular lymphomagenesis [136]. These transgenic mice were made on B6;129 and C57BL/6 backgrounds, respectively, so should be susceptible to diet-induced obesity.

Alterations in calorie intake can significantly alter lymphoma development in mice. A 40 % restriction in calorie intake in C57BL/6 mice reduced lymphoma incidence and prolonged onset in females [137]. Barbara Shields and colleagues put AKR mice onto a calorie-restricted diet, and showed that this prolonged their lifespan substantially [138]. We randomized male AKR mice at weaning to a high-fat (60 %) or low-fat (10 %) diet, and followed for onset of thymoma. High fat fed AKR mice were heavier than control mice at all timepoints, and developed thymomas significantly earlier than control mice (237 vs. 310 days, $p < 0.05$) [73].

There is increasing awareness that lymphoma cells may depend on free fatty acid (FFA) for energy and other cellular functions. Given the abundance of adipose tissue and elevated lipids in circulation and potentially in tumor microenvironments in obese patients, this could explain the clinically observed links between obesity and lymphoma. Based on this potential sensitivity, studies have been done to test strategies to block FFA metabolism in lymphoma cells. PPAR-alpha agonists, such as fenofibrate, stimulate FFA metabolism and are widely used to treat hypercholesterolemia. In one study, C57BL/6J mice transplanted with BCR/ABL transformed B cells exhibited lymphoma growth, which was associated with adipose tissue depletion, FFA release, and hepatic FFA uptake. Fenofibrate treatment decreased tumor growth rate, along with hepatic beta-oxidation and clearance of FFA [139]. Orlistat is an orally administered anti-obesity drug with minimal systemic absorption, which inhibits lipases in the intestine, and promotes body fat loss by reducing intestinal fat absorption. This drug can also inhibit fatty acid synthase, and so has been given by injection in mouse models to test against cancer cell lipogenesis. Orlistat injections reduced tumor growth, increased chemosensitivity and prolonged mouse survival in BALB/c mice transplanted with T cell lymphoma [140].

Many murine studies examining lymphoma and diet have centered on the relationship between particular dietary fatty acids and lymphoma progression. Counter to what one might expect based on the clinical data, a diet high in polyunsaturated fatty acids was associated with an *increased* incidence of spontaneous lymphoma in SJL/J mice [141]. In addition, AKR mice that fed on fish oil (which is high in omega-3 polyunsaturated fatty acids) exhibited faster growth of implanted syngeneic RDM-4 lymphoma cells than mice on normal chow or beef tallow (high in saturated fat) [142, 143]. On the other hand, a diet rich in marine fish was found to reduce the incidence of spontaneous lymphomas in Swiss-Webster mice, from 30 % on the control diet to 5 % [144]. Thus, the role of polyunsaturated fatty acids in lymphoma development remains unclear.

Recently, researchers have tested the activities of diabetes drugs against lymphoma cells. PPAR-gamma agonists like rosiglitazone are widely used to improve insulin sensitivity in patients with type II diabetes mellitus. These drugs can induce normal B lymphocyte and lymphoma apoptosis in vitro [145–148], though they can also stimulate T cell lymphoma growth in serum-starved conditions [149]. To our knowledge, these drugs have not been tested against lymphoma in vivo.

Hodgkin's Lymphoma (HL)

HL is a particular subset of B cell lymphoma typically characterized histologically by the presence of Reed–Sternberg cells in lymph node germinal or post-germinal centers. Incidence is more common in adolescents and the elderly, though overall it comprises only 10 % of all new lymphoma cases [150]. Unlike NHL and leukemia, there is sparse clinical data to support an association between obesity and HL. In one British case-control study, obese subjects were 2.2 times more likely to be diagnosed with HL, yet this association was only significant in men and those above 35 years of age [151]. The retrospective cohort analysis of white and African-African male veterans did not identify a significant risk for HL in obese individuals [98]. Given the clinical data, the research into any association between obesity and HL is lacking. Murine modeling of HL is limited to xenografting of human HL cells into Nude [152], SCID [153], and NSG mice [154].

Multiple Myeloma (MM)

MM is a malignancy of the most mature B cell type, the plasma cell, and is second in incidence only to NHL [155]. MM is primarily a disease of the elderly, with MM cells proliferating within the fatty bone marrow cavity. MM incidence is known to increase with age and have a greater occurrence in men and African-Americans. Through multiple clinical studies, an association between obesity and MM is evident. A meta-analysis done by researchers at the Karolinska Institute compiled data from 11 cohort and 4 case-control studies analyzing the association between obesity and MM. Overweight individuals were 12–43 % and obese individuals were 27–82 % more likely to be diagnosed with MM than normal weight individuals [6].

Some have hypothesized that obesity increases risk of MM through alterations in circulating adipokines. IL-6, which circulates at higher levels in obese individuals [156], has been shown to be potentially related to MM risk. Cozen et al. identified an association between MM diagnosis and IL-6 and IL-6 receptor polymorphisms [157]. IL-6-null BALB/c mice injected with pristane developed fewer plasma cell neoplasms than heterozygote or wild-type mice [158]. Two in vitro studies suggest that the hallmark decrease in adiponectin and rise in leptin levels in obese individuals would support an environment rich for MM development. Adiponectin stimulates AMP-activated protein kinase activation and subsequent MM cell cycle arrest and apoptosis [159]. Alternatively, murine and human MM cells express the leptin receptor and display a moderate increase in proliferation in a leptin-rich environment [160]. Like NHL, MM cells are also sensitive to the diabetes drug and PPAR-gamma agonist, rosiglitazone [161]. Rosiglitazone induces cell cycle arrest and differentiation of MM cells, further supporting the use of anti-diabetes and anti-obesity medication as adjuvants to chemotherapy. Unfortunately, supporting studies in murine models have not, to our knowledge, been done.

Two recent in vitro studies have demonstrated that MM cells are reliant on cholesterol and FFA metabolism for proliferation and survival. Low-density lipoprotein cholesterol (LDL-c) free media induced apoptosis in three human MM cell lines, while addition of exogenous LDL-c improved cell survival [162]. In a similar study, etomoxir, an inhibitor of beta-oxidation, and orlistat, (used as an inhibitor of fatty acid synthesis), induced MM cell cycle arrest and apoptosis [163].

The relative lack of clinically relevant MM murine models has limited the depth of in vivo studies. Only one mouse strain produces a spontaneous MM-like disease.

The C57BL/KaLwRij substrain of mice spontaneously develop MM after 2 years of age, but at a rather low incidence of 0.5 % [164]. These mice display monoclonal expansion of myeloma cells with accompanying lytic lesions in the long bones. Harvesting of these myeloma cells and transplantation into donor mice maintains these myeloma lines and is the foundation for the 5TMM model. Interestingly, these cells do not develop into myeloma when implanted in the parent C57BL6 strain. However, when C57BL6 mice are put on a high fat diet prior to implantation, they develop a myeloma-like condition, reminiscent of the clinical disorder which precedes myeloma, monoclonal gammopathy of undetermined significance (MGUS) [165].

Alternatively, two syngeneic models have been utilized to recapitulated MM-like disease. In one model, *Myc* transgenic mice on a B6;129 background were crossed with Bcl-X_L transgenic FVB/N mice. These mice develop plasma cell tumors within an average of 135 days [166]. BALB/c mice can also be exposed to pristane, developing a plasmacytosis similar to multiple myeloma only in the proliferation of plasma cells [167]. This is the most common model of MM.

Summary

The rising incidence of obesity throughout the world has spurred research into its effects on other diseases, including hematologic malignancies. Recent meta-analyses have demonstrated significant clinical associations between obesity and the incidence of each of the hematologic malignancy subtypes presented here. Further clinical studies have shown associations between obesity and treatment-related toxicity and mortality. More directed studies have sought to further understand these associations, illustrating increase in cancer risk with higher leptin, IL-6, FFAs, and calorie intake, and lower physical activity and adiponectin.

Advances in genetics and mouse husbandry have led to a variety of murine models available to test these clinical associations. Transgenic expression of gene fusions have allowed for the recapitulation of syngeneic hematologic malignancies on mice susceptible to diet-induced obesity. Selective culling and high-fat feeding have been utilized to create obese immune-competent and deficient mice. Obese athymic, SCID, and NSG mice have been developed and allow the exploration of the effect of obesity on human-derived blood cancers. Experiments using these models have shown that obese mice have a more rapid incidence of and mortality from ALL, AML, and NHL. These models have also been used to explore how FFAs stimulate cancer progression and whether current obesity and diabetes drugs, like fibrates, thiazolidinediones, and lipase inhibitors, can reverse the negative effect of obesity on hematologic malignancies.

Here, we reviewed how murine models can and have been adapted to study the effect of obesity on hematologic malignancies. Despite the wide research presented in this chapter, much is still unknown about the mechanisms behind this effect and the ways to effectively combat it.

References

1. Howlader N, et al. SEER Cancer Statistics Review, 1975–2011. 2014. http://seer.cancer.gov/csr/1975_2011/. Accessed 24 July 2014.
2. United States Cancer Statistics. 1999–2011. Incidence and mortality Web-based report. 2014. www.cdc.gov/uscs. Accessed 24 July 2014.
3. Mittelman SD, Berger NA. Energy balance and hematologic malignancies. New York: Springer; 2012. p. 184.
4. Larsson SC, Wolk A. Overweight and obesity and incidence of leukemia: a meta-analysis of cohort studies. Int J Cancer. 2008;122(6):1418–21.
5. Larsson SC, Wolk A. Obesity and risk of non-Hodgkin's lymphoma: a meta-analysis. Int J Cancer. 2007;121(7):1564–70.
6. Larsson SC, Wolk A. Body mass index and risk of multiple myeloma: a meta-analysis. Int J Cancer. 2007;121(11):2512–6.
7. Orgel E, et al. Obesity is associated with residual leukemia following induction therapy for childhood B-precursor acute lymphoblastic leukemia. Blood. 2014;124(26):3932–8.
8. Butturini AM, et al. Obesity and outcome in pediatric acute lymphoblastic leukemia. J Clin Oncol. 2007;25(15):2063–9.
9. Gelelete CB, et al. Overweight as a prognostic factor in children with acute lymphoblastic leukemia. Obesity (Silver Spring). 2011;19(9):1908–11.
10. Orgel E, et al. Impact on survival and toxicity by duration of weight extremes during treatment for pediatric acute lymphoblastic leukemia: a report from the Children's Oncology Group. J Clin Oncol. 2014;32(13):1331–7.
11. Lin A, et al. Influence of obesity on efficacy and toxicity of induction chemotherapy in patients with newly diagnosed acute myeloid leukemia. Leuk Lymphoma. 2013;54(3):541–6.
12. Esbenshade AJ, et al. Body mass index and blood pressure changes over the course of treatment of pediatric acute lymphoblastic leukemia. Pediatr Blood Cancer. 2011;56(3):372–8.
13. Surwit RS, et al. Diet-induced type II diabetes in C57BL/6J mice. Diabetes. 1988;37(9):1163–7.
14. West DB, et al. Dietary obesity in nine inbred mouse strains. Am J Physiol. 1992;262(6 Pt 2):R1025–32.
15. Petro AE, et al. Fat, carbohydrate, and calories in the development of diabetes and obesity in the C57BL/6J mouse. Metabolism. 2004;53(4):454–7.
16. Nishina PM, et al. Atherosclerosis and plasma and liver lipids in nine inbred strains of mice. Lipids. 1993;28(7):599–605.
17. Mills E, et al. Hypertension in CB57BL/6J mouse model of non-insulin-dependent diabetes mellitus. Am J Physiol. 1993;264(1 Pt 2):R73–8.
18. Pagliassotti MJ, et al. Changes in insulin action, triglycerides, and lipid composition during sucrose feeding in rats. Am J Physiol. 1996;271(5 Pt 2):R1319–26.
19. Sumiyoshi M, Sakanaka M, Kimura Y. Chronic intake of high-fat and high-sucrose diets differentially affects glucose intolerance in mice. J Nutr. 2006;136(3):582–7.
20. Sleder J, et al. Hyperinsulinemia in fructose-induced hypertriglyceridemia in the rat. Metabolism. 1980;29(4):303–5.
21. Barbosa CR, et al. Opposite lipemic response of Wistar rats and C57BL/6 mice to dietary glucose or fructose supplementation. Braz J Med Biol Res. 2007;40(3):323–31.
22. Koch CE, et al. High-fat diet induces leptin resistance in leptin-deficient mice. J Neuroendocrinol. 2014;26(2):58–67.
23. Pelleymounter MA, et al. Effects of the obese gene product on body weight regulation in ob/ob mice. Science. 1995;269(5223):540–3.
24. Kobayashi K, et al. The db/db mouse, a model for diabetic dyslipidemia: molecular characterization and effects of Western diet feeding. Metabolism. 2000;49(1):22–31.
25. Bates SH, et al. STAT3 signalling is required for leptin regulation of energy balance but not reproduction. Nature. 2003;421(6925):856–9.

26. Seeley RJ, et al. Melanocortin receptors in leptin effects. Nature. 1997;390(6658):349.
27. Bultman SJ, Michaud EJ, Woychik RP. Molecular characterization of the mouse agouti locus. Cell. 1992;71(7):1195–204.
28. Graham M, et al. Overexpression of Agrt leads to obesity in transgenic mice. Nat Genet. 1997;17(3):273–4.
29. Frayling TM, et al. A common variant in the FTO gene is associated with body mass index and predisposes to childhood and adult obesity. Science. 2007;316(5826):889–94.
30. Berulava T, Horsthemke B. The obesity-associated SNPs in intron 1 of the FTO gene affect primary transcript levels. Eur J Hum Genet. 2010;18(9):1054–6.
31. Church C, et al. Overexpression of Fto leads to increased food intake and results in obesity. Nat Genet. 2010;42(12):1086–92.
32. Stenzel-Poore MP, et al. Development of Cushing's syndrome in corticotropin-releasing factor transgenic mice. Endocrinology. 1992;130(6):3378–86.
33. Shepherd PR, et al. Adipose cell hyperplasia and enhanced glucose disposal in transgenic mice overexpressing GLUT4 selectively in adipose tissue. J Biol Chem. 1993;268(30):22243–6.
34. Kushi A, et al. Obesity and mild hyperinsulinemia found in neuropeptide Y-Y1 receptor-deficient mice. Proc Natl Acad Sci U S A. 1998;95(26):15659–64.
35. Naveilhan P, et al. Normal feeding behavior, body weight and leptin response require the neuropeptide Y Y2 receptor. Nat Med. 1999;5(10):1188–93.
36. Man TY, et al. Dietary manipulation reveals an unexpected inverse relationship between fat mass and adipose 11beta-hydroxysteroid dehydrogenase type 1. Am J Physiol Endocrinol Metab. 2011;300(6):E1076–84.
37. Paterson JM, et al. Metabolic syndrome without obesity: Hepatic overexpression of 11beta-hydroxysteroid dehydrogenase type 1 in transgenic mice. Proc Natl Acad Sci U S A. 2004;101(18):7088–93.
38. Coleman DL, Eicher EM. Fat (fat) and tubby (tub): two autosomal recessive mutations causing obesity syndromes in the mouse. J Hered. 1990;81(6):424–7.
39. Kim JY, et al. Obesity-associated improvements in metabolic profile through expansion of adipose tissue. J Clin Invest. 2007;117(9):2621–37.
40. Hoy AJ, et al. Adipose triglyceride lipase-null mice are resistant to high-fat diet-induced insulin resistance despite reduced energy expenditure and ectopic lipid accumulation. Endocrinology. 2011;152(1):48–58.
41. Reece EA, Homko CJ. Infant of the diabetic mother. Semin Perinatol. 1994;18(5):459–69.
42. Silverman BL, et al. Long-term effects of the intrauterine environment. The Northwestern University Diabetes in Pregnancy Center. Diabetes Care. 1998;21 Suppl 2:B142–9.
43. Nivoit P, et al. Established diet-induced obesity in female rats leads to offspring hyperphagia, adiposity and insulin resistance. Diabetologia. 2009;52(6):1133–42.
44. Gregorio BM, et al. Maternal high-fat intake predisposes nonalcoholic fatty liver disease in C57BL/6 offspring. Am J Obstet Gynecol. 2010:203(5):495.e1–8.
45. Krasnow SM, Nguyen ML, Marks DL. Increased maternal fat consumption during pregnancy alters body composition in neonatal mice. Am J Physiol Endocrinol Metab. 2011;301(6):E1243–53.
46. Stettler N, et al. Weight gain in the first week of life and overweight in adulthood: a cohort study of European American subjects fed infant formula. Circulation. 2005;111(15):1897–903.
47. Ibanez L, et al. Early development of adiposity and insulin resistance after catch-up weight gain in small-for-gestational-age children. J Clin Endocrinol Metab. 2006;91(6):2153–8.
48. Plagemann A, et al. Obesity and enhanced diabetes and cardiovascular risk in adult rats due to early postnatal overfeeding. Exp Clin Endocrinol. 1992;99(3):154–8.
49. Glavas MM, et al. Early overnutrition results in early-onset arcuate leptin resistance and increased sensitivity to high-fat diet. Endocrinology. 2010;151(4):1598–610.
50. Yura S, et al. Role of premature leptin surge in obesity resulting from intrauterine undernutrition. Cell Metab. 2005;1(6):371–8.

51. Bol VV, et al. Forced catch-up growth after fetal protein restriction alters the adipose tissue gene expression program leading to obesity in adult mice. Am J Physiol Regul Integr Comp Physiol. 2009;297(2):R291–9.
52. Szabo P, et al. Maturation of B cell precursors is impaired in thymic-deprived nude and old mice. J Immunol. 1998;161(5):2248–53.
53. Moiola CP, et al. Prostate tumor growth is impaired by CtBP1 depletion in high-fat diet-fed mice. Clin Cancer Res. 2014;20(15):4086–95.
54. Lamas B, et al. Dietary fat without body weight gain increases in vivo MCF-7 human breast cancer cell growth and decreases natural killer cell cytotoxicity. Mol Carcinog. 2013;54:58–71.
55. Nkhata KJ, et al. Mammary tumor development from T47-D human breast cancer cells in obese ovariectomized mice with and without estradiol supplements. Breast Cancer Res Treat. 2009;114(1):71–83.
56. Debons AF, et al. Gold thioglucose-induced hypothalamic damage, hyperphagia, and obesity: dependence on the adrenal gland. Endocrinology. 1982;110(6):2024–9.
57. Mombaerts P, et al. RAG-1-deficient mice have no mature B and T lymphocytes. Cell. 1992;68(5):869–77.
58. Winer S, et al. Normalization of obesity-associated insulin resistance through immunotherapy. Nat Med. 2009;15(8):921–9.
59. Weitman ES, et al. Obesity impairs lymphatic fluid transport and dendritic cell migration to lymph nodes. PLoS ONE. 2013;8(8):e70703.
60. Flores MB, et al. Obesity-induced increase in tumor necrosis factor-alpha leads to development of colon cancer in mice. Gastroenterology. 2012;143(3):741–53.e1–4.
61. Behan JW, et al. Activation of adipose tissue macrophages in obese mice does not require lymphocytes. Obesity (Silver Spring). 2013;21(7):1380–8.
62. Agliano A, et al. Human acute leukemia cells injected in NOD/LtSz-scid/IL-2Rgamma null mice generate a faster and more efficient disease compared to other NOD/scid-related strains. Int J Cancer. 2008;123(9):2222–7.
63. Dazzi F, et al. Normal and chronic phase CML hematopoietic cells repopulate NOD/SCID bone marrow with different kinetics and cell lineage representation. Hematol J. 2000;1(5):307–15.
64. Durig J, et al. A novel nonobese diabetic/severe combined immunodeficient xenograft model for chronic lymphocytic leukemia reflects important clinical characteristics of the disease. Cancer Res. 2007;67(18):8653–61.
65. Hauer J, et al. Loss of p19Arf in a Rag1(−/−) B-cell precursor population initiates acute B-lymphoblastic leukemia. Blood. 2011;118(3):544–53.
66. Lin S, et al. DDX5 is a positive regulator of oncogenic NOTCH1 signaling in T cell acute lymphoblastic leukemia. Oncogene. 2013;32(40):4845–53.
67. Lock RB, et al. The nonobese diabetic/severe combined immunodeficient (NOD/SCID) mouse model of childhood acute lymphoblastic leukemia reveals intrinsic differences in biologic characteristics at diagnosis and relapse. Blood. 2002;99(11):4100–8.
68. Loisel S, et al. Establishment of a novel human B-CLL-like xenograft model in nude mouse. Leuk Res. 2005;29(11):1347–52.
69. Lozzio BB, Lozzi CB, Machado E. Human myelogenous (Ph+) leukemia cell line: transplantation into athymic mice. J Natl Cancer Inst. 1976;56(3):627–9.
70. Skorski T, et al. Suppression of Philadelphia1 leukemia cell growth in mice by BCR-ABL antisense oligodeoxynucleotide. Proc Natl Acad Sci U S A. 1994;91(10):4504–8.
71. Yaccoby S, Barlogie B, Epstein J. Primary myeloma cells growing in SCID-hu mice: a model for studying the biology and treatment of myeloma and its manifestations. Blood. 1998;92(8):2908–13.
72. Heisterkamp N, et al. Acute leukaemia in bcr/abl transgenic mice. Nature. 1990;344(6263):251–3.
73. Yun JP, et al. Diet-induced obesity accelerates acute lymphoblastic leukemia progression in two murine models. Cancer Prev Res (Phila). 2010;3(10):1259–64.

74. Romana SP, et al. The t(12;21) of acute lymphoblastic leukemia results in a tel-AML1 gene fusion. Blood. 1995;85(12):3662–70.
75. Golub TR, et al. Fusion of the TEL gene on 12p13 to the AML1 gene on 21q22 in acute lymphoblastic leukemia. Proc Natl Acad Sci U S A. 1995;92(11):4917–21.
76. Bernardin F, et al. TEL-AML1, expressed from t(12;21) in human acute lymphocytic leukemia, induces acute leukemia in mice. Cancer Res. 2002;62(14):3904–8.
77. Neale GA, Rehg JE, Goorha RM. Ectopic expression of rhombotin-2 causes selective expansion of CD4-CD8- lymphocytes in the thymus and T-cell tumors in transgenic mice. Blood. 1995;86(8):3060–71.
78. Larson RC, et al. T cell tumours of disparate phenotype in mice transgenic for Rbtn-2. Oncogene. 1994;9(12):3675–81.
79. Weng AP, et al. Activating mutations of NOTCH1 in human T cell acute lymphoblastic leukemia. Science. 2004;306(5694):269–71.
80. Kong G, et al. NOTCH1 gene mutations target KRAS G12D-expressing CD8+ cells and contribute to their leukemogenic transformation. J Biol Chem. 2013;288(25):18219–27.
81. Kindler T, et al. K-RasG12D-induced T-cell lymphoblastic lymphoma/leukemias harbor Notch1 mutations and are sensitive to gamma-secretase inhibitors. Blood. 2008;112(8):3373–82.
82. Pastore F, et al. Combined molecular and clinical prognostic index for relapse and survival in cytogenetically normal acute myeloid leukemia. J Clin Oncol. 2014;32(15):1586–94.
83. Schlenk RF, et al. Mutations and treatment outcome in cytogenetically normal acute myeloid leukemia. N Engl J Med. 2008;358(18):1909–18.
84. Lange BJ, et al. Mortality in overweight and underweight children with acute myeloid leukemia. JAMA. 2005;293(2):203–11.
85. Castillo JJ, et al. Obesity but not overweight increases the incidence and mortality of leukemia in adults: a meta-analysis of prospective cohort studies. Leuk Res. 2012;36(7):868–75.
86. Estey E, et al. Association between increased body mass index and a diagnosis of acute promyelocytic leukemia in patients with acute myeloid leukemia. Leukemia. 1997;11(10):1661–4.
87. Breccia M, et al. Increased BMI correlates with higher risk of disease relapse and differentiation syndrome in patients with acute promyelocytic leukemia treated with the AIDA protocols. Blood. 2012;119(1):49–54.
88. Mazzarella L, Durfort T, Pelicci PG. Modelling the influence of diet on APL identifies insulin-growth factor 1 as a central mediator and provides a mechanistic rationale for therapeutic weight loss in American Society of Hematology Annual Meeting and Exposition. 2013. New Orleans, LA.
89. Thirman MJ, et al. Rearrangement of the MLL gene in acute lymphoblastic and acute myeloid leukemias with 11q23 chromosomal translocations. N Engl J Med. 1993;329(13):909–14.
90. Krivtsov AV, et al. H3K79 methylation profiles define murine and human MLL-AF4 leukemias. Cancer Cell. 2008;14(5):355–68.
91. Krivtsov AV, et al. Transformation from committed progenitor to leukaemia stem cell initiated by MLL-AF9. Nature. 2006;442(7104):818–22.
92. Corral J, et al. An Mll-AF9 fusion gene made by homologous recombination causes acute leukemia in chimeric mice: a method to create fusion oncogenes. Cell. 1996;85(6):853–61.
93. Horton SJ, et al. Acute myeloid leukemia induced by MLL-ENL is cured by oncogene ablation despite acquisition of complex genetic abnormalities. Blood. 2009;113(20):4922–9.
94. Lavau C, et al. Immortalization and leukemic transformation of a myelomonocytic precursor by retrovirally transduced HRX-ENL. Embo J. 1997;16(14):4226–37.
95. DiMartino JF, et al. The AF10 leucine zipper is required for leukemic transformation of myeloid progenitors by MLL-AF10. Blood. 2002;99(10):3780–5.
96. Fu JF, Hsu CL, Shih LY. MLL/AF10(OM-LZ)-immortalized cells expressed cytokines and induced host cell proliferation in a mouse bone marrow transplantation model. Int J Cancer. 2010;126(7):1621–9.

97. Fu JF, et al. Involvement of Gpr125 in the myeloid sarcoma formation induced by cooperating MLL/AF10(OM-LZ) and oncogenic KRAS in a mouse bone marrow transplantation model. Int J Cancer. 2013;133(8):1792–802.
98. Samanic C, et al. Obesity and cancer risk among white and black United States veterans. Cancer Causes Control. 2004;15(1):35–43.
99. Avcu F, et al. Association of plasma adiponectin concentrations with chronic lymphocytic leukemia and myeloproliferative diseases. Int J Hematol. 2006;83(3):254–8.
100. Tung S, et al. PPARalpha and fatty acid oxidation mediate glucocorticoid resistance in chronic lymphocytic leukemia. Blood. 2013;122(6):969–80.
101. Raveche ES, et al. Abnormal microRNA-16 locus with synteny to human 13q14 linked to CLL in NZB mice. Blood. 2007;109(12):5079–86.
102. Phillips JA, et al. The NZB mouse as a model for chronic lymphocytic leukemia. Cancer Res. 1992;52(2):437–43.
103. Klaus S, et al. Epigallocatechin gallate attenuates diet-induced obesity in mice by decreasing energy absorption and increasing fat oxidation. Int J Obes (Lond). 2005;29(6):615–23.
104. Haskell BD, et al. The diabetes-prone NZO/HlLt strain. I. Immunophenotypic comparison to the related NZB/BlNJ and NZW/LacJ strains. Lab Invest. 2002;82(7):833–42.
105. Narducci MG, et al. Regulation of TCL1 expression in B- and T-cell lymphomas and reactive lymphoid tissues. Cancer Res. 2000;60(8):2095–100.
106. Bichi R, et al. Human chronic lymphocytic leukemia modeled in mouse by targeted TCL1 expression. Proc Natl Acad Sci U S A. 2002;99(10):6955–60.
107. Munzert G, et al. Tumor necrosis factor receptor-associated factor 1 gene overexpression in B-cell chronic lymphocytic leukemia: analysis of NF-kappa B/Rel-regulated inhibitors of apoptosis. Blood. 2002;100(10):3749–56.
108. Aguilar-Santelises M, et al. Bcl-2, Bax and p53 expression in B-CLL in relation to in vitro survival and clinical progression. Int J Cancer. 1996;69(2):114–9.
109. Zapata JM, et al. TNF receptor-associated factor (TRAF) domain and Bcl-2 cooperate to induce small B cell lymphoma/chronic lymphocytic leukemia in transgenic mice. Proc Natl Acad Sci U S A. 2004;101(47):16600–5.
110. Montgomery MK, et al. Mouse strain-dependent variation in obesity and glucose homeostasis in response to high-fat feeding. Diabetologia. 2013;56(5):1129–39.
111. Chronic Myelogenous Leukemia. 2014. http://www.cancer.org/cancer/leukemia-chronic-myeloidcml/. Accessed 28 July 2014.
112. Strom SS, et al. Obesity, weight gain, and risk of chronic myeloid leukemia. Cancer Epidemiol Biomarkers Prev. 2009;18(5):1501–6.
113. Nakao T, et al. Expression of the leptin receptor in human leukaemic blast cells. Br J Haematol. 1998;102(3):740–5.
114. Hino M, et al. Leptin receptor and leukemia. Leuk Lymphoma. 2000;36(5/6):457–61.
115. Li S, et al. The P190, P210, and P230 forms of the BCR/ABL oncogene induce a similar chronic myeloid leukemia-like syndrome in mice but have different lymphoid leukemogenic activity. J Exp Med. 1999;189(9):1399–412.
116. Zhang X, Ren R. Bcr-Abl efficiently induces a myeloproliferative disease and production of excess interleukin-3 and granulocyte-macrophage colony-stimulating factor in mice: a novel model for chronic myelogenous leukemia. Blood. 1998;92(10):3829–40.
117. Neubauer A, et al. Expression of axl, a transforming receptor tyrosine kinase, in normal and malignant hematopoiesis. Blood. 1994;84(6):1931–41.
118. Augustine KA, et al. Noninsulin-dependent diabetes mellitus occurs in mice ectopically expressing the human Axl tyrosine kinase receptor. J Cell Physiol. 1999;181(3):433–47.
119. Non-Hodgkin Lymphoma. 2014. http://www.cancer.org/cancer/non-hodgkinlymphoma/index. Accessed 28 July 2014.
120. Chiu BC, et al. Obesity and risk of non-Hodgkin lymphoma (United States). Cancer Causes Control. 2007;18(6):677–85.

121. Pan SY, Mao Y, Ugnat AM. Physical activity, obesity, energy intake, and the risk of non-Hodgkin's lymphoma: a population-based case-control study. Am J Epidemiol. 2005;162(12):1162–73.
122. Teras LR, et al. Recreational physical activity, leisure sitting time and risk of non-Hodgkin lymphoid neoplasms in the American Cancer Society Cancer Prevention Study II Cohort. Int J Cancer. 2012;131(8):1912–20.
123. Chiu BC, et al. Diet and risk of non-Hodgkin lymphoma in older women. JAMA. 1996;275(17):1315–21.
124. Purdue MP, et al. Dietary factors and risk of non-Hodgkin lymphoma by histologic subtype: a case-control analysis. Cancer Epidemiol Biomarkers Prev. 2004;13(10):1665–76.
125. Zhang S, et al. Dietary fat and protein in relation to risk of non-Hodgkin's lymphoma among women. J Natl Cancer Inst. 1999;91(20):1751–8.
126. Zheng T, et al. Diet and nutrient intakes and risk of non-Hodgkin's lymphoma in Connecticut women. Am J Epidemiol. 2004;159(5):454–66.
127. Charbonneau B, et al. Trans fatty acid intake is associated with increased risk and n3 fatty acid intake with reduced risk of non-Hodgkin lymphoma. J Nutr. 2013;143(5):672–81.
128. Skibola CF, et al. Body mass index, leptin and leptin receptor polymorphisms, and non-hodgkin lymphoma. Cancer Epidemiol Biomarkers Prev. 2004;13(5):779–86.
129. Skibola DR, et al. Polymorphisms in ghrelin and neuropeptide Y genes are associated with non-Hodgkin lymphoma. Cancer Epidemiol Biomarkers Prev. 2005;14(5):1251–6.
130. Festing MF, Blackmore DK. Life span of specified-pathogen-free (MRC category 4) mice and rats. Lab Anim. 1971;5(2):179–92.
131. Frith CH, et al. Spontaneous lesions in virgin and retired breeder BALB/c and C57BL/6 mice. Lab Anim Sci. 1983;33(3):273–86.
132. Shimada Y, et al. Genetic susceptibility to thymic lymphomas and K-ras gene mutation in mice after exposure to X-rays and N-ethyl-N-nitrosourea. Int J Radiat Biol. 2003;79(6):423–30.
133. Ward JM. Lymphomas and leukemias in mice. Exp Toxicol Pathol. 2006;57(5–6):377–81.
134. Cloyd MW, Chattopadhyay SK. A new class of retrovirus present in many murine leukemia systems. Virology. 1986;151(1):31–40.
135. Mori S, et al. Utilization of pathway signatures to reveal distinct types of B lymphoma in the Emicro-Myc model and human diffuse large B-cell lymphoma. Cancer Res. 2008;68(20):8525–34.
136. Egle A, et al. VavP-Bcl2 transgenic mice develop follicular lymphoma preceded by germinal center hyperplasia. Blood. 2004;103(6):2276–83.
137. Blackwell BN, et al. Longevity, body weight, and neoplasia in ad libitum-fed and diet-restricted C57BL6 mice fed NIH-31 open formula diet. Toxicol Pathol. 1995;23(5):570–82.
138. Shields BA, et al. Calorie restriction suppresses subgenomic mink cytopathic focus-forming murine leukemia virus transcription and frequency of genomic expression while impairing lymphoma formation. Proc Natl Acad Sci U S A. 1991;88(24):11138–42.
139. Huang J, et al. The PPARalpha agonist fenofibrate suppresses B-cell lymphoma in mice by modulating lipid metabolism. Biochim Biophys Acta. 2013;1831(10):1555–65.
140. Kant S, Kumar A, Singh SM. Tumor growth retardation and chemosensitizing action of fatty acid synthase inhibitor orlistat on T cell lymphoma: implication of reconstituted tumor microenvironment and multidrug resistance phenotype. Biochim Biophys Acta. 2014;1840(1):294–302.
141. Cameron RG, et al. Changes in lymphoma development in female SJL/J mice as a function of the ratio in low polyunsaturated/high polyunsaturated fat diet. Cancer Lett. 1986;30(2):175–80.
142. el Ayachi N, et al. Susceptibility of RDM4 lymphoma cells to LAK-mediated lysis is decreased in tumor bearers fed fish oil high fat regimen. Cancer Lett. 1990;49(3):217–24.
143. Oth D, et al. Modulation of CD4 expression on lymphoma cells transplanted to mice fed (n-3) polyunsaturated fatty acids. Biochim Biophys Acta. 1990;1027(1):47–52.

144. Somers CM, Kwiecien JM, Quinn JS. A marine fish diet reduces spontaneous lymphoma in outbred Swiss-Webster mice. Leuk Lymphoma. 2005;46(12):1797–800.
145. Padilla J, Kaur K, Phipps RP. Effects of 15-deoxy-delta 12,14-PGJ2 on B lineage cells. Ann N Y Acad Sci. 2000;905:322–5.
146. Konopleva M, et al. Peroxisome proliferator-activated receptor gamma and retinoid X receptor ligands are potent inducers of differentiation and apoptosis in leukemias. Mol Cancer Ther. 2004;3(10):1249–62.
147. Ray DM, Akbiyik F, Phipps RP. The peroxisome proliferator-activated receptor gamma (PPARgamma) ligands 15-deoxy-delta12,14-prostaglandin J2 and ciglitazone induce human B lymphocyte and B cell lymphoma apoptosis by PPARgamma-independent mechanisms. J Immunol. 2006;177(8):5068–76.
148. Eucker J, et al. Peroxisome proliferator-activated receptor-gamma ligands inhibit proliferation and induce apoptosis in mantle cell lymphoma. Anticancer Drugs. 2006;17(7):763–9.
149. Jo SH, et al. Peroxisome proliferator-activated receptor gamma promotes lymphocyte survival through its actions on cellular metabolic activities. J Immunol. 2006;177(6):3737–45.
150. Hodgkin Disease. 2014. www.cancer.org/cancer/hodgkindisease/.
151. Willett EV, Roman E. Obesity and the risk of Hodgkin lymphoma (United Kingdom). Cancer Causes Control. 2006;17(8):1103–6.
152. Zamecnik PC, Long JC. Growth of cultured cells from patients with Hodgkin's disease and transplantation into nude mice. Proc Natl Acad Sci U S A. 1977;74(2):754–8.
153. Kapp U, et al. Preliminary report: growth of Hodgkin's lymphoma derived cells in immune compromised mice. Ann Oncol. 1992;3 Suppl 4:21–3.
154. Dewan MZ, et al. Hodgkin's lymphoma cells are efficiently engrafted and tumor marker CD30 is expressed with constitutive nuclear factor-kappaB activity in unconditioned NOD/SCID/gammac(null) mice. Cancer Sci. 2005;96(8):466–73.
155. Multiple Myeloma. 2014. http://www.cancer.org/cancer/multiplemyeloma/.
156. Kern PA, et al. Adipose tissue tumor necrosis factor and interleukin-6 expression in human obesity and insulin resistance. Am J Physiol Endocrinol Metab. 2001;280(5):E745–51.
157. Cozen W, et al. Interleukin-6-related genotypes, body mass index, and risk of multiple myeloma and plasmacytoma. Cancer Epidemiol Biomarkers Prev. 2006;15(11):2285–91.
158. Hilbert DM, et al. Interleukin 6 is essential for in vivo development of B lineage neoplasms. J Exp Med. 1995;182(1):243–8.
159. Medina EA, et al. PKA/AMPK signaling in relation to adiponectin's antiproliferative effect on multiple myeloma cells. Leukemia. 2014;28(10):2080–9
160. Caers J, et al. Neighboring adipocytes participate in the bone marrow microenvironment of multiple myeloma cells. Leukemia. 2007;21(7):1580–4.
161. Huang H, et al. Multiple myeloma cells undergo differentiation upon exposure to rosiglitazone and all-trans retinoic acid. Leuk Lymphoma. 2009;50(6):966–73.
162. Tirado-Velez JM, et al. Low-density lipoprotein cholesterol suppresses apoptosis in human multiple myeloma cells. Ann Hematol. 2012;91(1):83–8.
163. Tirado-Velez JM, et al. Inhibition of fatty acid metabolism reduces human myeloma cells proliferation. PLoS ONE. 2012;7(9):e46484.
164. Radl J, et al. Animal model of human disease. Multiple myeloma. Am J Pathol. 1988;132(3):593–7.
165. Lwin ST, et al. Diet-induced obesity promotes a myeloma-like condition in vivo. Leukemia. 2015;29(2):507–10.
166. Cheung WC, et al. Novel targeted deregulation of c-Myc cooperates with Bcl-X(L) to cause plasma cell neoplasms in mice. J Clin Invest. 2004;113(12):1763–73.
167. Potter M, Wax JS. Peritoneal plasmacytomagenesis in mice: comparison of different pristane dose regimens. J Natl Cancer Inst. 1983;71(2):391–5.

Chapter 6
Energy Balance, IGF-1, and Cancer: Causal Lessons from Genetically Engineered Mice

Stephen D. Hursting, Emily L. Rossi, Laura W. Bowers and Laura M. Lashinger

Abstract This chapter summarizes key findings, particularly from genetically engineered mouse models with alterations in serum levels of insulin-like growth factor (IGF)-1 on the biological mechanisms underlying many of the anticancer effects of calorie restriction (CR), with a particular focus on the role of IGF-1, its related downstream signaling pathways, and its interactions with other energy balance-related hormones, growth factors, and cytokines. It also describes some of the epidemiological and experimental evidence linking IGF-1, energy balance, and cancer and the emerging opportunities for investigation that will facilitate the translation of preclinical research on energy balance and cancer into effective strategies to prevent and control human cancer.

Keywords Insulin-like growth factor-1 · Obesity · Cancer · Calorie restriction · Energy balance · Adiposity · Mammalian target of rapamycin

S. D. Hursting (✉) · E. L. Rossi · L. W. Bowers . L. M. Lashinger
Department of Nutrition, University of North Carolina at Chapel Hill, 135 Dauer Drive, MJHRC2109, Chapel Hill, NC 27599, USA
e-mail: hursting@email.unc.edu

E. L. Rossi
e-mail: emilylro@live.unc.edu

L. W. Bowers
e-mail: laurawbowers@gmail.com

L. M. Lashinger
Department of Nutritional Sciences, The University of Texas at Austin, Austin, TX, USA
e-mail: lalash@austin.utexas.edu

N. A. Berger (ed.), *Murine Models, Energy Balance, and Cancer*,
Energy Balance and Cancer, DOI 10.1007/978-3-319-16733-6_6

Introduction

The Roman philosopher Titus Lucretius Carus (99–55 BC) is considered the first to publish a statement about the potential negative impact of the overconsumption of food on risk of chronic diseases such as cancer [1]. This possible link between excess energy intake and cancer developed into a working hypothesis espoused by John Hughes Bennett [2] and William Lambe [3] in the mid-1800s. Over the past century, animal models have played key roles in defining the calorie–cancer relationship. The first known experimental tests of the hypothesis that a low-energy calorie restriction (CR) dietary regimen can suppress tumors in animal models were reported in the early 1900s by Moreschi [4], Sweet, Corson-White and Saxon [5], and Rous [6]. These investigators showed that restricting food intake, relative to ad libitum (AL)-fed controls, inhibited the growth of transplanted tumors in mice. Intense interest in the comparison of CR versus AL-fed animals developed in the 1930s, when McCay et al. showed that reduced energy intake also increased lifespan in rodents [7]. CR research was further catalyzed by Albert Tannenbaum et al., who established that the incidence of tumors induced by various means (e.g., spontaneous, transplanted, or after carcinogen exposure) in mice, decreased when food intake was reduced [8].

As we and others have previously reviewed, CR has become the most widely studied and effective experimental strategy for the increasing median and maximal lifespan in rodents [9, 10]. CR is also the most potent, broadly acting dietary intervention for suppressing cancer development or progression in experimental models [10]. Studies of CR in rhesus monkeys and humans indicate extended lifespan and delayed tumor development in response to CR, suggesting that the anticancer effects of CR reported in rodent models may extend to humans. The monkey studies involved two cohorts of rhesus macaques; one led by Weindruch et al. in Madison, Wisconsin [11] and the other by Mattison et al. at the National Institute of Aging (NIA) in Baltimore, MD [12]. Both studies showed consistent anticancer effects of CR when begun in young adults [11, 12]. However, in the Mattison et al. study [12], there was no anticancer effect of CR when begun in older adults, and there was no effect of CR, regardless of age of onset, on overall survival. This is in contrast with the earlier report by Weindruch et al. [11] that shows both antiaging and anticancer effects of CR. Several differences between the studies may account for their differential findings. The Weindruch group, relative to the Mattison group, used a more purified, energy-dense diet that was ~30 % sucrose (versus 4 % sucrose in the Mattison study). Thus, the Weindruch group's controls, relative to the Mattison group's controls, were more obese and less healthy, and hence their CR monkeys had a greater difference in weight and metabolic parameters. The diets fed to the monkeys in the Mattison study also contained fish oil and were likely higher in phytochemicals, which probably contributed further to their monkeys being healthier and more metabolically similar regardless of caloric intake. Differences in genetics or microbiota may also have contributed to the observed differences, since the rhesus macaques used in the two studies came from different countries. Nonetheless, taken together there is evidence that CR can prevent cancer and improve metabolic tone in monkeys, with the magnitude of the effect dependent on several factors, including age of onset, nutritional quality of the diet, and genetic susceptibility. These are important and encouraging findings that suggest that the mechanisms characterized in animal model studies, and their translation into intervention targets and strategies, will have relevance to the prevention and treatment of cancers (particularly those related to obesity) in humans.

Several human studies have addressed the question of whether the observed health benefits of CR in rodents and nonhuman primates translate to human beings. This includes the Comprehensive Assessment of Long-Term Effects of Reducing Intake of Energy (CALERIE) study that evaluates the effects of a 2-year CR regimen (25 % less energy than controls) in healthy, nonobese individuals [13]. Preliminary reports on CALERIE indicate that many of the same metabolic and endocrine changes observed in rodents and monkeys are also occurring in human beings in response to CR [13]. These findings are consistent with studies on women at high risk for breast cancer, showing that inflammatory and growth factor signaling pathways are reduced by total CR or 2 days/week of restricted carbohydrate calories [14–16]. The observed metabolic effects of 2 days/week of restricted carbohydrate calories are of particular interest [14] since it is easier for most people to restrict a single macronutrient, such as carbohydrates, for a specific time period than to restrict total energy chronically. Unfortunately, despite more than a century of work on this topic (primarily in animals and more recently in humans), the mechanisms underlying this calorie–cancer connection remain unclear.

Findings from Genetically Engineered Mouse Models (GEMMs) About the Causal Links Between Energy Balance, IGF-1, and Cancer

Increased Cancer Risk in the Fatless A-Zip/F1 Mouse Model

A-Zip/F1 transgenic mice lack white adipose tissue (WAT) and have undetectable serum levels of leptin, adiponectin, and other adipokines, but they develop alipotrophic diabetes at a young age and have elevated insulin, IGF-1, inflammatory cytokines, and other metabolic perturbations often associated with obesity [17]. We showed that these mice are more susceptible to papilloma formation in a classical two-stage skin carcinogenesis experiment [17], a finding confirmed by others in a separate report [18]. Furthermore, when A-Zip/F1 mice were crossed with C3(1)/T-antigen transgenic mice (which spontaneously develop basal-like mammary tumors), the F1 female mice developed larger and earlier mammary tumors than their mothers [19]. Taken together, these findings (illustrated in Fig. 6.1) suggest that factors other than adipokines associated with obesity may be required for the obesity-associated enhancement of tumor development.

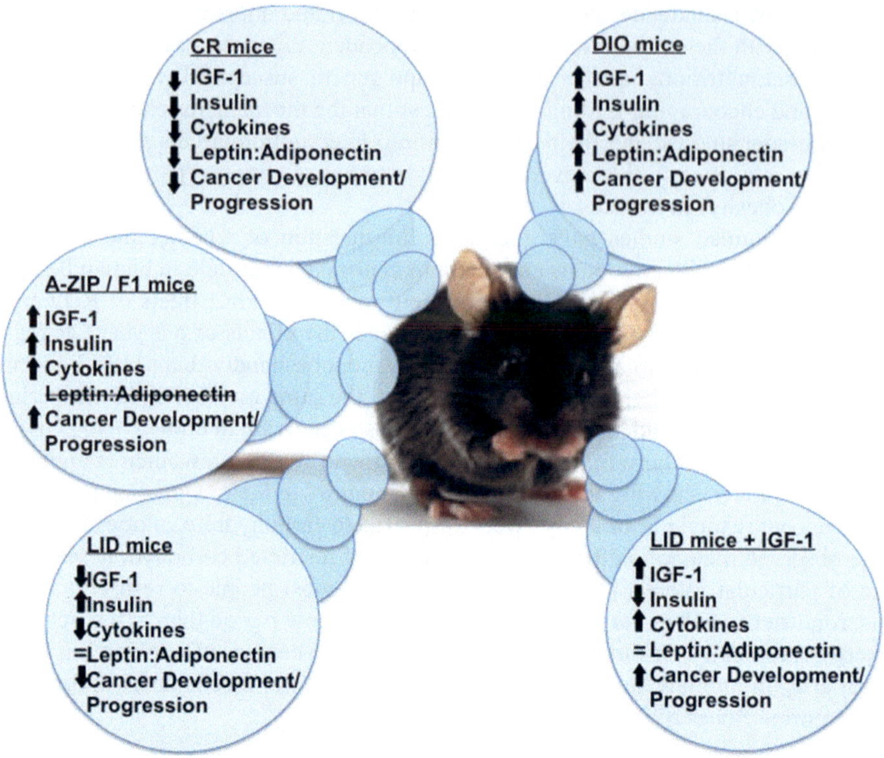

Fig. 6.1 Comparison of serum levels of insulin-like growth factor (*IGF*)-1, insulin, cytokines, leptin:adiponectin ratio, and cancer development or progression in diet-induced obese (*DIO*) mice, 30 % calorie restricted (*CR*) mice, A-Zip/F1 fatless mice, liver-specific IGF-1 deficient (*LID*) mice, and LID mice with exogenous IGF-1 infusion. Arrows indicate direction of significant change relative to wild-type mice on a control diet as reported in multiple mouse models in references [17, 19, 20–23, 27–29]. = indicates no change from wild-type control, cross-out of leptin:adiponectin (~~leptin:adiponectin~~) indicates no detectable levels of either factor in A-Zip/F1 mice. Cytokines refers to a panel of 6 cytokines (including IL-1, 2, 6, 10, TNF-α, and interferon-γ)

What may underlie the increased susceptibility to cancer in A-Zip/F1 mice? As mentioned, the A-Zip/F1 mice have high circulating levels of insulin, cytokines, and IGF-1, as well as activation of several carcinogenesis-related signaling pathways, particularly those downstream of the insulin and IGF-1 receptors, despite their lack of WAT [17]. This includes the Akt/mammalian target of rapamycin pathway (mTOR; discussed in more detail below and illustrated in Fig. 6.2), which we found was activated in the A-Zip/F1 mice [17]. It is possible that activation of these mitogenic and pro-inflammatory pathways, in the absence of obesity, underlies the observed accelerated tumor development in A-Zip/F1 mice. Thus, the studies on A-Zip/F1 fatless mice dissociate the effects of obesity-associated growth factor dysregluation from the myriad effects of increased adiposity. Deconvoluting these links will enhance our understanding of the relationships between obesity and different

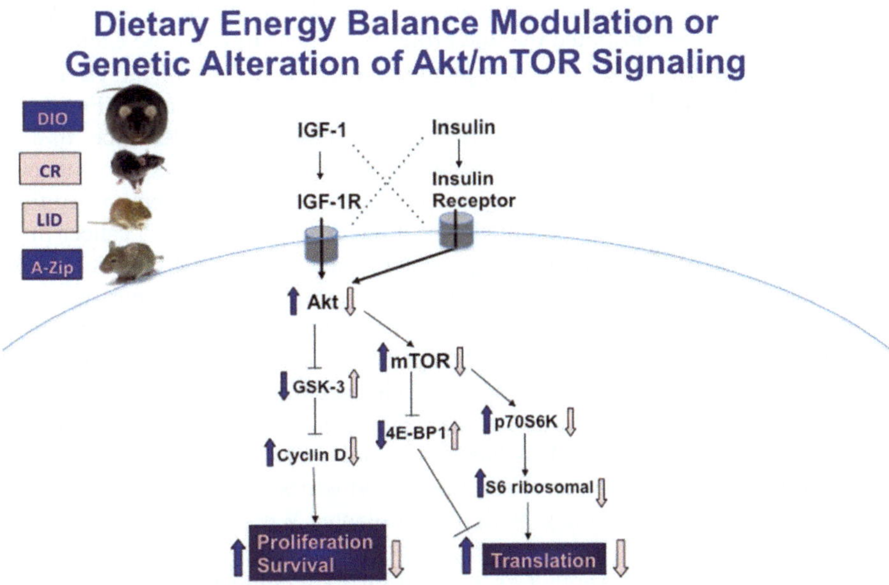

Fig. 6.2 The effect of dietary energy balance modulation or genetic alteration of adiposity or circulating insulin-like growth factor (*IGF*)-1 levels on signaling through the PI3K/Akt/mammalian target of rapamycin pathway (*mTOR*). *A-Zip/F1* fatless mice mirror diet-induced obese (*DIO*) mice, whereas liver-specific IGF-1 deficient (*LID*) mice mirror calorie restricted (*CR*) mice in their effects on phosphorylation (activation) of PI3K, Akt, mTOR, and their downstream targets. The *arrows* represent increased or decreased phosphorylation of each kinase relative to control levels as determined by Western blot analyses and immunohistochemistry as reported in multiple mouse models in references [17, 19, 20–23, 27–29]

cancer types and offer an opportunity to identify potential therapeutic and prevention targets.

Genetic Modulation of IGF-1 or mTOR Alters Tumorigenesis

Liver-specific IGF-1-deficient (LID) mice, which have a deletion in hepatic IGF-1 and consequently have reduced circulating IGF-1 levels, have been very useful in demonstrating that IGF-1 is an important tumor growth factor in the response to energy balance modulation (Figs. 6.1 and 6.2). Like CR, genetic reduction of IGF-1 in LID mice is associated with decreased mammary [19], colon [20], skin [21], and pancreatic tumor development and/or growth [22, 23]. Also similar to CR, the reduction in skin [21], mammary [19], and pancreatic tumors [22, 23] was associated with reduced steady-state signaling through the Akt/mTOR pathway (Fig. 6.2). Furthermore, restoration of serum IGF-1 levels in LID mice by infusion of recombinant IGF-1 restored pancreatic tumor growth and pancreatic mTOR signaling [22, 23]. Additional support for a causal relationship between energy balance, circulating IGF-1 levels, and tumor growth comes from the studies of Ford et al. [19]. Using a polyomavirus middle-T antigen transgenic mouse model of luminal-type mammary cancer, this report demonstrated that the protective effects of CR were exclusively IGF-1-dependent, and IGF-1 levels explained most, but not all, of the effects of a diet-induced obese (DIO) regimen on mammary tumor development (Fig. 6.2). In contrast to the reduction in tumor development observed in LID mice, tissue-specific overexpression of IGF-1 via the keratin 5 promoter increases spontaneous tumor development and susceptibility to carcinogens [24–26], providing further support that IGF-1 is an important mediator of the energy balance and cancer link. However, the role of IGF-1 in the energy balance–cancer connection is likely to be complex. Consistent with this inherent complexity are two other findings from our lab. For example, we observed that long-term exposure to a DIO regimen followed by weight loss later in life resulted in only a partial restoration of the metabolic dysregulation associated with the obese state and elevated tumor susceptibility, despite normalization of weight [27]. Specifically, 20 weeks of a DIO regimen followed by 8 weeks of gradual weight loss back to the level of control mice normalized leptin and insulin levels, but IGF-1 and several cytokines remained elevated in the "formerly obese mice," and the growth of transplanted Wnt-1 mammary tumors in these formerly obese mice was comparable to that of mice maintained on the DIO diet throughout the study. Another example of inherent complexity in the associations between energy balance, IGF-1, and cancer involves nature versus nurture, that is, the contributions of systemic factors such as IGF-1 in the context of cell autonomous effects. We found that constitutive activation of the mTOR pathway in Wnt-1 tumor cells prior to transplantation makes the resulting tumors insensitive to CR [28]. Kalaany and Sabatini [29] similarly reported that kidney cancer cells with constitutively activated PI3K mutations are proliferative in vitro in the absence of insulin or IGF-1 and form CR-resistant tumors in vivo. Taken together, these findings suggest that cell autonomous alterations, such as activating PI3K mutations, may influence the response of cells to alterations in energy balance and/or circulating levels of energy balance-related growth factors such as IGF-1.

Downstream Targets of IGF-1

Components of the PI3K/Akt and mTOR pathways are critical signal transducers of IGF-1 responses regulating cellular growth and metabolism [30]. The importance of the PI3K/Akt pathway in human cancers is evidenced by the observation that it is one of the most commonly altered pathways in human tumors [31–33]. Aberrations in Akt signaling, which commonly occur in human cancers, are frequently associated with elevations in mTORC1 signaling. As illustrated in Fig. 6.2, together these pathways form a sophisticated system that integrates cellular responses and environmental cues (mTORC1, a highly conserved serine/threonine protein kinase, acts as a sensor linking growth factor signals and energy status to translational control of new proteins [34]. mTORC1 regulates cell growth, proliferation, protein translation, and autophagy via serine/threonine kinase activity on downstream targets, most notably p70S6K and 4E-BP1 [35]. Nutrient deprivation conditions, such as those achieved during CR, inhibit mTORC1, leading to cell growth arrest, inhibition of protein translation, and induction of autophagy. We have demonstrated that modulation of Akt/mTORC1 signaling induced by dietary energy balance manipulation is due, in part, to altered signaling through the IGF-1R and modulation of IGF-1R/EGFR crosstalk [35].

Conclusions

Based on lessons learned from genetically engineered mouse models, such as A-Zip/F1 mice and LID mice, components of the IGF-1/Akt/mTOR pathway are emerging as plausible targets for breaking the obesity–cancer link. Clearly, no single pathway accounts for all of the procancer effects of obesity or the anticancer effects of CR (although as discussed, our findings suggest that most of the anticancer effects of CR in several GEMM models are largely attributable to the reduced IGF-1 levels). As with most chronic disease intervention strategies, combination approaches that target multiple pathways (and that maximize efficacy and minimize adverse effects) will likely be most successful for preventing cancer. Future studies that exploit the emerging mechanistic information, including from GEMMs, to target energy balance–responsive pathways through combinations of lifestyle (particularly diet and physical activity) and pharmacologic approaches should facilitate the translation of this research into more effective cancer prevention and treatment strategies.

References

1. Carus TL. The nature of things. New York: Penguin Group Incorporated; 2007.
2. Bennett JH. On cancerous and cancroid growths. Sutherland and Knox: BiblioLife; 1849.
3. Lambe W. Water and vegetable diet in consumption, scrofula, cancer, asthma, and other chronic diseases. New York: Fowlers & Wells; 1850.
4. Moreschi C. Beziehungen zwischen Ernahrung und Tumorwachstum. Z Immun. 1909;2:661–75.
5. Sweet JE, Corson-White EP, Saxon GJ. The relation of diets and of castration to the transmissible tumors of rats and mice. J Biol Chem. 1913;15:181–91.
6. Rous P. The influence of diet on transplanted and spontaneous mouse tumors. J Exp Med. 1914;20:433–51.
7. McCay CM, Crowell MF. Prolonging the life span. Sci Mon. 1934;39:405–14.
8. Tannenbaum A. The dependence of the genesis of induced skin tumors on the caloric intake during different stages of carcinogenesis. Cancer Res. 1944;4:673–79.
9. Chung KW, Kim DH, Park MH, Choi YJ, Kim ND, Lee J, et al. Recent advances in calorie restriction research on aging. Exp Gerontol. 2013;48:1049–53.
10. Hursting SD, Smith SM, Lashinger LM, Harvey AE, Perkins SN. Calories and carcinogenesis: lessons learned from 30 years of calorie restriction research. Carcinogenesis. 2010;31:83–9.
11. Colman RJ, Anderson RM, Johnson SC, Kastman EK, Kosmatka KJ, Beasley TM, et al. Caloric restriction delays disease onset and mortality in rhesus monkeys. Science. 2009;325:201–4.

12. Mattison JA, Roth GS, Beasley TM, Tilmont EM, Handy AM, Herbert RL, et al. Impact of caloric restriction on health and survival in rhesus monkeys from the NIA study. Nature. 2012;489:318–21.
13. Heilbronn LK, de Jonge L, Frisard MI, DeLany JP, Larson-Meyer DE, Rood J, et al. Effect of 6-month calorie restriction on biomarkers of longevity, metabolic adaptation, and oxidative stress in overweight individuals: a randomized controlled trial. JAMA. 2006;295:1539–48.
14. Harvie M, Wright C, Pegington M, McMullan D, Mitchell E, Martin B, et al. The effect of intermittent energy and carbohydrate restriction v. daily energy restriction on weight loss and metabolic disease risk markers in overweight women. Br J Nutr. 2013;110:1534–47.
15. Imayama I, Ulrich CM, Alfano CM, Wang C, Xiao L, Wener MH, et al. Effects of a caloric restriction weight loss diet and exercise on inflammatory biomarkers in overweight/obese postmenopausal women: a randomized controlled trial. Cancer Res. 2012;72:2314–26.
16. Fabian CJ, Kimler BF, Donnelly JE, Sullivan DK, Klemp JR, Petroff BK, et al. Favorable modulation of benign breast tissue and serum risk biomarkers is associated with >10% weight loss in postmenopausal women. Breast Cancer Res Treat. 2013;142:119–32.
17. Nunez NP, Oh WJ, Rozenberg J, Perella C, Anver M, Barrett JC, et al. Accelerated tumor formation in a fatless mouse with type 2 diabetes and inflammation. Cancer Res. 2006;66:5469–76.
18. Ablamunits V, Cohen Y, Brazee IB, Gaetz HP, Vinson C, Klebanov S. Susceptibility to induced and spontaneous carcinogenesis is increased in fatless A-ZIP/F-1 but not in obese ob/ob mice. Cancer Res. 2006;66:8897–902.
19. Ford NA, Nunez NP, Holcomb VB, Hursting SD. IGF1 dependence of dietary energy balance effects on murine Met1 mammary tumor progression, epithelial-to-mesenchymal transition, and chemokine expression. Endocr Relat Cancer. 2013;20:39–51.
20. Olivo-Marston SE, Hursting SD, Lavigne J, Perkins SN, Maarouf RS, Yakar S, et al. Genetic reduction of circulating insulin-like growth factor-1 inhibits azoxymethane-induced colon tumorigenesis in mice. Mol Carcinog. 2009;48:1071–6.
21. Moore T, Carbajal S, Beltran L, Perkins SN, Yakar S, Leroith D, et al. Reduced susceptibility to two-stage skin carcinogenesis in mice with low circulating insulin-like growth factor I levels. Cancer Res. 2008;68:3680–8.
22. Lashinger LM, Harrison LM, Rasmussen AJ, Logsdon CD, Fischer SM, McArthur MJ, et al. Dietary energy balance modulation of Kras- and Ink4a/Arf + /–driven pancreatic cancer: the role of insulin-like growth factor-I. Cancer Prev Res (Phila). 2013;6:1046–55.
23. Lashinger LM, Malone LM, McArthur MJ, Goldberg JA, Daniels EA, Pavone A, et al. Genetic reduction of insulin-like growth factor-1 mimics the anticancer effects of calorie restriction on cyclooxygenase-2-driven pancreatic neoplasia. Cancer Prev Res (Phila). 2011;4:1030–40.
24. Hursting SD, Perkins SN, Lavigne JA, Beltran L, Haines DC, Hill HL, et al. Urothelial overexpression of insulin-like growth factor-1 increases susceptibility to p-cresidine-induced bladder carcinogenesis in transgenic mice. Mol Carcinog. 2009;48:671–7.
25. DiGiovanni J, Bol DK, Wilker E, Beltran L, Carbajal S, Moats S, et al. Constitutive expression of insulin-like growth factor-1 in epidermal basal cells of transgenic mice leads to spontaneous tumor promotion. Cancer Res. 2000;60:1561–70.
26. DiGiovanni J, Kiguchi K, Frijhoff A, Wilker E, Bol DK, Beltran L, et al. Deregulated expression of insulin-like growth factor 1 in prostate epithelium leads to neoplasia in transgenic mice. Proc Natl Acad Sci U S A. 2000;97:3455–60.
27. De Angel RE, Conti CJ, Wheatley KE, Brenner AJ, Otto G, Degraffenried LA, et al. The enhancing effects of obesity on mammary tumor growth and Akt/mTOR pathway activation persist after weight loss and are reversed by RAD001. Mol Carcinog. 2013;52:446–58.
28. Nogueira LM, Dunlap SM, Ford NA, Hursting SD. Calorie restriction and rapamycin inhibit MMTV-Wnt-1 mammary tumor growth in a mouse model of postmenopausal obesity. Endocr Relat Cancer. 2012;19:57–68.
29. Kalaany NY, Sabatini DM. Tumours with PI3K activation are resistant to dietary restriction. Nature. 2009;458:725–31.
30. Pollak M. The insulin and insulin-like growth factor receptor family in neoplasia: an update. Nat Rev Cancer. 2012;12:159–69.

31. Engelman JA, Luo J, Cantley LC. The evolution of phosphatidylinositol 3-kinases as regulators of growth and metabolism. Nat Rev Genet. 2006;7:606–19.
32. Shaw RJ, Cantley LC. Ras, PI(3)K and mTOR signalling controls tumour cell growth. Nature. 2006;441:424–30.
33. Yuan TL, Cantley LC. PI3K pathway alterations in cancer: variations on a theme. Oncogene. 2008;27:5497–510.
34. Kim SG, Buel GR, Blenis J. Nutrient regulation of the mTOR complex 1 signaling pathway. Mol Cells. 2013;35:463–73.
35. Moore T, Beltran L, Carbajal S, Hursting SD, DiGiovanni J. Energy balance modulates mouse skin tumor promotion through altered IGF-1R and EGFR crosstalk. Cancer Prev Res (Phila). 2012;5:1236–46.

Chapter 7
Mouse Models to Study Leptin in Breast Cancer Stem Cells

Praveena S. Thiagarajan and Ofer Reizes

Abstract Leptin is a hormone originally identified as the gene mutated in the *obese* mouse and its receptor (LEPR) was identified as the gene mutated in the *diabetic* mouse. Both the *obese* and *diabetic* mice exhibit early onset obesity and have been extensively studied to delineate mechanisms of obesity and associated morbidities. Leptin/LEPR signaling regulate body fat via activation of brain pathways and deficiency in either gene is sufficient to induce obesity. Leptin/LEPR represent members of the druggable genome and over the past decade have been proposed as a link between obesity and breast cancer with studies focused on disrupting this pathway to inhibit breast cancer progression. This chapter reviews the roles of leptin and LEPR in breast cancer with a focus on tumor initiating or cancer stem cells (CSCs) in tumor progression. CSCs are cancer cells that exhibit resistance to chemo- and radio-therapy and are associated with recurrence and metastasis. Recently, leptin/LEPR have been shown to be necessary for survival of breast CSCs via induction in expression of the embryonic stem cell transcription factors *NANOG*, *SOX2*, and *OCT4*. These transcription factors are necessary for maintaining CSC self-renewal and pluripotency. Emerging studies strongly suggest that targeting CSC survival could inhibit cancer progression. The review will provide evidence for leptin/LEPR as a therapeutic target for inhibition of breast cancer progression and explore the mouse models to study this process.

O. Reizes (✉) · P. S. Thiagarajan
Department of Cellular & Molecular Medicine, Lerner Research Institute, Cleveland Clinic Foundation, 9500 Euclid Avenue, NC10, 44195 Cleveland, OH, USA
e-mail: thiagap@ccf.org

O. Reizes
e-mail: reizeso@ccf.org

N. A. Berger (ed.), *Murine Models, Energy Balance, and Cancer,*
Energy Balance and Cancer, DOI 10.1007/978-3-319-16733-6_7

Keywords Leptin receptor · Cancer stem cells · Transcription factors · NANOG · SOX 2 · OCT 4 · Janus kinase · Signal transduction and activator of transcription · ob/ob obese mice · db/db diabetic mice

Abbreviations

LEPR.	Leptin receptor
LEPRb.	Leptin receptor isoform b
CSC.	Cancer stem cell
JAK.	Janus kinases
STAT.	Signal transduction and activators of transcription
ER.	Estrogen receptor
ER+.	Estrogen receptor-positive
ER−.	Estrogen receptor-negative
CNS.	Central nervous system
ARN.	Arcuate nucleus
VMN.	Ventromedial nucleus
IGF.	Insulin-like growth factors
ERK.	Extracellular signal-regulated kinases
MAPK.	Mitogen-activated protein kinases
ob/ob.	Obese mice
db/db.	Diabetic mice
SH2.	Src homology 2
mTOR.	Mammalian target of rapamycin
PI3K.	Phosphatidylinositol-4,5-bisphosphate 3-kinase
SOCS3.	Suppressor of cytokine signaling 3
JNK.	c-Jun N-terminal kinase
TNF-α.	Tumor necrosis factor-α
IL-1.	Interleukin-1
AMPK.	AMP-Activated protein kinase
PKC.	Protein kinase C
TGF-α.	Transforming growth factor-α
MMTV.	Mouse mammary tumor virus
VEGF.	Vascular endothelial growth factor
VEGF-R.	Vascular endothelial growth factor receptor
LPrA2.	Leptin receptor antagonist
DMBA.	7,12-Dimethylbenz(a)anthracene
TNM.	Tumor, node, metastasis
TNBC.	Triple negative breast cancer
IHC.	Immunohistochemistry
IRS	Insulin receptor substrate

Introduction

Today, over 25 % of the US population is either obese or morbidly obese, a condition of excess adipose tissue and fat [1–5]. Obesity results primarily from an energy imbalance of increased calorie intake to energy expenditure [6]. Associated with obesity is a significantly increased risk in development of multiple diseases, including diabetes, cardiovascular disease, and of relevance to this book, cancer [7, 8]. Breast cancer is the most common malignancy and the leading cause of cancer death among women worldwide [9–11] with incidence more common among postmenopausal than premenopausal women [12–16]. Calle and colleagues provided the first compelling evidence for a link between obesity and breast cancer in postmenopausal women. The findings indicated both an increase in incidence and mortality from obesity in this patient population [7, 17]. Subsequent studies have provided clear evidence that obesity is associated with poor prognosis not only in postmenopausal women but also in premenopausal women [18]. The link to premenopausal women is particularly alarming as these women present with a more aggressive disease that is resistant to chemotherapy and highly metastatic [19]. The mechanisms underlying the link between obesity and breast cancer remain rudimentary with limited therapeutic options to disrupt the link [19–21].

Accumulated evidence suggests that adipose tissue is an important mediator of the breast cancer-obesity link [20–23]. Obesity is associated with changes within the breast as well as induction of a chronic low-grade inflammation and activation of pro-inflammatory conditions involved in the pathogenesis of obesity-related insulin resistance and type-2 diabetes [8, 24]. Indicators of obesity including body mass index, weight gain, and waist to hip ratio have provided clear evidence of a positive correlation with the risk of developing breast cancer [25]. Prognostic indicators of breast cancer including recurrence, metastases, and mortality are higher in obese women. Adipose tissue, the fat storing organ in the body, functions as a major endocrine organ and is also a source of cytokines that communicate with other organs on the levels of fat reserves and provide inflammatory, and immunological cues [26]. Expression of several adipocyte-derived factors is elevated in obesity and some of these have been implicated in cancer development. Among these adipose factors, the most extensively studied is leptin [27].

Leptin is a metabolic hormone primarily secreted by fat cells [28]. Its "textbook" physiological function is to regulate feeding behavior, metabolism, and body weight by binding to its specific receptor, (LEPR) [29, 30]. The primary physiological function of leptin is to convey the status of the energy or adipose stores to the central nervous system (CNS) to inhibit food intake and promote energy expenditure [6, 31, 32] (Fig. 7.1). This is achieved by leptin binding and activation of LEPR in the brain. Deficiency in either leptin or the LEPR leads to morbid obesity as a consequence of increased food intake and reduced energy metabolism [33]. LEPR has also been detected in peripheral tissues and is highly expressed in multiple tumors [34–37]. In breast cancer, leptin is implicated as a promoter of cell proliferation, migration, and induction of angiogenesis [38–40]. Many of these activities are due to LEPR activation of signaling pathways implicated in cancer [30, 41–44].

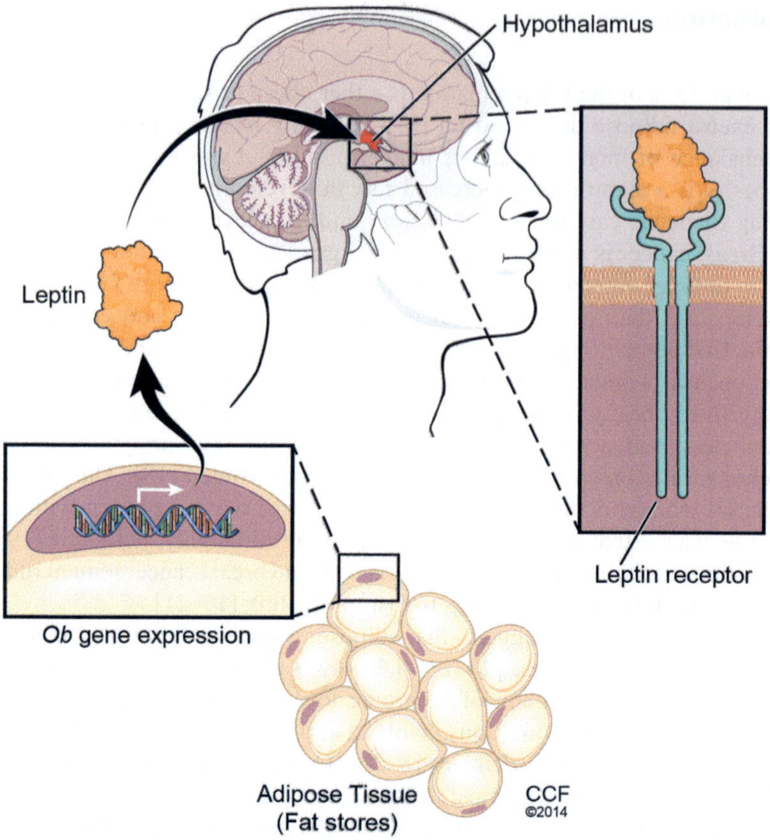

Fig. 7.1 Regulation of energy balance by leptin. Leptin circulates in the blood and activates neurons in the hypothalamus which in-turn regulate appetite and energy balance. Leptin secretion by adipose tissue is maintained by a negative feedback loop released in proportion to adipose mass. Secreted leptin binds primarily to its receptor in the hypothalamus inducing downstream signaling functions. Excess adipose tissue leads to increased leptin secretion resulting in suppression of appetite and increased energy expenditure inducing weight loss until the adipose tissue mass is restored. Leptin regulation maintains the homeostatic control of adipose tissue mass in the body

Leptin functions in conjunction with its cognate plasma membrane receptor LEPR to activate the Janus Kinase JAK2 that in turn activates the signal transducer and activator of transcription (STAT3/5) proteins, resulting in transcriptional regulation of its target genes [43, 45]. Notably, among STAT family members, STAT3, STAT5a, and STAT5b are known to play a role in cancer [46]. These proteins enhance cell cycle progression, angiogenesis, and survival, and are considered to be oncogenes [38, 40, 47, 48]. STAT target genes include the cell cycle regulators *cyclin D1* and *cyclin D3*, the oncogene *c-Myc*, the growth factor vascular endothelial growth factor (*VEGF*), genes involved in migration and invasion such as *MMP-2* and *MMP-9*, and anti-apoptotic genes including *survivin*, *Mcl-1*, and *Bcl-XL* [49,

50]. Moreover, JAK2/STAT signaling is closely involved in cancer stem cell renewal, a process thought to critically underlie tumor development discussed later in the chapter [51, 52]. Accordingly, it has been reported that leptin via LEPR potently promotes tumor development [53–56]. This chapter will focus on leptin/LEPR and models to study their role in breast cancer.

Crosstalk Between Leptin and Estrogen

Estrogen dependent breast cancer was one of the first cancer subtypes implicated in obesity [57]. Ovaries are the main site of estrogen synthesis in premenopausal women [58]. However, in postmenopausal women, circulating estrogens are produced by the adipose tissue [59]. Adipose tissue is a well-established source of high levels of circulating estrogens in postmenopausal women and has been associated with breast cancer. In support of the link, circulating estrogen is proportional to the amount of adipose tissue, making it an important source of estrogens. Circulating levels of estrogen correlates positively with obesity providing evidence for the role of estrogen in postmenopausal breast cancer pathogenesis [60]. Estrogen receptor-positive (ER+) breast cancer patients, primarily associated with obesity, have an overall better prognosis than estrogen receptor-negative (ER−) breast cancer patients [61]. However, ER+ breast cancer patients have a poorer prognosis if obese. Various mechanisms have been proposed linking postmenopausal obesity with breast cancer including increased estrogen synthesis by adipose tissue via aromatase (CYP19) [62, 63]. Aromatase is a critical enzyme in the conversion of androgens into estrogens in adipose tissue [63]. The high estrogen levels and impairment of cellular immunity in obese postmenopausal women induces development of metastatic disease [64–67].

Leptin can stimulate estrogen production through an increase of aromatase expression in postmenopausal women [68]. Early studies indicated that cytokines, which leptin is a member, secreted by macrophages regulate the synthesis of adipose tissue-specific promoters of stromal cells aromatase [63, 69]. Further, leptin can increase aromatase activity in human luteinized granulosa cells, adipose stromal cells, and human breast cancer cells [63, 70–73]. The increase in peripheral aromatization in obese postmenopausal women is hypothesized due to elevated circulating leptin [74, 75]. Androgen aromatization is not limited to fat tissue but breast tumor tissue has also been shown to possess aromatase activity influencing breast cancer progression in an autocrine fashion [27, 36, 63]. Estrogen could then influence breast cancer progression in a paracrine manner by interacting with the estrogen receptor (ER) on neighboring cancer cells [48, 76, 77].

LEPR and ER are co-expressed in mammary tumors and breast cancer cells [78] including the ER+MCF7 and T47D cells [79–81]. As expected, in human ER+ breast cancer cells, leptin increases cell viability and proliferation [39, 82]. These data suggest that leptin and estrogen might act cooperatively to promote tumor growth in estrogen-dependent breast cancers.

Adding to the cross talk between leptin and estrogen are rodent studies where the levels of estrogen are manipulated. In 8-week-old female rats, the relative expression of LEPR in the brain is altered by either estrogen treatment or ovariectomy. Expression of LEPR is decreased by estradiol treatment, while the expression is increased by ovariectomy in the arcuate nucleus (ARN) and the ventromedial nucleus (VMN) of the hypothalamus, regions necessary for regulation of body weight [83]. Consistent with these studies, ovariectomy in young adult female rats results in an initial lower serum leptin levels characterized by weight gain than leptin levels detected in ovariectomized mice treated with estradiol. The leptin levels in these two groups showed similar levels after 13 weeks of surgery [84–86]. These studies point to regulation of leptin expression by estrogen.

Leptin and Regulation of Body Weight

To understand leptin and utility of mouse models of breast cancer, it is necessary to first consider leptin's role in energy homeostasis [87, 88]. Leptin is the product of *ob* locus, identified by Friedman and colleagues in 1994, in the search for the gene mutated in obese (*ob/ob)* mice [28, 89, 90]. *ob/ob* mice exhibit early onset obesity and importantly lack any functional leptin. The mice provide a unique model for studying the role of leptin in various contexts including breast cancer as discussed later in the chapter. Leptin is highly expressed in the adipose tissue, though low levels can be detected in the skeletal muscle, placenta, and the brain [91]. Leptin is synthesized and secreted by adipocytes and preadipocytes as well as mammary epithelia and other cell types [37]. Leptin is a cytokine, based on its structural relationship to the long-chain four α-helical bundle, a characteristic of this family of proteins [92]. Leptin is considered an adipokine because it is a fat secreted cytokine [93]. Leptin is primarily considered a hormone due to its role at communication to the brain on the body's energy status [94]. As body fat increases, circulating leptin levels increase in the serum. Leptin is anorexigenic and functions primarily by activating the hypothalamic ARN nuclei neurons [95–97].

Leptin regulates energy balance through its effects on satiety and energy expenditure via LEPRs in the hypothalamus. In rodents, primates, and humans [98], leptin controls and regulates body fat stores by coordinating feeding behavior, metabolism, neuroendocrine responses, and the autonomic nervous system [87, 96]. Many studies have indicated that leptin regulates other physiological processes including fetal development, glycemic control, sexual maturation, reproduction, lactation, hematopoiesis, angiogenesis, immunity, bone development, and wound healing [38, 40, 67, 99–104].

The pleiotropic functions of leptin on multiple organ systems and the widespread expression of the LEPR reflect its diverse functions (Fig. 7.2). In addition to its central regulation of body weight, leptin also regulates glucose and lipid metabolism [105]. Leptin is also involved in the modulation of inflammatory response and

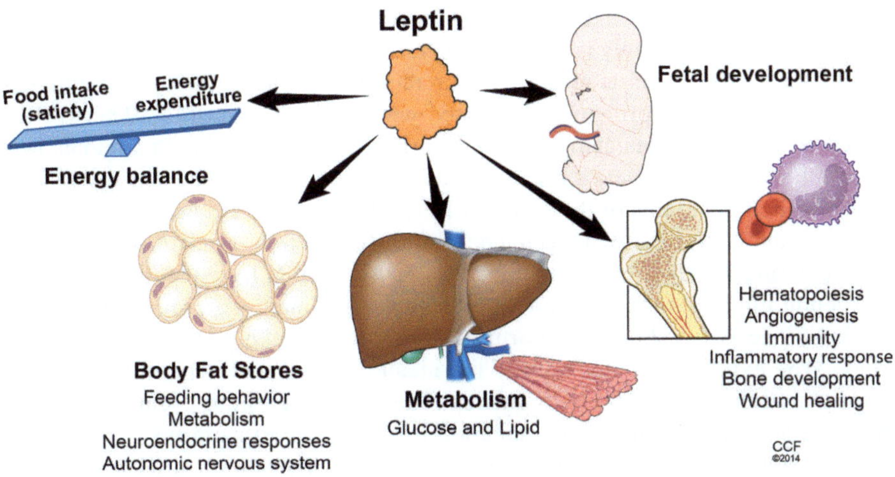

Fig. 7.2 Pleiotrophic functions of leptin. Circulating leptin is a multifunctional adipokine. It regulates energy balance and maintains body fat. Leptin also regulates the glucose and lipid metabolism, fetal development, hematopoiesis, angiogenesis, inflammatory and immune responses, bone development, and wound healing

immune system. Indeed, in adipose tissue, leptin expression is potently increased in pro-inflammatory conditions and down regulated by factors including exposure to cooler temperatures, adrenergic stimulation, and growth hormone [37]. Further, inflammatory stimuli including LPS and interleukin-1 (IL-1) acutely induce an increase in leptin levels indicating that leptin may participate in the host response to acute infection [97].

Structure and Functions of LEPR and Its Isoforms

Within a year of the discovery of leptin as the mutated gene in *ob*/ob mice, the receptor was identified as the gene mutated in the diabetic (*db/db)* mice. These mice exhibit early onset obesity, and are a phenocopy of the *ob/ob* mice [106]. Unlike the *ob/ob* mice, which are leptin-deficient, *db/db* mice are leptin resistant, thus have high levels of circulating leptin in the serum [96]. This distinction provides a unique model system to define the role of leptin in various physiological and pathophysiological processes including cancer.

Cloning of LEPR identified the receptor as a single membrane-spanning receptor of the class I cytokine family [107]. This is highly consistent with the structure of leptin as a cytokine. LEPRs are widely expressed in areas such as the cortex, cerebellum, brainstem, basal ganglia, and hippocampus throughout the brain [108]. Further, LEPRs are expressed in many other tissues, including placenta, pancreas, stomach, adrenal gland, hematopoietic cells, liver, lung, and heart [37, 91, 101].

LEPR is expressed in as many as six isoforms (LEPRa-f), formed by alternative splicing or proteolysis, only two of the splice isoforms are well studied. Leptin acts through the long form of its receptor, leptin receptor isoform b (LEPRb) mediating intracellular signaling primarily associated with JAK2 activity [44]. The LEPRb isoform is the isoform mutated in the *db/db* mice [87, 106]. The long functional isoform LEPRb has a complete intracellular domain with docking sites for the JAK2.

Four splicing variants of the LEPR (LEPRa, LEPRc, LEPRd and LEPRf) share the membrane-spanning region sequence but differ in the 3–11 additional residues in the C-termini at their cytosolic domain. LEPRa–LEPRf have identical extracellular leptin binding domains. LEPRa is the most abundant isoform of the LEPRs [109]. LEPRe is the shortest isoform of the receptor lacking the transmembrane region. The extracellular domains of the membrane-bound receptors can also be released into the circulation by proteases. The soluble receptors include both free and protein-bound leptin in the circulation [110]. The short isoforms of the LEPR have not been shown to elicit intracellular signaling cascades. LEPRa contains a truncated intracellular domain, and is unable to activate the STAT pathway but may still transduce signals by activating JAK2, insulin receptor substrate-1 (IRS-1c or extracellular factor-regulated kinases (ERKS) including mitogen-activated protein kinases (MAPKs) [44]. Presence of LEPR mRNA in the meninges and the microcirculation suggest that the shorter isoforms facilitate leptin uptake or efflux from cerebrospinal fluid in the receptor-mediated transport of leptin across the blood brain barrier [111]. Only LEPRb has an extended intracellular domain with the typical cytokine receptors structural elements [112].

LEPR Signaling

Upon leptin binding, LEPRb initiates signaling via multiple intracellular signal transduction pathways. LEPRb has been shown to activate JAK2, STAT3 and 5, IRS-1, and MAPK pathways (Fig. 7.3). The best-characterized pathway in leptin signaling is the JAK/STAT pathway [42–44, 113]. The binding of leptin to its receptor induces conformational changes leading to homo-oligomerization and recruitment and binding of JAK2 forming the LEPR/JAK2 complex followed by cross-phosphorylation of the receptor on Tyr985, Tyr1077, and Tyr1138. JAK2 autophosphorylation on LEPRb is independent of tyrosine phosphorylation sites and can activate multiple pathways [114]. Phosphorylation of the LEPRb on these unique tyrosines leads to recruitment and binding of SH2 (*src* homology 2) containing signal transducers and in vivo studies have shown signaling through STAT3 [42, 43]. Binding of the SH2 domain containing transducers to the receptor leads to phosphorylation and activation of STAT family transcription factors, extracellular signal-regulated kinases (ERK), phosphoinositol-3 kinase (PI3K), and mammalian target of rapamycin (mTOR) [44, 72]. One of these effects is to phosphorylate c/PI3K signaling and activate ERK pathways activating downstream signals. Tyr1138

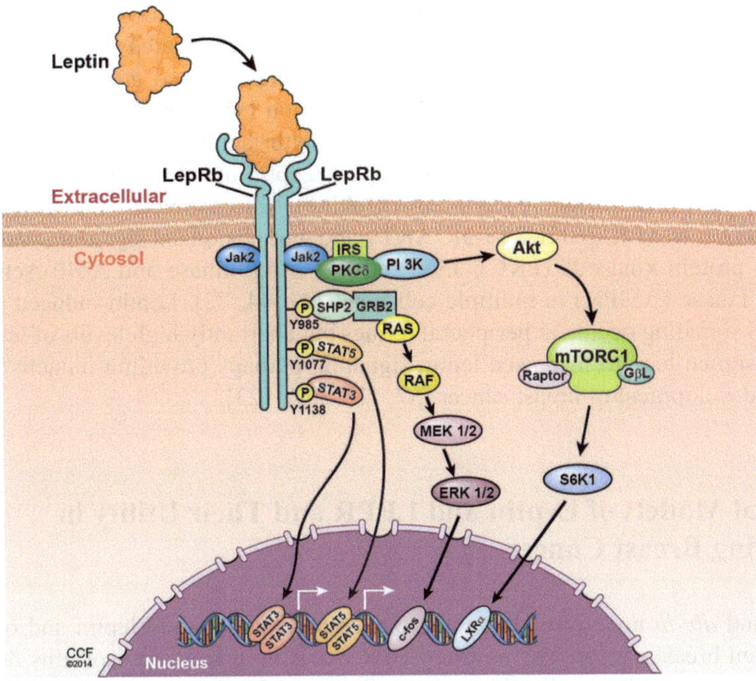

Fig. 7.3 Leptin signaling. Schematic diagram of signal transduction pathways induced by leptin. Leptin binding to LEPRb leads to activation of the associated JAK2 and phosphorylation of the intracellular domain of LEPR. The phosphorylation leads to binding and activation of STAT, PI3K and ERK pathways. Phosphorylated STAT3 (pSTAT3) translocates to the nucleus altering the expression of several genes including SOCS3. SOCS3 negatively regulates JAK2 phosphorylation. LEPRb signaling mediated by STAT3 primarily regulates energy homeostasis. PI3K metabolic pathway is activated by IRS which phosphorylates AKT (pAKT). pAKT induces activation of mammalian target of rapamycin complex 1 (mTORC1) substrate S6 kinase (S6K) which induces the expression of liver X receptors (LXR) [193]. The mitogenic pathway is activated (via the RAS/RAF/MEK/ERK) inducing c-FOS gene expression in hypothalamic neurons

phosphorylation on LEPRb is crucial for STAT3 activation. Upon phosphorylation, tyrosine phosphorylated STAT3 dimerizes and translocates to the nucleus activating transcription of target genes including the gene for a member of the suppressors of the cytokine signaling family (SOCS3) [115]. Tyr1138 phosphorylation stimulates SOCS3 expression negatively inhibiting leptin signaling primarily via binding to Tyr985. This mediates negative feedback on leptin signaling.

JAK2 phosphorylation of Tyr985 leads to phosphorylation of the SH2 domain of the tyrosine phosphatase SHP-2 (*src* homology 2-containing tyrosine phosphatase) with subsequent activation of the ERK signaling [72]. As a cytokine, leptin activates the stress-activated protein kinase c-Jun N-terminal kinase (JNK). A downstream target for p38 and the JNK ERK pathways is Nuclear factor kappaB (NF-kB) because this transcription factor is essential in the transcriptional regulation of proinflammatory cytokines such as tumor necrosis factor-α (TNF-α) and interleukin IL-1. Other leptin signaling pathways are still being elucidated owing to the diverse tissue specific expression [37]. Multiple oncogenes and growth factors like HER2 [116] and AKT [117] including transcription factors such as STAT3 [118], NF-κB [119], Notch [120], and insulin-like growth factors (IGF) [121] cross-talk with leptin inducing a multitude of effects [52]. Leptin-induced signals via multiple pathways commonly triggered by many cytokines. The canonical pathways include JAK2/STAT, MAPK/ERK1/2 PI-3K/AKT1 and, non-canonical signaling pathways include protein kinase C (PKC), JNK and p38 MAP kinase and AMP-Activated protein kinase (AMPK) in multiple cell types [42–44, 72]. Leptin-induced physiological signaling occurs in peripheral tissues but aberrantly high levels of leptin in obese women lead to increased leptin signaling perhaps providing inductive cues for the development of breast cancer [37, 112, 122, 123].

Animal Models of Leptin and LEPR and Their Utility in Studying Breast Cancer

ob/ob and *db/db* mice provide unique mouse mutants to study leptin and obesity effects on breast cancer. The murine LEPR exists in at least six isoforms derived from a single *LEPR* gene by alternative splicing. LEPR functional mutations result in morbid obesity. All aspects of epithelial development and mammary gland morphogenesis are impaired in *ob/ob* and *db/db* mice [124]. Further, defects in leptin production, in *ob/ob*, or in the LEPR (*db/db* mice and *fa/fa* rats) are characterized by infertility leading to the suggestion that leptin signaling is necessary for mammary gland biology and breast cancer [125].

The first in vivo role of leptin in mammary tumor development is based on the observation that when transgenic leptin-deficient (*ob/*ob) mice were bred with mice that overexpress transforming growth factor-α (TGF-α), no mammary tumors developed (0 of 59 mice). Wild type transgenic TGF-α mice develop tumors by 15 months of age with a 30 % incidence of mammary tumors [126]. Wild type TGF-α mice have circulating leptin that is twice the concentration of heterozygote ob/+TGF-α mice. In agreement with these studies, genetically obese *db/db* mice do not form mammary tumors when bred with TGF-α mice [127, 128]. Though these studies provide compelling evidence, the studies must be interpreted carefully since both the *ob/ob* and *db/db* lack epithelial mammary gland development [124]. In whole-mount preparations of the mammary gland, morphogenesis was found to be impaired in both genetically obese *ob/ob* and *db/db* mice [124]. This has significant ramifications on the use of these models in genetic breeding studies because the cells sensitive to oncogenic transformation are never generated. However, these mouse models have great utility in orthotopic cancer studies. Leptin was shown to be responsible for the increase in expression of cyclin D1, a cell-cycle control protein necessary for mammary gland development, in orthotopically injected murine mammary tumor virus-Wnt-1 mammary cancer cells [50, 129].

Evidence for a LEPR role in mammary carcinogenesis is further supported in carcinogen-induced models. When genetically obese Zucker (*fa/fa*) rats, which also have a LEPR defect, were administered the chemical carcinogen methylnitrosourea, only a small percentage of carcinomas developed in obese compared with lean rats [130]. Mammary tumors that developed were palpable and were comparable between the lean and obese rats but only a small percentage of those were carcinomas in the obese rats supporting the role of LEPR in mammary tumorigenesis [136]. Similarly, obesity increased the susceptibility of female *fa/fa* rats to 7,12-dimethylbenz(a)anthracene (DMBA)-induced mammary tumors suggesting that obesity and its downstream effector molecules play a significant role in carcinogenesis [131]. Obese Zucker rats had greater susceptibility to this carcinogen than did lean rats attributing to different degrees of LEPR defects in different substrains of the Zucker rats.

Diet-Induced Obesity Provides Evidence for a Role for Leptin in Mammary Gland Oncogenesis

Obesity-prone mice show a higher incidence of high-grade adenocarcinomas than obesity-resistant mice [127]. TGF-α wild type mice fed a high-fat diet become obese and develop mammary tumors at a younger age than do transgenic obesity-resistant TGF-α mice fed the same high-fat diet. TGF-α obesity-prone mice have 3-fold higher serum leptin levels compared to obesity-resistant mice and 5-fold higher level than mice fed a low fat diet. The dietary effect on serum leptin, body fat, and subsequent mammary tumor development provide more direct evidence for the role of leptin because these mice exhibit normal mammary gland development. These studies raise the question of the pattern of leptin and LEPR expression in the mammary gland and in breast cancer.

Expression of Leptin and LEPR in Breast Cancer

Leptin and LEPR expression is increased in cancer versus noncancerous mammary tissues. Analyses of tissue biopsies reveals that, similar to the LEPR, leptin is overexpressed in breast tumor tissue compared to noncancer breast epithelium [33, 37]. Patients with LEPR-positive tumors also overexpress leptin and are associated with high rates of metastases compared to patients with LEPR-negative tumors [132]. Increased *LEPR* mRNA expression with increased serum leptin levels is associated with poor prognosis [33]. Further, leptin and LEPR expression, based on immunohistochemistry (IHC), is increased in 85 and 75 % of breast tumors and correlated with the size of the tumor [133].

As indicated above, leptin may promote the development of ER+ breast cancer by potentiating the proliferation of cancer cells [112]. Notably, early studies indicated that LEPR is expressed at higher levels in ER positive compared with ER negative cells, though subsequent studies found that it is also expressed in triple negative breast cancers (TNBC). LEPR is expressed in 92% and leptin in 86% of TNBC [134]. This breast cancer subtype lacks the expression of the estrogen and progesterone receptors and does not exhibit *ERBB2* overexpression [135]. With regards to the role of LEPR in estrogen receptor positive breast cancer, several mechanisms have been proposed to explain the apparent leptin-estrogen crosstalk including leptin prevention of ER degradation, enhancement of ER stability, and increased ER transcriptional potential in the presence of anti-estrogens [79]. Further, it has been shown that leptin may compete with breast cancer therapeutics including anti-estrogens.[37, 63].

The use of leptin as a biomarker of breast cancer disease and progression has been evaluated and remains unresolved [91, 136–138]. Some, but not all, studies indicate that serum leptin levels between breast cancer patients and controls are significantly elevated in postmenopausal women [122]. In postmenopausal women, leptin is also found to be an independent predictor of tumor classification and Tumor, Node, Metastasis (TNM) stage [139]. Further, serum leptin levels were higher in breast cancer patients than in healthy women or patients with benign breast disease [33, 140]. However, contradicting results of serum leptin levels in subjects with breast cancer were found by some groups while other studies could not confirm this pattern [136, 137, 141–144]. Based on these studies, leptin may impact breast cancer development due to tumor subtype and its microenvironment. Indeed, obesity increases the risk of ER− breast cancer, which is more aggressive and difficult to treat due to limited clinical therapeutic options. Serum leptin levels are elevated in ER- breast cancer patients suggesting reducing circulating leptin may lead to reduced breast cancer in these patients [139].

Autocrine and Paracrine Roles of Leptin in Breast Cancer

Breast cancer proliferation may be induced and/or enhanced by overabundance of locally produced, paracrine or autocrine leptin rather than an endocrine source (Fig. 7.4). Indeed, the source of leptin influencing breast cancer cells is an area of intense investigation as it may impact the design of therapeutics [145]. Increased leptin and LEPR expression in breast cancer suggests that leptin can act as an autocrine factor on cancer cells inducing growth of the tumor [27, 133].

Evidence for Autocrine Leptin Leptin expression in breast cancer is associated with high proliferative rate. In intraductal in situ proliferative breast tumors, leptin expression is high, leading to suggestions that leptin may also contribute to breast cancer development in an autocrine manner [27, 146]. Multiple studies clearly demonstrate that breast cancer cells in response to various obesity-related conditions including hypoxia, insulin, IGF-1, and estrogen can induce an increase in leptin expression [147]. This has led to the speculation that leptin can promote breast cancer progression in an autocrine fashion [27, 113, 146].

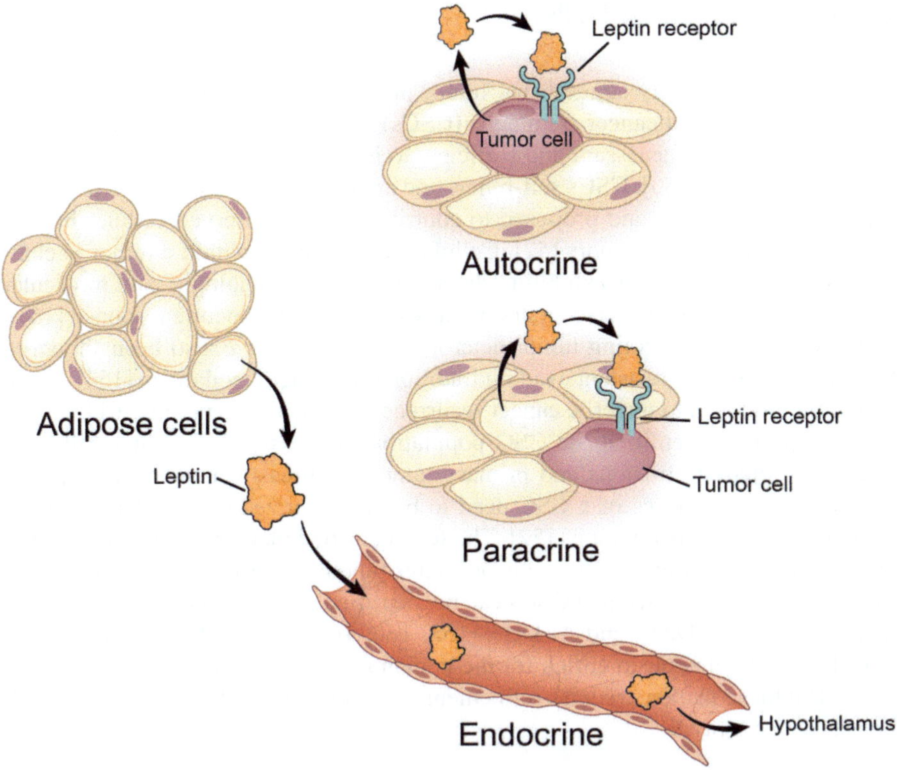

Fig. 7.4 Autocrine, paracrine, and endocrine leptin can promote breast cancer. Adipocytes serve as the major source of endocrine and paracrine leptin for binding to LEPR on breast cancer cells. Because breast cancer cells express and synthesize leptin they may be an alternate source of leptin for direct induction via an autocrine manner to promote breast cancer progression and metastases

Evidence for Paracrine Leptin. Leptin produced in a paracrine fashion by mammary gland adipocytes can enhance proliferation and invasion of human breast cancer cell lines and play a role in breast cancer progression [37, 48, 76, 77]. Leptin and LEPR were significantly related to invasiveness of breast cancers, especially in high grade tumors [27, 33, 37, 148]. Expression of Leptin and LEPR in breast cancer tissues in both primary tumors and metastastatic sites implicates that the autocrine and paracrine axes may be operative in breast cancer progression [27, 37, 77].

While most leptin functions are attributed to its role as a hormone secreted into the circulation and activating LEPR at distant sites, these studies point to leptin activating LEPR in a more localized manner via paracrine and autocrine activities. This provides a complexity in associating circulating levels of leptin with cancer risk and mortality.

Breast Cancer Stem Cells

Recent studies strongly suggest that tumor propagation relies on a population of stem-like cells called cancer stem cells (CSCs) [154]. Functional dysregulation of these cancer cells is necessary and sufficient to initiate cancer and promote tumorigenic progression [150]. CSCs self-renew and are pluripotent, thus can differentiate into diverse cell types and proposed to contribute to the observed cellular heterogeneity found in many tumors [52]. Cellular heterogeneity within a tumor is characteristic of breast cancer exhibiting diverse genetic, morphologic, and molecular profiles, as well as cell surface marker expression, and therapeutic response [151–153]. Apart from promoting tumor heterogeneity, CSCs have also been shown to be chemo- and radio-resistant evading traditional treatment modalities [154]. Thus, CSCs are proposed to underlie cancer recurrence and metastasis [155–161].

Al-Hajj et al. identified the CSC population in breast cancer using cells isolated from pleural effusions and primary tumors of breast cancer patients [162]. The phenotypic features of a CSC need to be characterized to elucidate the CSC subpopulation and its functions [52]. The few identified markers are predominantly expressed on the surface of the cells or confer functional properties for stem cells. Many groups have shown that CSCs can be identified and enriched using molecular markers CD44+/CD24− and ALDH+ [153, 160, 163–166]. CD49f^{hi} has been established as the marker to enrich for the human breast cancer stem cell population [167]. But these markers require refinement as the phenotype does not extend to all breast cancer subtypes.

Mouse Models to Study Breast Cancer Stem Cells

CSCs are implicated to have a central role in breast cancer progression and metastases [168, 169]. Stem cells are characterized by their ability to self-renew and maintain pluripotency [170]. In order to maintain this pluripotent state, the transcription factors NANOG, SOX2, and OCT4 have been identified as master regulators of self-renewal [171]. Recent studies indicate that CSCs utilize the same transcriptional machinery to maintain their stem-like state [52]. Consistent with these observations, *NANOG* has emerged as a key regulator of CSC self-renewal and is considered pro-carcinogenic [172]. Importantly, NANOG is overexpressed in poorly differentiated breast cancers and correlated with poor prognosis [173]. NANOG inhibition correlates with differentiation and subsequent loss of pluripotency and self-renewal. *NANOG* transcription is regulated by the SOX2 and OCT4 binding to the *NANOG* promoter [174]. Current studies suggest these factors are regulated by distinct signaling pathways but converge to regulate common targets and cooperate to maintain stem cell in a pluripotent state and self-renewal. As expression of these factors is inhibited the cells differentiate [175].

The discovery and development of specific therapies that target CSCs has the potential to revolutionize the treatment of these tumors. Elucidation of specific mechanisms by which CSCs survive chemotherapy, regulate self-renewal and interact with their primary and metastatic niches will be useful for the design of new therapeutic alternatives. Such approaches may become the basis for the next generation of effective and clinically applicable therapies that prevent disease relapse, metastasis, and enhance patient survival. Studies on CSCs have been largely performed using immunocompromised mice. Thus, syngeneic mouse models with an intact immune system are needed to validate the findings.

The merits of using mouse models to study mammary gland stem cells are that the isolation and characterization of stem cells from the mammary glands from mice are relatively easy [176]. Furthermore, orthotopic transplantation into the mammary fat pad is a well-developed standard procedure in studying mammary gland biology [176, 177]. The entire mammary gland can be effectively regenerated by serial transplantation of mammary stem cells orthotopically following which limiting dilution assays can be performed to calculate their self-renewing potential [176].

Different models of mammary carcinogenesis exist to study the CSC subpopulation in mice by use of diverse cell surface markers [178–180]. Because most CSC studies utilize immunodeficient mice, Clarke and colleagues set out to define a mammary CSC mouse model to study breast cancer [181]. Mammary tumors from MMTV-Wnt-1 mice were used for this purpose. Using the MMTV-Wnt-1 mouse mutant, a CSC population was identified based on the expression of CD24 and CD29 markers and account for 5–10 % of the total mammary epithelial cell population. Further, studies have been performed using in vitro limiting dilution mammosphere assays wherein the spheres were larger compared to other cell surface marker combinations [180]. The investigators harvested and dissociated MMTV-Wnt-1 breast tumors followed by flow cytometry based on Thy1, CD24, and CD45. The sorted cells were then injected into recipient background FVB/NJ female syngeneic mice. Thy1+CD24+ cancer cells constituted approximately 1 %–4 % of tumor cells and were highly enriched for cancer stem cells in 6 out of 7 tumors. Phenotypic diversity of the developed tumors showed high similarity to the original tumor. Microarray analysis comparing Thy1+CD24+ tumor cells to Thy1-CD24– cells identified a bunch of differentially expressed genes. Survival of human breast cancer patients from two different study groups was predicted based on the differentially expressed genes. This model strongly suggests that the CSC population in the MMTV-Wnt-1 murine breast tumor can be employed to study cancer stem cells in breast cancer [182]. Indeed, Zheng and colleagues used the same model to identify the role of leptin in CSCs [56].

The cell surface markers CD29, CD24, and CD61 exhibit similar utility for enriching CSCs capable of forming MMTV–Wnt-1 mammary tumors [179]. Further, expression and increased activity of aldehyde dehydrogenase (ALDH) has also been used to enrich for CSCs. In breast carcinomas, cell fractions possessing high ALDH activity can self-renew contributing to cellular heterogeneity [166]. The limitation with using mouse models for studying CSCs in mammary tumor is the difficulty in the identification of the mammary stem cell niche location in the gland.

The merits and demerits of the mouse models to study breast CSCs have been reviewed. Future challenges in CSC research include establishing the hierarchy of mammary stem and progenitor cells for future studies of mammary carcinogenesis [183, 184]. Despite extensive efforts to identify and isolate CSCs in human cancers, the molecular mechanisms of CSCs and their role in the tumor initiation and progression are yet to be addressed.

LEPR and Breast Cancer Stem Cell Maintenance

Leptin is a procarcinogenic adipokine that can promote proliferation, inhibit apoptosis, induce angiogenesis, and is pro-inflammatory. All of these actions are associated with cancer stem cell phenotype [52]. A critical barrier to the identification of CSC regulatory mechanisms is experimental systems that enable the reliable enrichment of CSCs from non-CSCs for comparative and functional analysis. To address this limitation and to interrogate the stem cell state in real time, a novel reporter system was developed in which a GFP reporter was placed under the control of the *NANOG* promoter. Using this reporter, breast cancer cells were sorted for GFP+(CSC) or GFP−(nonCSC). GFP+ cells exhibit CSC characteristics including expression of *NANOG*, *OCT4*, and *SOX2*, high tumorsphere formation capacity, and high frequency of tumor initiation in mice [185]. LEPR was enriched in the GFP+, CSC population [186]. The studies suggest that inhibition of LEPR may be a promising therapeutic approach to inhibit NANOG and thereby neutralizeing CSC functions (Fig. 7.5). Complete characterization of CSCs based on markers on the cell surface and molecular targets at the functional level are essential to design novel treatment strategies [152, 187–189]. Therapeutic targeting of CSCs will be developed with the aim to specifically target breast cancer initiation and development, resistance, recurrence and metastasis.

Endocrine leptin deficiency leads to functional depletion of CSCs in Murine Mammary Tumor Virus-Wnt-1 (MMTV-Wnt-1) transgenic mice [56]. Zheng and colleagues transplanted isolated cancer cells from spontaneous MMTV-Wnt-1 tumors into *ob*/ob and *db/db* mice and determined that leptin was necessary for maintaining CSCs using limiting dilution analysis [56]. Limiting dilution analysis of cancer cells derived from MMTV-Wnt-1 into wild type mice yields a tumor initiating or stem cell frequency of 1 in 2000 cells. However, when MMTV-Wnt-1 cancer cells were first injected in *ob/ob* mice, the tumors that develop exhibit a reduced initiating frequency when serially transplanted into wild type mice with a frequency of 1 in 30,000 cells. This study provided the first definitive evidence for a role of leptin in maintaining CSCs in tumors [56]. Based on these studies, Zheng et al. tested the hypothesis that the LEPR expressed in breast cancer cells is critical for maintaining CSC-like and metastatic properties [186]. *LEPR* was silenced via shRNA lentivirus transduction and determined that expression of stem cell self-renewal transcription factors NANOG, SOX2, and OCT4 was inhibited. The LEPR-NANOG signaling pathway is conserved between species because *NANOG* expression was rescued in human LEPR-silenced cells with the mouse LEPR. These studies show that LEPR expression is necessary to maintain CSCs in culture and LEPR silenced cells fail

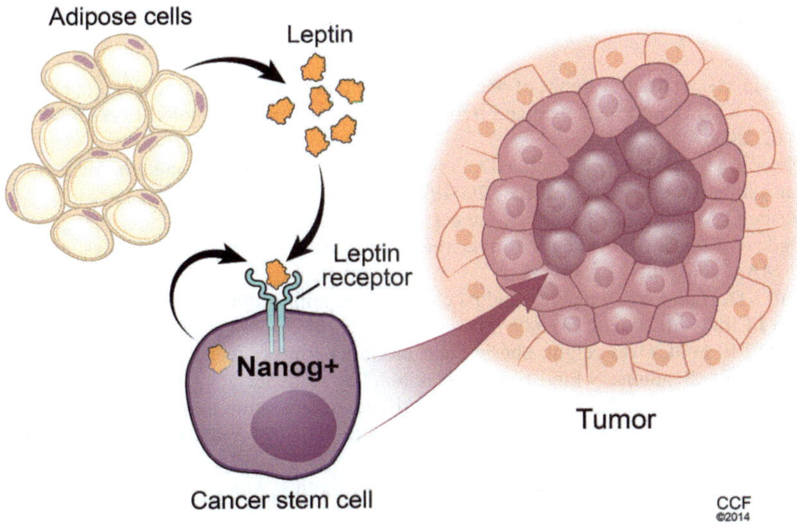

Fig. 7.5 Role of leptin in breast cancer stem cells. In obesity, the levels of leptin are elevated due to increased adipocytes. These adipocytes serve as a source of endocrine/paracrine leptin to promote breast CSC self-renewal and maintain pluripotency. Breast CSCs can provide an independent pool of leptin that may signal in an autocrine mannter. All sources of leptin can bind to LEPR expressed on *NANOG* expressing breast CSCs leading to maintenance of pluripotency and self-renewal contributing to tumor progression

to form tumors in mice. Consistent with these results, Feldman et al. observed that in response to leptin, CSCs induced phosphorylation and activation of STAT3 with subsequent activation of the transcription factors *OCT4* and *SOX2* [190]. Cross talk between leptin signaling and CSC strongly suggest leptin as a potential therapeutic target in breast cancer [52, 186, 190].

Summary

Leptin and LEPR are members of the druggable genome [34]. As such, they represent ligand-receptor targets for development of therapeutics to inhibit breast cancer [37, 123, 191]. While systemic inhibition of leptin could induce obesity, a risk factor for breast cancer initiation and progression, targeting LEPR on breast CSCs in the periphery represents a viable approach to block cancer progression [52]. In fact, therapeutics that do not cross the blood-brain barrier can be readily designed [192], so the strategy holds promise for breast cancer therapy. The discovery that LEPR is expressed on cancer stem cells and is necessary and sufficient to drive the expression of the self-renewal proteins *NANOG*, *SOX2*, and *OCT4* may be useful for development of therapeutics to block cancer recurrence and metastasis.

References

1. Baskin ML, et al. Prevalence of obesity in the United States. Obes Rev. 2005;6(1):5–7.
2. Flegal KM, et al. Prevalence and trends in obesity among US adults, 1999–2000. JAMA. 2002;288(14):1723–7.
3. Flegal KM, Ogden CL, Carroll MD. Prevalence and trends in overweight in Mexican-american adults and children. Nutr Rev. 2004;62(7 Pt 2):S144–8.
4. Haslam DW, James WP. Obesity. Lancet. 2005;366(9492):1197–209.
5. Ogden CL, et al. Prevalence of overweight and obesity in the United States, 1999–2004. JAMA. 2006;295(13):1549–55.
6. Hall KD, et al. Energy balance and its components: implications for body weight regulation. Am J Clin Nutr. 2012;95(4):989–94.
7. Calle EE, et al. Overweight, obesity, and mortality from cancer in a prospectively studied cohort of U.S. adults. N Engl J Med. 2003;348(17):1625–38.
8. Goran MI, Alderete TL. Targeting adipose tissue inflammation to treat the underlying basis of the metabolic complications of obesity. Nestle Nutr Inst Workshop Ser. 2012;73:49–60 (discussion p61–6).
9. American Cancer Society. Breast cancer facts & figures. American Cancer Society, 2011–2012.
10. Li CI, Daling JR, Malone KE. Age-specific incidence rates of in situ breast carcinomas by histologic type, 1980 to 2001. Cancer Epidemiol Biomark Prev. 2005;14(4):1008–11.
11. Chia KS, et al. Profound changes in breast cancer incidence may reflect changes into a Westernized lifestyle: a comparative population-based study in Singapore and Sweden. Int J Cancer. 2005;113(2):302–6.
12. Majed B, et al. Increased risk of contralateral breast cancers among overweight and obese women: a time-dependent association. Breast Cancer Res Treat. 2011;126(3):729–38.
13. Cheraghi Z, et al. Effect of body mass index on breast cancer during premenopausal and postmenopausal periods: a meta-analysis. PLoS ONE. 2012;7(12):e51446.
14. Key TJ, et al. Body mass index, serum sex hormones, and breast cancer risk in postmenopausal women. J Natl Cancer Inst. 2003;95(16):1218–26.
15. Sweeney C, et al. Risk factors for breast cancer in elderly women. Am J Epidemiol. 2004;160(9):868–75.
16. Cleary MP, Maihle NJ. The role of body mass index in the relative risk of developing premenopausal versus postmenopausal breast cancer. Proc Soc Exp Biol Med. 1997;216(1):28–43.
17. Calle EE, Thun MJ. Obesity and cancer. Oncogene. 2004;23(38):6365–78.
18. Ligibel J. Obesity and breast cancer. Oncology (Williston Park). 2011;25(11):994–1000.
19. Carmichael AR. Obesity and prognosis of breast cancer. Obes Rev. 2006;7(4):333–40.
20. Harvie M, Hooper L, Howell AH. Central obesity and breast cancer risk: a systematic review. Obes Rev. 2003;4(3):157–73.
21. Pierobon M, Frankenfeld CL. Obesity as a risk factor for triple-negative breast cancers: a systematic review and meta-analysis. Breast Cancer Res Treat. 2013;137(1):307–14.
22. Donohoe CL, Doyle SL, Reynolds JV. Visceral adiposity, insulin resistance and cancer risk. Diabetol Metab Syndr. 2011;3:12.
23. Esser N, et al. Inflammation as a link between obesity, metabolic syndrome and type 2 diabetes. Diabetes Res Clin Pract. 2014;105(2):141–50.
24. Coleman DL, Obese anddiabetes. Two mutant genes causing diabetes-obesity syndromes in mice. Diabetologia. 1978;14(3):141–8.
25. Borugian MJ, et al. Waist-to-hip ratio and breast cancer mortality. Am J Epidemiol. 2003;158(10):963–8.
26. Calabro P, Yeh ET. Obesity, inflammation, and vascular disease: the role of the adipose tissue as an endocrine organ. Subcell Biochem. 2007;42:63–91.

27. Ishikawa M, Kitayama J, Nagawa H. Enhanced expression of leptin and leptin receptor (OB-R) in human breast cancer. Clin Cancer Res. 2004;10(13):4325–31.
28. Zhang Y, et al. Positional cloning of the mouse obese gene and its human homologue. Nature. 1994;372(6505):425–32.
29. Friedman JM. Leptin and the regulation of body weight. Harvey Lect. 1999;95:107–36.
30. Myers MG, Cowley MA, Munzberg H. Mechanisms of leptin action and leptin resistance. Annu Rev Physiol. 2008;70:537–56.
31. Heymsfield SB, et al. Recombinant leptin for weight loss in obese and lean adults: a randomized, controlled, dose-escalation trial. JAMA. 1999;282(16):1568–75.
32. Blundell JE, Goodson S, Halford JC. Regulation of appetite: role of leptin in signalling systems for drive and satiety. Int J Obes Relat Metab Disord. 2001;25 Suppl 1:S29–34.
33. Miyoshi Y, et al. High expression of leptin receptor mRNA in breast cancer tissue predicts poor prognosis for patients with high, but not low, serum leptin levels. Int J Cancer. 2006;118(6):1414–9.
34. Garofalo C, Surmacz E. Leptin and cancer. J Cell Physiol. 2006;207(1):12–22.
35. Surmacz E. Obesity hormone leptin: a new target in breast cancer? Breast Cancer Res. 2007;9(1):301.
36. Vona-Davis L, Rose DP. Adipokines as endocrine, paracrine, and autocrine factors in breast cancer risk and progression. Endocr Relat Cancer. 2007;14(2):189–206.
37. Garofalo C, et al. Increased expression of leptin and the leptin receptor as a marker of breast cancer progression: possible role of obesity-related stimuli. Clin Cancer Res. 2006;12(5):1447–53.
38. Cao R, et al. Leptin induces vascular permeability and synergistically stimulates angiogenesis with FGF-2 and VEGF. Proc Natl Acad Sci U S A. 2001;98(11):6390–5.
39. Dieudonne MN, et al. Leptin mediates a proliferative response in human MCF7 breast cancer cells. Biochem Biophys Res Commun. 2002;293(1):622–8.
40. Hausman GJ, Richardson RL. Adipose tissue angiogenesis. J Anim Sci. 2004;82(3):925–34.
41. Myers MG, Jr. Leptin receptor signaling and the regulation of mammalian physiology. Recent Prog Horm Res. 2004;59:287–304.
42. Yin N, et al. Molecular mechanisms involved in the growth stimulation of breast cancer cells by leptin. Cancer Res. 2004;64(16):5870–5.
43. Jiang H, et al. Upregulation of survivin by leptin/STAT3 signaling in MCF-7 cells. Biochem Biophys Res Commun. 2008;368(1):1–5.
44. Fruhbeck G. Intracellular signalling pathways activated by leptin. Biochem J. 2006;393(1):7–20.
45. Bates SH, et al. STAT3 signalling is required for leptin regulation of energy balance but not reproduction. Nature. 2003;421(6925):856–9.
46. Buettner R, Mora LB, Jove R. Activated STAT signaling in human tumors provides novel molecular targets for therapeutic intervention. Clin Cancer Res. 2002;8(4):945–54.
47. Quesnelle KM, Boehm AL, Grandis JR. STAT-mediated EGFR signaling in cancer. J Cell Biochem. 2007;102(2):311–9.
48. Perera CN, et al. Leptin-regulated gene expression in MCF-7 breast cancer cells: mechanistic insights into leptin-regulated mammary tumor growth and progression. J Endocrinol. 2008;199(2):221–33.
49. Leeman RJ, Lui VW, Grandis JR. STAT3 as a therapeutic target in head and neck cancer. Expert Opin Biol Ther. 2006;6(3):231–41.
50. Saxena NK, et al. Leptin-induced growth stimulation of breast cancer cells involves recruitment of histone acetyltransferases and mediator complex to CYCLIN D1 promoter via activation of Stat3. J Biol Chem. 2007;282(18):13316–25.
51. Dreesen O, Brivanlou AH. Signaling pathways in cancer and embryonic stem cells. Stem Cell Rev. 2007;3(1):7–17.
52. Guo S, et al. Oncogenic role and therapeutic target of leptin signaling in breast cancer and cancer stem cells. Biochim Biophys Acta. 2012;1825(2):207–22.

53. Renehan AG, Roberts DL, Dive C. Obesity and cancer: pathophysiological and biological mechanisms. Arch Physiol Biochem. 2008;114(1):71–83.
54. Park J, Euhus DM, Scherer PE. Paracrine and endocrine effects of adipose tissue on cancer development and progression. Endocr Rev. 2011;32(4):550–70.
55. Hoda MR, et al. The adipocyte-derived hormone leptin has proliferative actions on androgen-resistant prostate cancer cells linking obesity to advanced stages of prostate cancer. J Oncol. 2012;2012:280386.
56. Zheng Q, et al. Leptin deficiency suppresses MMTV-Wnt-1 mammary tumor growth in obese mice and abrogates tumor initiating cell survival. Endocr Relat Cancer. 2011;18(4):491–503.
57. Enger SM, et al. Body size, physical activity, and breast cancer hormone receptor status: results from two case-control studies. Cancer Epidemiol Biomark Prev. 2000;9(7):681–7.
58. Miller WR. Aromatase inhibitors: mechanism of action and role in the treatment of breast cancer. Semin Oncol. 2003;30(4 Suppl 14):3–11.
59. Simpson ER, et al. Regulation of estrogen biosynthesis by human adipose cells. Endocr Rev. 1989;10(2):136–48.
60. Rose DP, Komninou D, Stephenson GD. Obesity, adipocytokines, and insulin resistance in breast cancer. Obes Rev. 2004;5(3):153–65.
61. Rochefort H, et al. How to target estrogen receptor-negative breast cancer? Endocr Relat Cancer. 2003;10(2):261–6.
62. Althuis MD, et al. Etiology of hormone receptor-defined breast cancer: a systematic review of the literature. Cancer Epidemiol Biomarkers Prev. 2004;13(10):1558–68.
63. Magoffin DA, et al. Leptin regulation of aromatase activity in adipose stromal cells from regularly cycling women. Ginekol Pol. 1999;70(1):1–7.
64. Purohit A, Reed MJ. Regulation of estrogen synthesis in postmenopausal women. Steroids. 2002;67(12):979–83.
65. Siiteri PK. Adipose tissue as a source of hormones. Am J Clin Nutr. 1987;45(1 Suppl):277–82.
66. Kershaw EE, Flier JS. Adipose tissue as an endocrine organ. J Clin Endocrinol Metab. 2004;89(6):2548–56.
67. Chandra RK. Cell-mediated immunity in genetically obese C57BL/6J ob/ob) mice. Am J Clin Nutr. 1980;33(1):13–6.
68. Geisler J, et al. Total body aromatization in postmenopausal breast cancer patients is strongly correlated to plasma leptin levels. J Steroid Biochem Mol Biol. 2007;104(1–2):27–34.
69. Zhao Y, et al. Aromatase P450 gene expression in human adipose tissue. Role of a Jak/STAT pathway in regulation of the adipose-specific promoter. J Biol Chem. 1995;270(27):16449–57.
70. Kitawaki J, et al. Leptin directly stimulates aromatase activity in human luteinized granulosa cells. Mol Hum Reprod. 1999;5(8):708–13.
71. Catalano S, et al. Leptin enhances, via AP-1, expression of aromatase in the MCF-7 cell line. J Biol Chem. 2003;278(31):28668–76.
72. Catalano S, et al. Leptin induces, via ERK1/ERK2 signal, functional activation of estrogen receptor alpha in MCF-7 cells. J Biol Chem. 2004;279(19):19908–15.
73. Ray A, Nkhata KJ, Cleary MP. Effects of leptin on human breast cancer cell lines in relationship to estrogen receptor and HER2 status. Int J Oncol. 2007;30(6):1499–509.
74. Maccio A, Madeddu C. Obesity, inflammation, and postmenopausal breast cancer: therapeutic implications. ScientificWorldJournal. 2011;11:2020–36.
75. Dubois V, et al. Leptin induces a proliferative response in breast cancer cells but not in normal breast cells. Nutr Cancer. 2014;66(4):645–55.
76. Manabe Y, et al. Mature adipocytes, but not preadipocytes, promote the growth of breast carcinoma cells in collagen gel matrix culture through cancer-stromal cell interactions. J Pathol. 2003;201(2):221–8.
77. Barone I, et al. Leptin mediates tumor-stromal interactions that promote the invasive growth of breast cancer cells. Cancer Res. 2012;72(6):1416–27.
78. Cleary MP, Grossmann ME. Minireview: obesity and breast cancer: the estrogen connection. Endocrinology. 2009;150(6):2537–42.

79. Garofalo C, Sisci D, Surmacz E. Leptin interferes with the effects of the antiestrogen ICI 182,780 in MCF-7 breast cancer cells. Clin Cancer Res. 2004;10(19):6466–75.
80. Laud K, et al. Identification of leptin receptors in human breast cancer: functional activity in the T47-D breast cancer cell line. Mol Cell Endocrinol. 2002;188(1–2):219–26.
81. Nkhata KJ, et al. Mammary tumor development from T47-D human breast cancer cells in obese ovariectomized mice with and without estradiol supplements. Breast Cancer Res Treat. 2009;114(1):71–83.
82. Fusco R, et al. Cellular and molecular crosstalk between leptin receptor and estrogen receptor-{alpha} in breast cancer: molecular basis for a novel therapeutic setting. Endocr Relat Cancer. 2010;17(2):373–82.
83. Bennett PA, et al. Differential expression and regulation of leptin receptor isoforms in the rat brain: effects of fasting and oestrogen. Neuroendocrinol. 1998;67(1):29–36.
84. Chu SC, et al. Fluctuation of serum leptin level in rats after ovariectomy and the influence of estrogen supplement. Life Sci. 1999;64(24):2299–306.
85. Fungfuang W, et al. Effects of estrogen on food intake, serum leptin levels and leptin mRNA expression in adipose tissue of female rats. Lab Anim Res. 2013;29(3):168–73.
86. Shimizu H, et al. Estrogen increases in vivo leptin production in rats and human subjects. J Endocrinol. 1997;154(2):285–92.
87. Jin L, et al. Leptin and leptin receptor expression in rat and mouse pituitary cells. Endocrinology. 2000;141(1):333–9.
88. Otvos L, Jr, et al. Efficacy of a leptin receptor antagonist peptide in a mouse model of triple-negative breast cancer. Eur J Cancer. 2011;47(10):1578–84.
89. Croker AK, Allan AL. Inhibition of aldehyde dehydrogenase (ALDH) activity reduces chemotherapy and radiation resistance of stem-like ALDHhiCD44(+) human breast cancer cells. Breast Cancer Res Treat. 2012;133(1):75–87.
90. Feldman DE, et al. Pluripotency factor-mediated expression of the leptin receptor (OB-R) links obesity to oncogenesis through tumor-initiating stem cells. Proc Natl Acad Sci U S A. 2012;109(3):829–34.
91. Artac M, Altundag K. Leptin and breast cancer: an overview. Med Oncol. 2012;29(3):1510–4.
92. Eric Hu X, et al. Small-molecule melanin-concentrating hormone-1 receptor antagonists require brain penetration for inhibition of food intake and reduction in body weight. J Pharmacol Exp Ther. 2008;324(1):206–13.
93. Steffensen KR, Gustafsson JA. Putative metabolic effects of the liver X receptor (LXR). Diabetes. 2004;53 Suppl 1:S36–42.
94. Ahima RS. Central actions of adipocyte hormones. Trends Endocrinol Metab. 2005;16(7):307–13.
95. Jarde T, et al. Involvement of adiponectin and leptin in breast cancer: clinical and in vitro studies. Endocr Relat Cancer. 2009;16(4):1197–210.
96. Gunter MJ, et al. Insulin, insulin-like growth factor-I, and risk of breast cancer in postmenopausal women. J Natl Cancer Inst. 2009;101(1):48–60.
97. Ouchi N, et al. Reciprocal association of C-reactive protein with adiponectin in blood stream and adipose tissue. Circulation. 2003;107(5):671–4.
98. Mantzoros C, et al. Adiponectin and breast cancer risk. J Clin Endocrinol Metab. 2004;89(3):1102–7.
99. Matsuda M, et al. Role of adiponectin in preventing vascular stenosis. The missing link of adipo-vascular axis. J Biol Chem. 2002;277(40):37487–91.
100. Okamoto Y, et al. An adipocyte-derived plasma protein, adiponectin, adheres to injured vascular walls. Horm Metab Res. 2000;32(2):47–50.
101. Campfield LA, et al. Recombinant mouse OB protein: evidence for a peripheral signal linking adiposity and central neural networks. Science. 1995;269(5223):546–9.
102. Jequier E. Leptin signaling, adiposity, and energy balance. Ann N Y Acad Sci. 2002;967:379–88.

103. Frederich RC, et al. Expression of ob mRNA and its encoded protein in rodents. Impact of nutrition and obesity. J Clin Invest. 1995;96(3):1658–63.
104. Ingalls AM, Dickie MM, Snell GD. Obese, a new mutation in the house mouse. J Hered. 1950;41(12):317–8.
105. Hoggard N, et al. Leptin expression in placental and fetal tissues: does leptin have a functional role? Biochem Soc Trans. 2001;29(2):57–63.
106. Zhang F, et al. Crystal structure of the obese protein leptin-E100. Nature. 1997;387(6629):206–9.
107. Santos-Alvarez J, Goberna R, Sanchez-Margalet V. Human leptin stimulates proliferation and activation of human circulating monocytes. Cell Immunol. 1999;194(1):6–11.
108. Campfield LA, Smith FJ, Burn P. The OB protein (leptin) pathway–a link between adipose tissue mass and central neural networks. Horm Metab Res. 1996;28(12):619–32.
109. Glaum SR, et al. Leptin, the obese gene product, rapidly modulates synaptic transmission in the hypothalamus. Mol Pharmacol. 1996;50(2):230–5.
110. Maffei M, et al. Leptin levels in human and rodent: measurement of plasma leptin and ob RNA in obese and weight-reduced subjects. Nat Med. 1995;1(11):1155–61.
111. Sachot C, Poole S, Luheshi GN. Circulating leptin mediates lipopolysaccharide-induced anorexia and fever in rats. J Physiol. 2004;561(1):263–72.
112. Murad A, et al. Leptin is an autocrine/paracrine regulator of wound healing. FASEB J. 2003;17(13):1895–7.
113. Cella F, Giordano G, Cordera R. Serum leptin concentrations during the menstrual cycle in normal-weight women: effects of an oral triphasic estrogen-progestin medication. Eur J Endocrinol. 2000;142(2):174–8.
114. Fantuzzi G, Faggioni R. Leptin in the regulation of immunity, inflammation, and hematopoiesis. J Leukoc Biol. 2000;68(4):437–46.
115. Cento RM, et al. Leptin levels in menopause: effect of estrogen replacement therapy. Horm Res. 1999;52(6):269–73.
116. Cymbaluk A, Chudecka-Glaz A, Rzepka-Gorska I. Leptin levels in serum depending on Body Mass Index in patients with endometrial hyperplasia and cancer. Eur J Obstet Gynecol Reprod Biol. 2008;136(1):74–7.
117. Geisthovel F, et al. Serum pattern of circulating free leptin, bound leptin, and soluble leptin receptor in the physiological menstrual cycle. Fertil Steril. 2004;81(2):398–402.
118. Pelleymounter MA, et al. Effects of the obese gene product on body weight regulation in ob/ob mice. Science. 1995;269(5223):540–3.
119. Faggioni R, et al. IL-1 beta mediates leptin induction during inflammation. Am J Physiol. 1998;274(1 Pt 2):R204–8.
120. Fei H, et al. Anatomic localization of alternatively spliced leptin receptors (Ob-R) in mouse brain and other tissues. Proc Natl Acad Sci U S A. 1997;94(13):7001–5.
121. Murakami T, et al. A short form of leptin receptor performs signal transduction. Biochem Biophys Res Commun. 1997;231(1):26–9.
122. Devos R, et al. Ligand-independent dimerization of the extracellular domain of the leptin receptor and determination of the stoichiometry of leptin binding. J Biol Chem. 1997;272(29):18304–10.
123. Golden PL, Maccagnan TJ, Pardridge WM. Human blood-brain barrier leptin receptor. Binding and endocytosis in isolated human brain microvessels. J Clin Invest. 1997;99(1):14–8.
124. Frankenberry KA, et al. Leptin receptor expression and cell signaling in breast cancer. Int J Oncol. 2006;28(4):985–93.
125. Bartella V, et al. Insulin-dependent leptin expression in breast cancer cells. Cancer Res. 2008;68(12):4919–27.
126. Jiang L, Li Z, Rui L. Leptin stimulates both JAK2-dependent and JAK2-independent signaling pathways. J Biol Chem. 2008;283(42):28066–73.
127. Bjorbaek C, et al. The role of SOCS-3 in leptin signaling and leptin resistance. J Biol Chem. 1999;274(42):30059–65.

128. Korkaya H, et al. HER2 regulates the mammary stem/progenitor cell population driving tumorigenesis and invasion. Oncogene. 2008;27(47):6120–30.
129. Korkaya H, et al. Regulation of mammary stem/progenitor cells by PTEN/Akt/beta-catenin signaling. PLoS Biol. 2009;7(6):e1000121.
130. Zhou J, et al. Activation of the PTEN/mTOR/STAT3 pathway in breast cancer stem-like cells is required for viability and maintenance. Proc Natl Acad Sci U S A. 2007;104(41):16158–63.
131. Pratt MA, et al. The canonical NF-kappaB pathway is required for formation of luminal mammary neoplasias and is activated in the mammary progenitor population. Oncogene. 2009;28(30):2710–22.
132. Guo S, Gonzalez-Perez RR. Notch, IL-1 and leptin crosstalk outcome (NILCO) is critical for leptin-induced proliferation, migration and VEGF/VEGFR-2 expression in breast cancer. PLoS ONE. 2011;6(6):e21467.
133. Ozbay T, Nahta R. A novel unidirectional cross-talk from the insulin-like growth factor-I receptor to leptin receptor in human breast cancer cells. Mol Cancer Res. 2008;6(6):1052–8.
134. Cirillo D, et al. Leptin signaling in breast cancer: an overview. J Cell Biochem. 2008;105(4):956–64.
135. McCormack D, et al. The antiproliferative effects of pterostilbene on breast cancer in vitro are via inhibition of constitutive and leptin-induced Janus kinase/signal transducer and activator of transcription activation. Am J Surg. 2011;202(5):541–4.
136. Hu X, et al. Leptin–a growth factor in normal and malignant breast cells and for normal mammary gland development. J Natl Cancer Inst. 2002;94(22):1704–11.
137. Chehab FF, Lim ME, Lu R. Correction of the sterility defect in homozygous obese female mice by treatment with the human recombinant leptin. Nat Genet. 1996;12(3):318–20.
138. Cleary MP, et al. Genetically obese MMTV-TGF-alpha/Lep(ob)Lep(ob) female mice do not develop mammary tumors. Breast Cancer Res Treat. 2003;77(3):205–15.
139. Cleary MP, Grande JP, Maihle NJ. Effect of high fat diet on body weight and mammary tumor latency in MMTV-TGF-alpha mice. Int J Obes Relat Metab Disord. 2004;28(8):956–62.
140. Cleary MP, et al. Leptin receptor-deficient MMTV-TGF-alpha/Lepr(db)Lepr(db) female mice do not develop oncogene-induced mammary tumors. Exp Biol Med (Maywood). 2004;229(2):182–93.
141. Zheng Q, Hursting SD, Reizes O. Leptin regulates cyclin D1 in luminal epithelial cells of mouse MMTV-Wnt-1 mammary tumors. J Cancer Res Clin Oncol. 2012;138(9):1607–12.
142. Lee WM, et al. Susceptibility of lean and obese Zucker rats to tumorigenesis induced by N-methyl-N-nitrosourea. Cancer Lett. 2001;162(2):155–60.
143. Hakkak R, et al. Obesity promotes 7,12-dimethylbenz(a)anthracene-induced mammary tumor development in female zucker rats. Breast Cancer Res. 2005;7(5):R627–33.
144. Cleveland RJ, et al. Common genetic variations in the LEP and LEPR genes, obesity and breast cancer incidence and survival. Breast Cancer Res Treat. 2010;120(3):745–52.
145. Snoussi K, et al. Leptin and leptin receptor polymorphisms are associated with increased risk and poor prognosis of breast carcinoma. BMC Cancer. 2006;6:38.
146. Foulkes WD, Smith IE, Reis-Filho JS. Triple-negative breast cancer. N Engl J Med. 2010;363(20):1938–48.
147. Coskun U, et al. Serum leptin, prolactin and vascular endothelial growth factor (VEGF) levels in patients with breast cancer. Neoplasma. 2003;50(1):41–6.
148. Mantzoros CS, et al. Leptin in relation to carcinoma in situ of the breast: a study of pre-menopausal cases and controls. Int J Cancer. 1999;80(4):523–6.
149. Tessitore L, et al. Leptin expression in colorectal and breast cancer patients. Int J Mol Med. 2000;5(4):421–6.
150. Goodwin PJ, et al. Is leptin a mediator of adverse prognostic effects of obesity in breast cancer? J Clin Oncol. 2005;23(25):6037–42.

151. Sauter ER, et al. Leptin expression in breast nipple aspirate fluid (NAF) and serum is influenced by body mass index (BMI) but not by the presence of breast cancer. Horm Metab Res. 2004;36(5):336–40.
152. Petridou E, et al. Leptin and insulin growth factor I in relation to breast cancer (Greece). Cancer Causes Control. 2000;11(5):383–8.
153. Chen DC, et al. Serum adiponectin and leptin levels in Taiwanese breast cancer patients. Cancer Lett. 2006;237(1):109–14.
154. Ozet A, et al. Effects of tamoxifen on the serum leptin level in patients with breast cancer. Jpn J Clin Oncol. 2001;31(9):424–7.
155. O'Brien S N, Welter BH, Price TM. Presence of leptin in breast cell lines and breast tumors. Biochem Biophys Res Commun. 1999;259(3):695–8.
156. Cascio S, et al. Mechanism of leptin expression in breast cancer cells: role of hypoxia-inducible factor-1alpha. Oncogene. 2008;27(4):540–7.
157. Falk RT, et al. Interrelationships between serum leptin, IGF-1, IGFBP3, C-peptide and prolactin and breast cancer risk in young women. Breast Cancer Res Treat. 2006;98(2):157–65.
158. Maccio A, et al. Correlation of body mass index and leptin with tumor size and stage of disease in hormone-dependent postmenopausal breast cancer: preliminary results and therapeutic implications. J Mol Med (Berl). 2010;88(7):677–86.
159. Visvader JE, Lindeman GJ. Cancer stem cells in solid tumours: accumulating evidence and unresolved questions. Nat Rev Cancer. 2008;8(10):755–68.
160. Visvader JE, Lindeman GJ. Cancer stem cells: current status and evolving complexities. Cell Stem Cell. 2012;10(6):717–28.
161. Yoder BJ, Wilkinson EJ, Massoll NA. Molecular and morphologic distinctions between infiltrating ductal and lobular carcinoma of the breast. Breast J. 2007;13(2):172–9.
162. Fillmore CM, Kuperwasser C. Human breast cancer cell lines contain stem-like cells that self-renew, give rise to phenotypically diverse progeny and survive chemotherapy. Breast Cancer Res. 2008;10(2):R25.
163. Nakshatri H, Srour EF, Badve S. Breast cancer stem cells and intrinsic subtypes: controversies rage on. Curr Stem Cell Res Ther. 2009;4(1):50–60.
164. Chuthapisith S, et al. Breast cancer chemoresistance: emerging importance of cancer stem cells. Surg Oncol. 2010;19(1):27–32.
165. Sakariassen PO, Immervoll H, Chekenya M. Cancer stem cells as mediators of treatment resistance in brain tumors: status and controversies. Neoplasia. 2007;9(11):882–92.
166. Tang C, Chua CL, Ang BT. Insights into the cancer stem cell model of glioma tumorigenesis. Ann Acad Med Singapore. 2007;36(5):352–7.
167. Phillips TM, McBride WH, Pajonk F. The response of CD24(-/low)/CD44+ breast cancer-initiating cells to radiation. J Natl Cancer Inst. 2006;98(24):1777–85.
168. Li X, et al. Intrinsic resistance of tumorigenic breast cancer cells to chemotherapy. J Natl Cancer Inst. 2008;100(9):672–9.
169. Dave B, Chang J. Treatment resistance in stem cells and breast cancer. J Mammary Gland Biol Neoplasia. 2009;14(1):79–82.
170. Tanei T, et al. Association of breast cancer stem cells identified by aldehyde dehydrogenase 1 expression with resistance to sequential Paclitaxel and epirubicin-based chemotherapy for breast cancers. Clin Cancer Res. 2009;15(12):4234–41.
171. Ling LJ, et al. A novel mouse model of human breast cancer stem-like cells with high CD44+CD24-/lower phenotype metastasis to human bone. Chin Med J (Engl). 2008;121(20):1980–6.
172. Al-Hajj M, et al. Prospective identification of tumorigenic breast cancer cells. Proc Natl Acad Sci U S A. 2003;100(7):3983–8.
173. Liu R, et al. The prognostic role of a gene signature from tumorigenic breast-cancer cells. N Engl J Med. 2007;356(3):217–26.
174. Shipitsin M, et al. Molecular definition of breast tumor heterogeneity. Cancer Cell. 2007;11(3):259–73.

175. Ricardo S, et al. Breast cancer stem cell markers CD44, CD24 and ALDH1: expression distribution within intrinsic molecular subtype. J Clin Pathol. 2011;64(11):937–46.
176. Ginestier C, et al. ALDH1 is a marker of normal and malignant human mammary stem cells and a predictor of poor clinical outcome. Cell Stem Cell. 2007;1(5):555–67.
177. Cariati M, et al. Alpha-6 integrin is necessary for the tumourigenicity of a stem cell-like subpopulation within the MCF7 breast cancer cell line. Int J Cancer. 2008;122(2):298–304.
178. Charafe-Jauffret E, et al. Breast cancer cell lines contain functional cancer stem cells with metastatic capacity and a distinct molecular signature. Cancer Res. 2009;69(4):1302–13.
179. Sheridan C, et al. CD44+/CD24- breast cancer cells exhibit enhanced invasive properties: an early step necessary for metastasis. Breast Cancer Res. 2006;8(5):R59.
180. Takahashi K, Yamanaka S. Induction of pluripotent stem cells from mouse embryonic and adult fibroblast cultures by defined factors. Cell. 2006;126(4):663–76.
181. Boyer LA, et al. Core transcriptional regulatory circuitry in human embryonic stem cells. Cell. 2005;122(6):947–56.
182. Jeter CR, et al. Functional evidence that the self-renewal gene NANOG regulates human tumor development. Stem Cells. 2009;27(5):993–1005.
183. Nagata T, et al. Prognostic significance of NANOG and KLF4 for breast cancer. Breast Cancer. 2014;21(1):96–101.
184. Rodda DJ, et al. Transcriptional regulation of nanog by OCT4 and SOX2. J Biol Chem. 2005;280(26):24731–7.
185. Thiagarajan PS, Hitomi M, Hale JS, Alvarado AG, Otvos B, Sinyuk M, Stoltz K, Wiechert A, Mulkearns-Hubert E, Jarrar A, Zheng Q, Thomas D, Egelhoff T, Rich JN, Liu H, Lathia JD, Reizes O. Development of a fluorescent reporter system to delineate cancer stem cells in triple-negative breast cancer. Stem Cells. 2015. doi:10.1002/stem.2021.
186. Kashyap V, et al. Regulation of stem cell pluripotency and differentiation involves a mutual regulatory circuit of the NANOG, OCT4, and SOX2 pluripotency transcription factors with polycomb repressive complexes and stem cell microRNAs. Stem Cells Dev. 2009;18(7):1093–108.
187. Shackleton M, et al. Generation of a functional mammary gland from a single stem cell. Nature. 2006;439(7072):84–8.
188. Stingl J, et al. Purification and unique properties of mammary epithelial stem cells. Nature. 2006;439(7079):993–7.
189. Cho RW, Clarke MF. Recent advances in cancer stem cells. Curr Opin Genet Dev. 2008;18(1):48–53.
190. Vaillant F, et al. The mammary progenitor marker CD61/beta3 integrin identifies cancer stem cells in mouse models of mammary tumorigenesis. Cancer Res. 2008;68(19):7711–7.
191. Zhang M, et al. Identification of tumor-initiating cells in a p53-null mouse model of breast cancer. Cancer Res. 2008;68(12):4674–82.
192. Clarke MF, Fuller M. Stem cells and cancer: two faces of eve. Cell. 2006;124(6):1111–5.
193. Cho RW, et al. Isolation and molecular characterization of cancer stem cells in MMTV-Wnt-1 murine breast tumors. Stem Cells. 2008;26(2):364–71.
194. Visvader JE, Stingl J. Mammary stem cells and the differentiation hierarchy: current status and perspectives. Genes Dev. 2014;28(11):1143–58.
195. Stingl J, Caldas C. Molecular heterogeneity of breast carcinomas and the cancer stem cell hypothesis. Nat Rev Cancer. 2007;7(10):791–9.
196. Zheng Q, et al. Leptin receptor maintains cancer stem-like properties in triple negative breast cancer cells. Endocr Relat Cancer. 2013;20(6):797–808.
197. Garcia Bueno JM, et al. An update on the biology of cancer stem cells in breast cancer. Clin Transl Oncol. 2008;10(12):786–93.
198. Korkaya H, Wicha MS. Selective targeting of cancer stem cells: a new concept in cancer therapeutics. BioDrugs. 2007;21(5):299–310.

Chapter 8
Mouse Models Used to Study the Effects of Diabetes, Insulin, and IGFs on Cancer

Zara Zelenko, Derek LeRoith and Emily J. Gallagher

Abstract Type 2 diabetes (T2D) and obesity are epidemic problems worldwide. These disorders are linked to decreased life expectancy and increased incidence and mortality from cancer. Understanding the mechanisms of cancer progression in obesity and T2D will aid in discovering cancer prevention and treatment strategies for these patients. Hyperinsulinemia, insulin receptor, and insulin-like growth factor-1 (IGF-1) receptor signaling appear to be important factors linking diabetes, obesity, and cancer. A number of mouse models have been developed in order to accurately understand the roles of insulin, IGF-1, and their receptor signaling in obesity and diabetes-related cancer. This chapter summarizes many of the mouse models used to study T2D and cancer.

Keywords Type 2 diabetes · Obesity · Cancer · Mouse models · Insulin · Insulin-like growth factor · IGF-1 · Hyperinsulinemia · Hyperglycemia

E. J. Gallagher (✉) · Z. Zelenko · D. LeRoith
Division of Endocrinology, Diabetes and Bone Diseases,
Icahn School of Medicine at Mount Sinai, One Gustave L. Levy Place,
Box 1055, 10029 New York, NY, USA
e-mail: emily.gallagher@mssm.edu

Z. Zelenko
e-mail: Zara.Zelenko@mssm.edu

D. LeRoith
e-mail: derek.leroith@mssm.edu

E. J. Gallagher
e-mail: emily.gallagher@mssm.edu

N. A. Berger (ed.), *Murine Models, Energy Balance, and Cancer*,
Energy Balance and Cancer, DOI 10.1007/978-3-319-16733-6_8

Introduction

Obesity and type 2 diabetes (T2D) are major global health issues. The prevalence of both conditions has increased over the past. In 2008, the World Health Organization (WHO) reported that 10% of men and 14% of women worldwide were obese. As of 2010, it was estimated that there were approximately 285 million diabetic patients globally [1]. Obesity and T2D are both associated with reduced life expectancy and epidemiological evidence has linked both conditions to an increase in incidence and mortality from various cancers [2–6].

Different mechanisms that may link T2D and cancers include insulin resistance, hyperinsulinemia, increased systemic or tissue insulin-like growth factor-1 (IGF-1), hyperglycemia, and dyslipidemia [7–9]. Moreover, inflammatory cytokines and adipokines may contribute to metabolic dysfunction and may promote cancer development [10,11]. For tumor growth, invasion, and metastasis to occur, normal cells undergo various morphological changes. These changes can be influenced by IGF-1, hyperinsulinemia, hyperglycemia, and chronic inflammation [12]. It is believed that normal cells, which have developed different oncogenic mutations, are susceptible to these growth and metastasis-promoting changes associated with T2D and obesity, further promoting cancer progression. Both obesity and T2D have been associated with more advanced stages of a number of cancers at presentation, resistance to therapy, and recurrence; all of these are factors that contribute to greater cancer mortality [13–16].

Although human epidemiological studies have reported associations between circulating insulin, IGF-1, and their respective receptor expression on cancer incidence and mortality, these studies cannot determine the mechanisms through which insulin and IGF-1 may drive tumor progression, or if these factors are causative, or simply associated with cancer progression. In vitro studies have examined the direct effects of insulin, IGF-1, and their receptors on cancer cell proliferation, migration, and morphology; however, in vitro modeling does not recapitulate the complex systemic, microenvironment, and cell changes that occur in T2D. Therefore, it is vital to study the effects of obesity and T2D on the progression of cancer in vivo. This chapter will examine various mouse models that have been generated to study the pathophysiology of cancer in T2D and obesity. The mouse models have been generated either through transgene insertions and/or targeted gene-deletion approaches.

Insulin and IGF-1 Signaling

Studies have proposed various mechanisms to link T2D, obesity, and breast cancer. In humans, insulin resistance and hyperinsulinemia characterize pre-diabetes and early T2D. Insulin resistance in metabolic tissues and hyperinsulinemia due to β-cell compensation occur long before the development of hyperglycemia, the hallmark of T2D [17–20]. Hyperglycemia defines the clinical onset of diabetes [5]. Insulin is a well-recognized growth factor and hyperinsulinemia in patients has been

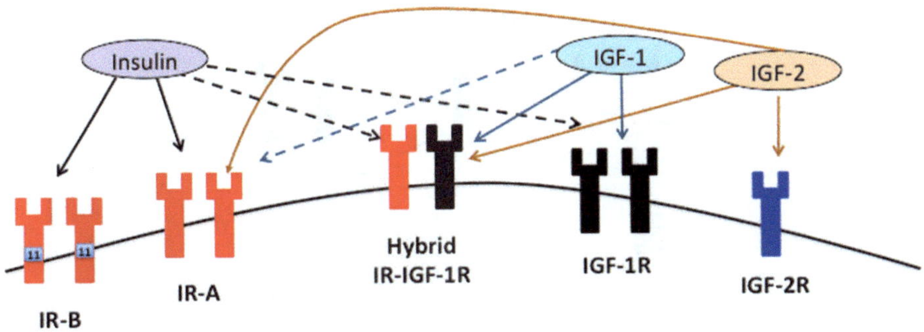

Fig. 8.1 Schematic of Insulin, IGF-1, and IGF-2 binding to target receptors. The IR has two isoforms, IR-A (lacks exon 11) and IR-B. The IR isoforms, the IGF receptors, and the hybrid IR-IGF-1R can be stimulated by the ligands shown. Insulin has a high affinity (*solid line*) for IR-A, IR-B, and low affinity (*dashed line*) to the IGF-1R and the hybrid receptor. IGF-1 can bind IGF-1R with a high affinity, the hybrid receptor and at lower affinities to IR-A. IGF-2 can bind to the IGF-2R and has high affinity for IR-A, thus allowing it to bind to the IR-A and the IR-IGF-1R hybrid receptor. *IGF-1* insulin-like growth factor 1, *IGF-2* insulin-like growth factor 2, *IR-A* insulin receptor A, *IR-B* insulin receptor B, *IGF-1R* insulin-like growth factor 1 receptor, *IGF-2R* insulin-like growth factor 2 receptor, *Solid line* represents high affinity for receptor, *Dashed line* represents low affinity for receptor

associated with decreased breast cancer survival and decreased recurrence free survival [21,22]. Therefore, hyperinsulinemia may promote the progression of breast cancer in patients. Studies have reported that the expression levels of the insulin receptor (IR) and the insulin-like growth factor-1 receptor (IGF-1R) are elevated in human breast cancer specimens and cancer cell lines [23–25]. Furthermore, an increase in IR/IGF-1R phosphorylation in pathological breast cancer specimens has been associated with a worse prognosis [26].

Hyperinsulinemia may be exerting direct effects of the IR on the tumor cells, promoting growth and progression. The IR is highly expressed in many tumors, particularly the mitogenic splice variant of the IR, IR-A. The IR-A isoform of the IR lacks exon 11, whereas IR-B the metabolic variant contains exon 11 (Fig. 8.1) [27,28]. IR-B is expressed in liver, muscle, and fat, but IR-A is expressed in fetal tissues, the spleen and is frequently found in cancers [29]. Therefore, insulin or hyperinsulinemia may stimulate tumor growth by acting directly on the IR, possibly by acting on the IR isoform A. Insulin may also have indirect effects on tumor growth through IGF-1. IGF-1 and insulin-like growth factor-binding proteins (IGFBPs) are synthesized and secreted by the liver due to the stimulation by growth hormone (GH) or insulin [30]. Circulating IGFBPs can limit the bioavailability of IGF-1. The detailed actions of IGFBPs are beyond the scope of this chapter, but have been recently reviewed by RC Baxter [31]. Insulin increases circulating IGF-1 levels and decreases IGFBP-1, which potentially leads to the increase of local "free" IGF-1 in the tissues [32,33]. IGF-1 can also be made in extrahepatic tissues [34]. The roles of IGF-1 include regulating cell proliferation, differentiation, and apoptosis [35]. In a muscle-specific growth hormone receptor (GHR) knockout mouse, there were no changes to IGF-1 levels suggesting that in some tissues, such as skeletal muscle,

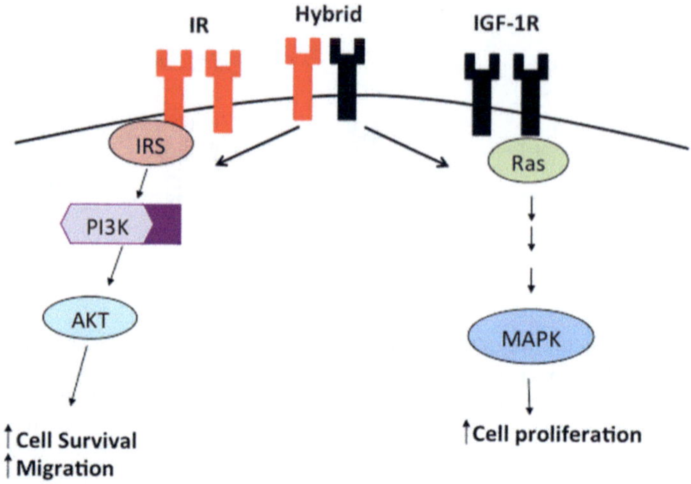

Fig. 8.2 Insulin and IGF-1 receptor signaling pathway. Binding of ligand to the IR, IGF-1R, or hybrid IR-IGF-1R initiates signaling cascades that involve multiple phosphorylation events. Upon ligand binding to the extracellular subunits of the receptors, there is cross-autophosphorylation of the receptor tyrosine domains. The IR or hybrid receptor phosphorylates and activates IRS, which triggers the phosphorylation of PI3K. The catalytic domain of PI3K (*light purple*) activates the serine/theronine kinase, AKT. AKT activation leads to various outcomes such as increased cell survival and migration. The IGF-1R signal transduction pathway goes through the Ras-GTPase that leads to the activation of MAPK. MAPK activation allows for increased cell proliferation. *IR* insulin receptor, *IGF-1R* insulin-like growth factor 1 receptor, *IRS* insulin receptor substrate, *PI3K* phosphoinositide 3-kinase, *MAPK* mitogen-activated protein kinase

IGF-1 may not be GH dependent [36]. This local IGF-1 may also affect tumor growth. Some studies have reported that individuals with higher serum levels of IGF-1 have an increased risk of developing various cancers, such as prostate, colon, cervical, ovarian, and breast [35,37].

Insulin and IGF-1 can bind avidly to the extracellular α-subunits of the IR and IGF-1R, respectively. Both the IR and IGF-1R are receptor tyrosine kinases (RTKs) that typically signal through phosphoinositide 3-kinase (PI3K) and mitogen-activated protein kinase (MAPK) pathways (Fig. 8.2) [38]. The IR and IGF-1R receptors can homo- or heterodimerize, which leads to a cross-autophosphorylation of the tyrosine kinase domain (Fig. 8.1). Active phosphorylated forms of IR, IGF-1R, or the heterodimerized hybrid IR-IGF-1R can phosphorylate and activate the insulin receptor substrate-1 (IRS-1). IRS-1 triggers the phosphorylation and activation of the PI3K catalytic subunit, which in turn activates a serine/threonine kinase, Akt [39,40]. MAPK activation leads to increased cell proliferation, whereas Akt activation leads to increased cell survival and migration [41]. The activation of the IGF-1R and IR receptor have been correlated to increased disease progression and poor prognosis in breast cancer [42,43].

It has been noted that Akt activation can promote cell cycle progression, cell survival, as well as tumor cell invasion [44,45]. Moreover, it has been shown that

Akt activation can induce epithelial to mesenchymal transition (EMT) by altering the expression of a number of factors including repressing the transcription of E-cadherin [44,46]. EMT is a process that leads to the loss of epithelial markers, and the acquisition of mesenchymal properties. This process has been reported as being important for tumor metastases and for the development of tumor cells with stem cell-like properties that may be responsible for tumor recurrence [47].

IGF-1 has been associated with breast/mammary cancer progression because of its mitogenic and anti-apoptotic effects on mammary epithelial cells [48]. Cells overexpressing the IGF-1R exhibit growth factor-independent proliferation, lack of contact inhibition, have anchorage-independent growth, and are tumorigenic in vivo [49]. They exhibit downregulation of epithelial markers such as E-cadherin and β-catenin and upregulation of classic mesenchymal markers such as vimentin, N-cadherin, and fibronectin. The in vitro studies demonstrate that IGF-1 and IGF-1R signaling contribute to EMT [49]. Therefore, it is possible that hyperinsulinemia may act through the IR, IGF-1R, or the hybrid receptor of cancer cells and promote downstream signaling and growth of these cells [24].

It is clear that the IR and IGF-1R may have important roles in the pathogenesis of cancer in the setting of T2D. Therefore, it was critical to establish mouse models to understand the effect of T2D on cancer progression.

Insulin Receptor Gene Deletion

In order to study the role of diabetes and insulin signaling on cancer progression, mouse models were created. In these animal models, complete disruption of the insulin receptor gene was generated by introducing mutations to the IR gene through homologous recombination [50,51]. Mice with complete IR deletion have growth defects as well as skeletal muscle hypotrophy [50,51]. Furthermore, the mice develop severe diabetes with hyperglycemia and hyperinsulinemia [50,51]. This ultimately leads to death within 7 days after birth [50,51]. Therefore, approaches to obtain tissue-specific IR gene deletion were established.

Tissue-Specific Insulin Receptor Gene Deletion

The most commonly used system employed to generate tissue-specific gene deletion is the Cre-lox system. The Cre recombinase recognizes and excises a pair loxP sites that flank the gene of interest [52]. The loxP sites can be inserted into noncoding regions around the exon of interest by homologous recombination. In order to generate a tissue-specific gene deletion, the Cre recombinase is expressed under a tissue-specific promoter. Therefore, the gene of interest will be deleted only in the tissue expressing the Cre (Fig. 8.3). This system is utilized to generate IR deletion in muscle, fat, liver, and β-cells to determine the tissue-specific effects of IR deletion

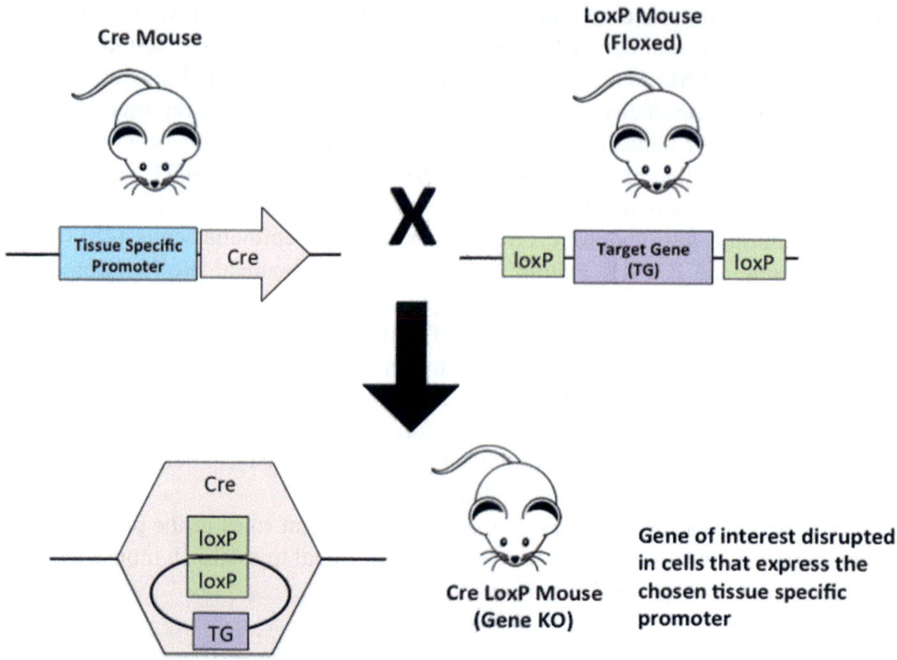

Fig. 8.3 Model of Cre function. In the Cre mouse, Cre recombinase is expressed under the control of a tissue-specific promoter. In the LoxP mouse, LoxP sites are inserted around a target gene. When the Cre mouse is crossed to the LoxP mouse, the Cre recombinase excises the loxP sites, disrupting the gene of interest in the specific tissue

on the development of T2D. These models have also been used to study the effects of insulin and IR signaling on cancer development.

Placing the Cre recombinase under the pancreatic β-cell-specific promoter generates mice lacking β-cell-insulin receptor (βIRKO) [53]. These mice lose the acute first-phase insulin response to glucose. The mice also develop glucose intolerance, assessed by glucose tolerance tests, which worsened as βIRKO mice aged. The βIRKO mice were crossed with a transgenic mouse model of neuroendocrine tumors of the pancreatic β-cells (RIP1-Tag2 transgenic mice) [54]. These RIP1-Tag2 mice express the Simian Vacuolating Virus 40 (SV40) large T-antigen 2 (Tag2) oncogene under the rat insulin promoter-1 (RIP-1) and develop pancreatic β-cell neuroendocrine tumors. The growth of these tumors is not significantly reduced by a monoclonal antibody targeting the IGF-1R (A12) that causes receptor internalization and degradation. However, the RIP1-Tag2/βIRKO mice exhibit a significant reduction of tumor volume compared to controls. The tumors from these mice have a 1.5-fold increase in the percentage of apoptotic cells compared to RIP1-Tag2 mice with functional β-cell IR. Moreover, treatment of the RIP1-Tag2/βIRKO with the IGF-1R inhibitor A12 leads to a significant decrease in tumor formation and proliferation. This demonstrates that in certain tumor types, targeting IGF-1R may not be effective, perhaps due to IR compensation. However, inhibiting the IR reduced the tumor growth. Furthermore, inhibiting both the IR and the IGF-1R produced an even greater reduction in tumor growth. Therefore, it may be necessary to target both receptors signaling within the tumor cells to prevent the formation of larger tumors.

Insulin Receptor/IGF-1R Signaling Proteins

Another approach at targeting the IR/IGF-1R signaling pathway includes targeting proteins downstream of the receptors, such as insulin receptor substrate (IRS)-1, IRS-2, and AKT. IRS-1 and IRS-2 are major substrates in the IR/IGF-1R-signaling pathway, which are rapidly phosphorylated on tyrosine residues after ligand stimulation of the receptors. Mice with deletion of the IRS-1 have severely stunted growth and reduced body weight compared to controls [55–57]. These mice demonstrate significantly higher serum insulin levels and have mild-moderate insulin resistance [55,57]. IRS-1 null and control mice were placed on a high-fat diet and then injected with diethylnitrosamine (DEN), an organic compound that is toxic and induces liver tumors [58]. DEN increases alanine transaminase activity while decreasing glutathione, glutathione peroxidase, and gluthatione-s-transferase activity in liver tissues, which contributes to the formation of tumors in the liver [59]. Furthermore, DEN has been linked to the activation of various proto-oncogenes like *c-Myc* and *N-ras*, and inhibiting tumor suppressor genes such as p53 and retinoblastoma (RB) [60]. IRS-1 null mice treated with DEN develop fewer liver tumors and were protected against hepatic steatosis [58]. This suggests that inhibiting the IR/IGF-1R signaling pathway by blocking IRS-1 reduces hepatic tumorigenesis [58]. In other tumor cell lines, including breast cancer cells higher IRS-1 expression has been linked to an increase in cell proliferation [61,62]. Human MCF7 cells, which are an estrogen receptor alpha (ERα)-positive breast cancer cell line, express higher levels of IRS-1 than MDA-MB-231, human triple negative breast cancer cells [63].

IRS-2 expression in human breast cancers has been associated with greater mortality [61]. Moreover, IRS-2 expression has been associated with high-grade invasive ductal breast carcinomas [63]. Deregulated IRS-2 expression has been positively correlated with the progression of colorectal cancer by activation of the PI3K pathway [64]. Global disruption of IRS-2 produces a phenotype in mice similar to T2D. The IRS-2 null mice develop insulin resistance due to multiorgan impairment of insulin signaling, causing hyperinsulinemia, followed by β-cell decompensation and failure, and finally hyperglycemia [65,66]. IRS-2 null mice were crossed with mice that express the polyoma virus middle T antigen (PyVmT) oncogene under the mouse mammary tumor virus (MMTV) promoter, a mouse model of luminal type B breast cancer. These IRS-2 null/PyVmT mice exhibit increased tumor latency and decreased incidence of pulmonary metastasis compared to control PyVmT mice [67]. These studies show that IRS-2 contributes to the progression of cancer. In the IRS-2 null/PyVmT model, mammary tumor cells that lacked IRS-2 were less invasive, again suggesting that inhibiting the IR/IGF-1R signaling pathway within the tumor results in decreased tumor growth. Furthermore, IRS-2 null tumors demonstrated a higher prevalence of apoptosis. The data suggest that IR/IGF-2R signaling through IRS-2 is important for mammary tumor cell invasion and survival.

As mentioned in the section "Insulin and IGF-1 Signaling," further downstream of the IR/IGF-1R pathway is the serine-threonine protein kinase, AKT. AKT1 and AKT2 genes have both been deleted in mice. AKT1 null mice do not exhibit a strong phenotype [68] AKT2 null mice exhibit mild growth retardation, glucose intolerance, insulin resistance, dyslipidemia, and hyperglycemia [69]. When AKT2 null mice were bred with transgenic mice that form mammary tumors (MMTV-PyVmT and MMTV-Neu), the AKT2 deletion increases tumor latency and decreases mammary tumor growth rate [70].

These mouse models provide insight into the potential mechanisms through which cancer can progress in the type 2 diabetic setting. Silencing components of the IR/IGF-1R signaling pathway within the tumors promotes an increase in tumor latency and decreases tumor growth rate in many of these mouse models. This suggests that the IR/IGF-1R signaling pathways are important in tumor growth and that hyperinsulinemia in the setting of T2D and obesity can promote tumor growth by stimulating the IR/IGF-1R pathway in tumors with functional IR and IGF-1R. This concept is addressed in the next section.

MKR Mouse

To establish a mouse model of human T2D, the MKR mice were created [71]. The LeRoith group overexpressed the kinase-dead IGF-1R in skeletal muscles under the control of muscle creatine kinase (MCK) promoter. The kinase-dead IGF-1R was established by a point mutation of a lysine (K) to arginine (R) residue in the ATP-binding domain, [72]. As mentioned previously, the IGF-1R and IR are known to heterodimerize to form hybrid IR-IGF-1R receptors. Through the formation of hybrid receptors between the endogenous IR and the dominant negative IGF-1R, both IGF-1R and IR signaling was inhibited in these MCK:K→R (MKR) mice (Fig. 8.4). Stimulation with IGF-1 or insulin should induce phosphorylation of tyrosine residues of receptors as well as the association of p85 with IRS1 [73,74].

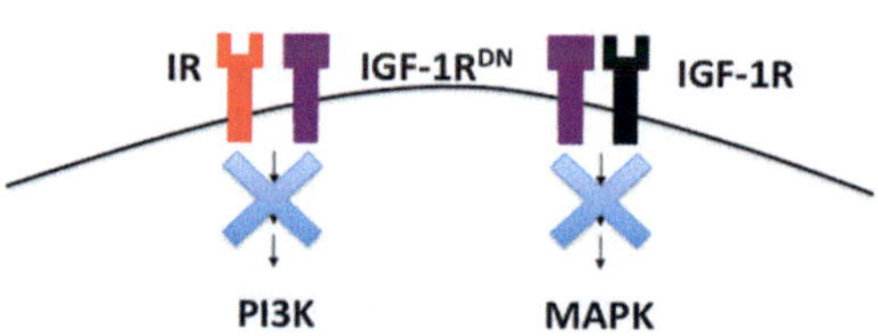

Fig. 8.4 Mode of IR and IGF-1R inhibition in MKR mouse. The IGF-1R^{DN} was established by a point mutation of a lysine to arginine residue in the ATP-binding domain of the IGF-1R. This IGF-1R^{DN} is able to homodimerize with endogenous IGF-1R and heterodimerize with endogenous IR, disrupting the downstream signaling cascades of both receptors. *IGF-1R* insulin-like growth factor 1 receptor, *IR* insulin receptor, *IGF-1R^{DN}* dominant negative insulin-like growth factor 1 receptor

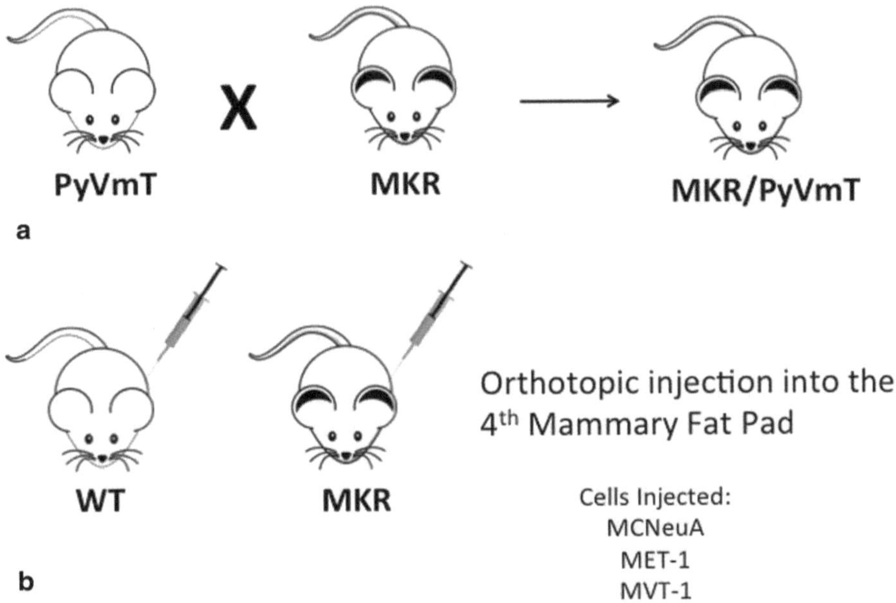

Fig. 8.5 Transgenic and orthotopic induction of mammary tumors. **a** The PyVmT mice contain the PyVmT oncogene under the MMTV promoter, which leads to the formation of mammary tumors in these mice. These mice were crossed with the type 2 diabetic MKR mice to produce a transgenic mouse model of breast cancer in type 2 diabetes. The MKR/PyVmT mice develop larger primary tumors with more pulmonary metastases compared to controls. **b** Orthotopic injections of various mouse-mammary cancer cells were injected into the 4th mammary fat pad of control and type 2 diabetic mice. MKR mice developed larger primary tumors compared to controls. The MVT-1 cell line had increased pulmonary metastases in the MKR mice compared to controls. *PyVmT* polyoma virus middle T antigen, *MMTV* mouse mammary tumor virus

In the MKR mice, there is no induction of tyrosine phosphorylation or association of p85 with IRS1 in the skeletal muscle upon injection with IGF-1 or insulin [72]. This demonstrates that the overexpression of dominant negative IGF-1R impairs both insulin and IGF-1 signaling in the skeletal muscle of MKR mice. Furthermore, there is impaired glucose uptake in the skeletal muscle of MKR mice compared to control mice following insulin stimulation. The male MKR mice exhibit significant insulin resistance, hyperinsulinemia, hyperglycemic, and hypertriglyceridemia, consistent with T2D. The female mice only demonstrate insulin resistance and hyperinsulinemia without hyperglycemia and hyperlipidemia, recapitulating a pre-diabetic phenotype in humans [75]. Both the males and females have reduced body fat establishing the MKR mouse as a nonobese model of T2D [75].

In the hyperinsulinemic female MKR mice, increased mammary tumor growth was established following carcinogen, transgenic, or orthotopic induction [75]. More extensive mammary epithelial cell development with more terminal end buds and epithelial branching is observed in the MKR mice compared to controls [75]. Analyzing the mammary epithelial cells revealed that there is a threefold increase in IR expression in the MKR mice compared to controls, with an increase in phosphorylation of the IR/IGF-1R and AKT. Breeding these MKR mice with the MMTV-PyVmT mice generated the MKR/PyVmT mice (Fig. 8.5a). The MKR/PyVmT mice develop larger primary tumors with more secondary tumor sites compared to controls [75]. Similar to the mammary epithelial cells from the MKR mice, tumors from the MKR/PyVmT mice have increased levels of phosphorylated IR/IGF-1R and AKT compared to tumors from control mice [76]. Orthotopically injecting mammary cancer cells derived from MMTV-Neu transgenic tumors (MC-NeuA cells), MMTV-PyVmT tumors (Met-1 cells), or MMTV-c-Myc/vegf tumors (MVT-1 cells), the female MKR mice develop larger primary tumors, assessed by tumor size and tumor weight (Fig. 8.5b) [75,76]. As with the MKR/PyVmT tumors, and mammary epithelial cells, there is a marked elevation of IR/IGF-1R phosphorylation and activation of the PI3K/Akt signaling pathway in all of the primary orthotopic tumors from the MKR mice compared to wild-type control mice. In addition, an increase in IR and IGF-1R protein expression is found in the MVT-1 tumors from the MKR mice. Utilizing the metastatic MVT-1 cell line, an increase in pulmonary metastases is observed in the MKR mice [76]. In order to assess if the increased pulmonary metastases resulted from larger primary tumors in MKR mice, intravenous injections of the MVT-1 cells were performed into the tail vein of control and MKR mice. The increase in lung metastases is also observed after this intravenous injection of the MVT-1 cells, suggesting that the hyperinsulinemia leads to increased survival or proliferation of the metastasizing cells [76]. To understand the mechanisms through which hyperinsulinemia and IR/IGF-1R/Akt signaling increase tumor metastases, primary tumors were analyzed for the expression of proteins related to tumor progression. Increased levels of c-Myc, matrix metalloprotease-9 (MMP9), and VEGF in tumors from the MKR mice [76]. In vitro analysis of the MVT-1 cells stimulated with physiological levels of insulin found that insulin increases expression of c-Myc, MMP9 and activates AKT [76]. The cell culture experiments recapitulate the results seen in vivo and suggest that insulin acting through the IR and perhaps the IGF-1R on the tumor cells can promote mammary tumor cell proliferation and metastasis [76].

Utilizing a small-molecule tyrosine kinase inhibitor, which blocks the ATP-binding domain of IR/IGF-1R significantly reduced the tumor growth in MKR mice compared to controls [75]. Furthermore, reducing insulin levels using an insulin-sensitizing agent, a β3-adrenoceptor agonist decreased circulating insulin levels in the MKR mice [77]. This CL-316243, insulin-sensitizing agent, significantly decreased the growth rate of the PyVmT tumors as well as the orthotopically injected Met-1, and MCNeuA in the MKR mice [77]. Administering the insulin-sensitizing agent to the mice after intravenous MVT-1 cell injection promotes a significant reduction of metastases in the MKR mice compared to the vehicle-treated MKR mice [76]. Inhibiting the downstream target of insulin signaling, the PI3K/Akt pathway with a pan-class I PI3K inhibitor leads to a reduction of tumor growth in the MKR mice compared to controls. The reduction in tumor growth after the administration of the inhibitor is attributed to a decrease in Akt phosphorylation, as well as a decrease in proliferation due to a reduction in 5-bromo-2'-deoxyuridine (BrdU) incorporation in the tumors from the MKR mice compared to controls. Inhibiting the mTOR pathway with rapamycin also results in reduced primary tumor growth [78]. Taken together, the data show that inhibiting the insulin signaling pathway and its downstream targets decreases tumor growth and pulmonary metastasis in the MKR mice [75–79].

In contrast to the previously discussed mouse models of T2D (the βIRKO, IRS-2 null, Akt2 null mice), the MKR mice only have a genetic abnormality of IR/IGF-1R signaling in the skeletal muscle. As demonstrated in these studies, the normal mammary epithelial cells and primary tumors in the MKR mice have functional insulin and IGF-1 receptors. Therefore, the MKR mouse is a model of hyperinsulinemia, demonstrating that hyperinsulinemia promotes the increased survival or proliferation of mammary tumors as well as pulmonary metastases by acting through the IR/IGF-1R signaling pathway. Inhibiting the IR/IGF-1R pathway with targeted therapies decreased tumor growth and the number of lung metastases in the hyperinsulinemic MKR mouse.

Overexpression of IGF-1

In this section, we will examine the mouse models with tissue-specific IGF-1 overexpression and its effect on cancer. In female mice, overexpression of IGF-1 under the control of the bovine keratin 5 (BK5) promoter allows for the overexpression of the IGF-1 transgene in the myoepithelial cells of the mouse mammary gland [80]. The mammary epithelial cells were exposed to high levels of IGF-1 by paracrine signaling, which is believed to model the local production of IGF-1 and paracrine signaling that occurs in human breast cancers [80]. Local IGF-1 overexpression in this model promotes mammary gland hyperplasia, spontaneous mammary tumorigenesis, and increased susceptibility to chemical carcinogens [80].

Male mice with IGF-1 overexpression in prostate epithelial cells under the BK5 promoter develop prostate cancer at an increased rate [81]. The mice develop prostatic adenocarcinomas through a stepwise progression of preneoplasia to full neoplasia [81]. The development proceeds through the preneoplastic changes that include atypical hyperplasia, prostatic intraepithelial neoplasia (PIN)-like lesions, and then proceed to adenocarcinomas that exhibit invasive properties, which is similar to the human progression of prostate cancer [82]. In male mice, the findings from the BK5.IGF1 transgenic mice are able to provide a link between IGF-1R signaling and prostate tumor development.

Another mouse model to study the effects of IGF-1 on the mammary gland used DES-IGF-1; an IGF-1 analog that has reduced affinity for IGFBP, meaning it has increased IGF-1R-binding capacity [83]. This analog was placed under the whey acidic promoter (WAP) [83]. The WAP promoter is hormonally regulated and its expression increases during pregnancy and peaks at lactation [84]. The WAP promoter can be used to drive the overexpression of genes in the mammary gland during development. Therefore, these WAP-DES mice exhibit an increase of IGF-1 in the mammary gland. This in turn leads to an increase in the development of mammary adenocarcinomas in WAP-DES mice and a reduction of apoptosis in the normal mammary gland [85]. IGF-1 analog overexpression in these mice increased the frequency of mammary interepithelial neoplasia [85]. These mouse studies demonstrate that the increase in local IGF-1 levels promotes the formation of tumors.

IGF-1R Overexpression

A mouse model was designed to express a CD8-IGF-1R fusion protein that was placed under the MMTV promoter [86]. These extracellular portions of the human T-cell antigen CD8-alpha can homodimerize, thus constitutively activating IGF1R signaling [86]. At 4 weeks of age, the mammary gland from the CD8-IGF1R mice was abnormal, with reduction in terminal end buds and aberrant side branching compared to nontransgenic controls [86]. The MMTV-CD8-IGF-1R transgenic mice developed mammary and salivary adenocarcinomas at an early age, as early at 6 weeks after birth compared to controls [86]. The tumors were palpable within 8 weeks of age compared to controls [86].

Another mouse model utilizing a tissue-specific IGF-1R overexpression is the MTB-IGF-1R transgenic mouse that is an inducible IGF-1R mouse model. The IGF-1R gene can be induced by doxycycline at 21 days of age to promote the development of mammary epithelial tumors [87]. Wild-type mice treated with doxycycline maintained low levels of IGF-1R in mammary ducts and terminal end buds [87]. MTB-IGF-1R mice showed significantly higher levels of IGF-1R in mammary ducts and terminal end buds post doxycycline treatment [87]. Mammary tumors from MTB-IGF-1R mice expressed high levels of IGF-1R and were highly proliferative [87]. Furthermore, the cytokeratin profile of the MTB-IGF-1R mammary tumors was altered. In normal tissue, luminal cells stain positive for cytokeratin 8 and 18 whereas cytokeratin 5 and 14 are associated with basal cells [87]. The tumors with solid sheets of cells were typically negative for cytokeratin 5 and 14 [87]. The tumors that were formed by the CD8-IGF-1R and MTB-IGF1R were primarily solid epithelial sheets with some of the tumor cells displaying a more mesenchymal phenotype [87]. Some of the CD8-IGF-1R and 40% of the MTB-IGF1R mice had increased lung metastasis [87].

In order to assess the importance of IGF-1R expression, doxycycline was used to induce tumors in the MTB-IGF-1R mice. After the development of mammary tumors, doxycycline administration was removed which promoted the regression of approximately 80% of the tumors [88]. Tumors showed three different fates: (1) regressed to a nonpalpable state and did not resume growth, (2) regressed to a nonpalpable state and then resumed growth, or (3) tumors regressed or did not change and then resumed growth [88]. Tumor regression was associated with a decrease in proliferation and an increase in apoptosis assessed by a reduction in PCNA staining and an increased TUNEL staining [88]. Only 13% of the mice had tumor recurrence within 21–83 days post doxycycline removal [88]. The tumors that recurred did not express IGF-1R and had a more mesenchymal phenotype based on the increased levels of Twist, Snail, Slug, and Zeb and the decreased levels of E-cadherin [88]. Histological sections of the mammary tumors resembled those with activated ErbB2 or Wnt signaling [88]. This suggests that mammary tumors may be able to activate these oncogenes and undergo EMT to recur. The mouse models in this section demonstrate that increased IGF-1R expression in the tumors promotes the formation of larger primary tumors and increases pulmonary metastases. Furthermore, these studies suggested that decreasing IGF-1R signaling within the tumors may be therapeutically beneficial. However, as discussed in the section "Tissue-Specific Insulin Receptor Gene Deletion" in certain tumors the IR may compensate for the loss of IGF-1R, and as discussed in the next section IGF-2 may signal through either the IGF-1R or IR to promote tumor growth. IGF-1R inhibition and reduction is evaluated in the section "IGF-1 Reduction and IGF-1R Inhibition."

IGF-2 Overexpression

IGF-2 is a growth factor that is involved in the control of cell proliferation and inhibition of apoptosis [89]. The IGF-2 gene is maternally imprinted and has homology with IGF-1 [90]. IGF-2 overexpression has been linked to various cancers such as adrenocortical carcinomas, lung cancer, and breast cancer [89,91–94]. It has been demonstrated that IGF-2 can bind IGF-1R, IR, or the hybrid IGF-1R/IR receptor to mediate its action (Fig. 8.1) [95]. Moreover, IGF-2 has a high affinity for the insulin receptor-A (IR-A) isoform [29]. Therefore, it is believed that in the presence of increased IGF-2, IR-A can be activated and contribute to mitogenesis. Human endometrial carcinoma cell lines treated with IGF-2 demonstrated an increase in cell cycle progression and a decrease in apoptosis [96]. Furthermore, stimulation with IGF-2 led to an increase in Akt phosphorylation. African-American women present with more advanced stages of breast cancer and have lower survival than white women [97]. Analyzing breast tissue from these two ethnic groups found that there is significantly higher IGF-2 expression in breast tissue and cell lines from African-American women compared to white women [97]. Tumors from African-American women expressed higher levels of pro-survival and anti-apoptotic proteins, survivin, and Bcl-2 [97]. Moreover, breast cancers from African-American women had a higher levels of IR-A compared to breast cancer tumors from white women [98]. Therefore, it is possible that the higher levels of IGF-2 acting through IR-A leads to the increase in cell cycle progression and decrease in apoptosis, which contributes to the more aggressive breast cancer phenotype that is observed in African-American women.

Mice with an overexpression of IGF-2 in the liver are larger than their normal littermates, whereas ones with inactive IGF-2 gene are 40 % smaller than controls [95]. Transgenic mice overexpressing IGF2 in the mammary gland were generated by placing the IGF2 gene under the MMTV promoter [99]. This allows for the overexpression of IGF-2 in mammary epithelial cells. The overexpression of IGF-2 in the mammary epithelium delayed postlactation mammary involution, decreased mammary epithelial apoptosis, and prolonged the expression of active phosphorylated Akt in the MMTV-IGF-2 mice compared to wild-type controls [99]. It is believed that this could be due to constitutive activation of the pathways downstream of IGF-1R driving more ERK1/2 and p38 MAPK signaling leading to activation of cyclic-AMP response element-binding protein (CREB), a transcription factor that has been found to be overexpressed or constitutively activated in various cancers, including lung, breast, and prostate [94,100]. IGF-2 reduced mammary epithelial apoptosis assessed by in situ labeling and detecting of DNA breaks [99]. Moreover, there is evidence that shows that IGF-2 promotes the translocation of β-catenin into the nucleus, thus activating the Wnt signaling pathway [101]. Through the constitutive activation of the MAPK or Wnt signaling pathways and inhibition of apoptosis, IGF-2 may promote tumor development. High IGF-2 levels in the serum have also been correlated with an increase in tumor incidence in older mice [92,99]. IGF-2 was placed under the phosphoenolpyruvate carboxykinase (PEPCK) promoter leading to overexpression of IGF-2 in the colon and various other tissues, such as the liver, intestine, and kidney [102,103]. IGF-2 serum levels were significantly elevated in the transgenic mice compared to controls [102]. The PEPCK-IGF-2 transgenic mice developed more colon tumors as well as an increase in tumor size [102].

The RIP1-Tag2 mouse model, previously mentioned in the section "Tissue-Specific Insulin Receptor Gene Deletion," has an overexpression of IGF-2 [104]. It was noted that IGF-2 is selectively upregulated in the tumors, contributing to the hyperproliferative β-cell lesions [104]. The tumors progress in a stepwise manner from hyperplastic/dysplastic islet cells, which undergo an angiogenic switch [104]. The solid tumors then transform into invasive carcinomas and display a loss of E-cadherin [105]. The effects of IGF-1R and IGF-2 overexpression in the RIP-Tag2 mouse model was evaluated by using double transgenic mouse models [105]. IGF-1R was placed under the rat insulin promoter 7 (RIP7) to generate the RIP7-IGF-1R mice [105]. These mice were crossed with RIP-Tag2 mice, which overexpress IGF-2 [105]. The overexpression of both IGF-2 and IGF-1R in the double transgenic mice promotes the acceleration of β-cell tumor formation [105]. Furthermore, many of the mice display lymph node metastases, which are not present in the single transgenic RIP1-Tag2 mouse model [105]. To investigate the importance of IGF-2 in this mouse model, RIP-Tag2 mice were crossed with IGF-2 null mice, which develop normally, even though they have a reduced body mass [105]. The RIP-Tag2/IGF-2$^{-/-}$ mice developed smaller tumors with increased apoptosis [105].

Furthermore, studies have shown that mammary tumor onset can be delayed by increasing the expression of the IGF-2R in the mammary gland [106]. One of the IGF-2R functions is to clear IGF-2 from circulation, thus reducing the availability of IGF-2 to bind to the IR and/or IGF-1R [107]. Transgenic mice with IGF-2 overexpression in the mammary gland were crossed with mice with ubiquitous IGF-2R overexpression [106]. As previously mentioned, mice with IGF-2 overexpression in the mammary gland promotes the formation of mammary tumors [93]. The presence of the IGF-2R delayed the onset of mammary tumors in the mice compared to IGF-2 overexpressing mice [106]. These mouse models demonstrate that deregulation of insulin, IGF-1, IGF-2, and their receptors plays an important role in cancer development and metastases.

IGF-1 Reduction and IGF-1R Inhibition

Using the Cre-loxP system described in the previous sections, the IGF-1 gene was deleted in a liver-specific manner in mice, by expressing Cre under the albumin promoter [108]. These liver IGF-1-deficient (LID) mice demonstrated normal growth even though they had a 75 % reduction in serum IGF-1 levels [108]. By 1 month of age, LID mice exhibit insulin resistance and demonstrate an age-dependent onset of hyperinsulinemia starting at 2.5 months of age [109]. Orthotopic transplantation of colon adenocarcinoma cells demonstrated that the incidence of colon tumor growth was significantly higher in controls compared to LID mice [110]. This demonstrates that IGF-1 promotes increased tumor development and more metastasis. The LID mice, having decreased circulating levels of IGF-1, therefore, had smaller primary tumors and less metastasis to the lungs compared to the control mice.

LID mice exposed to the carcinogen 7,12-dimethybenz (a)anthracene (DMBA) which induces tumors exhibited a delayed onset of mammary tumor formation [111]. In the same study, C3(1)/SV40-T-antigen transgenic mice were crossed with LID mice [111]. C3(1) is the 5′ flanking region of the rat prostate steroid-binding protein (PSBP) that has been used to target the expression of the SV40-large T-antigen to the epithelial cells of the prostate and mammary glands [112]. The female C3(1)-Tag mice develop invasive mammary carcinomas, with metastases to the lung [111]. The LID mice were crossed with the C3(1)-Tag transgenic mice to examine the effects of lower circulating levels of IGF-1 on genetic mammary tumor development [111]. There was an earlier tumor formation in the control mice compared to LID mice with the C(3)1-Tag transgene [111]. These studies demonstrated that the reduction in circulating IGF-1 allowed for a decrease in colon and mammary tumor growth and metastasis. This was not the case for murine osteosarcoma primary tumor growth. K7M2 murine osteosarcoma cells were injected intramuscularly into LID and control mice [113]. These cells express both IGF-1 and the IGF-1R [113]. There was no significant difference in primary tumor growth or pulmonary metastasis in the LID mice compared to controls [113]. The osteosarcoma primary tumors exhibited a similar tumor growth and equal time of onset [113]. This study demonstrated that the decrease in serum IGF-1 did not influence the tumor growth rate or ability of the osteosarcoma cell to metastasize [113]. This suggests that paracrine IGF-1 signaling is important in progression of cancer and that reducing circulating IGF-1 levels alone may not be sufficient in delaying tumor formation in certain cancers.

A reduction of circulating IGF-1 levels can prolong tumor latency and reduce tumor size [111]. However, studies silencing IGF-1R signaling, such as in the βIRKO mice (Tissue-Specific Insulin Receptor Gene Deletion), have not shown the predicted beneficial effects on tumor growth. A study combined dietary fat reduction with an IGF-1R blocking antibody to examine the effects on prostate cancer progression in a mouse model [114]. Although, previous in vitro studies found that blocking the IGF-1R inhibited cell growth and induced apoptosis, neither the dietary fat reduction nor the IGF-1R blocking therapy separately, or in combination affected the tumor weight or volume [114].

Various clinical trials tried to utilize specific anti-IGF-1R antibodies and have been relatively unsuccessful in reducing cancer growth [115–117]. Phase II studies on patients with squamous cell carcinoma of the head and neck (SCCHN) found that there was no clinically significant benefit of the anti-IGF-1R antibody [116]. The trial demonstrated that the anti-IGF-1R antibody was unable to significantly inhibit the PI3K/Akt and MAPK signaling pathways in patients with SCCHN [116]. A clinical trial in patients with advanced non-small cell lung cancer (NSCLC) and in patients with metastatic refractory colorectal cancer (CRC) also demonstrated little benefit [115,117]. It is believed that only a specific subset of cancers are responsive to IGF-1R targeted therapy and that other receptors and signaling pathways such as the insulin receptor signaling pathway may be compensating for the inhibition of IGF-1R signaling. A review by D. Yee examined current IGF-1R inhibitors in cancer therapy [118]. Inhibition of only the IGF-1R promotes the elevation of serum growth hormone and IGF-1, which then allows for serum insulin levels to increase. Patients with insulin resistance or hyperinsulinemia had increased toxicity associated with IGF-1R inhibitors. This is most likely due to the inability of IGF-1R inhibitors to block IR, thus leading to enhanced IR signaling. Thus, it is important to keep in mind possible compensation between the IR and the IGF-1R when developing pharmacological therapies.

Conclusions

Mouse studies have led to important discoveries regarding the effects of insulin and IGF-1 on cancer in the setting of T2D and obesity. Numerous epidemiological studies have connected T2D, the metabolic syndrome, and obesity with increased risk and mortality in many different cancers. As discussed in this chapter, there are various mouse models that can be appropriately used to study the mechanisms that may contribute to the progression of cancer in insulin resistant, diabetic, or obese individuals. It is believed that normal cells obtain oncogenic mutations that transform these cells into cancer cells. The growth of these mutated cells is then propagated in the setting of T2D and obesity. It is vital to understand the mechanisms involved in the link between T2D, obesity, and cancer. The mouse models highlighted in this chapter illustrate the importance of IR and IGF-1R signaling in the progression of cancer and metastases. Current studies focus on uncovering these mechanisms further in order to allow for therapeutic targeting.

Disclosure Authors have nothing to disclose.

References

1. Boyle P, Boniol M, Koechlin A, et al. Diabetes and breast cancer risk: a meta-analysis. Br J Cancer. 2012. doi:10.1038/bjc.2012.414.
2. Adami HO, McLaughlin J, Ekbom A, et al. Cancer risk in patients with diabetes mellitus. Cancer Causes Control. 1991;2(5):307–14.
3. Renehan AG, Tyson M, Egger M, et al. Body-mass index and incidence of cancer: a systematic review and meta-analysis of prospective observational studies. Lancet. 2008;371(9612):569–78. doi:10.1016/S0140-6736(08)60269-X.
4. Chari ST, Leibson CL, Rabe KG, et al. Probability of pancreatic cancer following diabetes: a population-based study. Gastroenterology. 2005;129(2):504–11. doi:10.1016/j.gastro.2005.05.007.
5. Onitilo AA, Engel JM, Glurich I, et al. Diabetes and cancer I: risk, survival, and implications for screening. Cancer Causes Control. 2012;23(6):967–81. doi:10.1007/s10552-012-9972-3.
6. Zhang PH, Chen ZW, Lv D, et al. Increased risk of cancer in patients with type 2 diabetes mellitus: a retrospective cohort study in China. BMC Public Health. 2012;12(1):567. doi:10.1186/1471-2458-12-567.
7. Redaniel MT, Jeffreys M, May MT, et al. Associations of type 2 diabetes and diabetes treatment with breast cancer risk and mortality: a population-based cohort study among British women. Cancer Causes Control. 2012;23(11):1785–95. doi:10.1007/s10552-012-0057-0.
8. Yang Y, Mauldin PD, Ebeling M, et al. Effect of metabolic syndrome and its components on recurrence and survival in colon cancer patients. Cancer. 2012. doi:10.1002/cncr.27923.
9. Lawlor DA, Smith GD, Ebrahim S. Hyperinsulinaemia and increased risk of breast cancer: findings from the British Women's Heart and Health Study. Cancer Causes Control. 2004;15(3):267–75. doi:10.1023/B:CACO.0000024225.14618.a8.
10. Zanders MM, Boll D, van Steenbergen LN, et al. Effect of diabetes on endometrial cancer recurrence and survival. Maturitas. 2013;74(1):37–43. doi:10.1016/j.maturitas.2012.10.007.
11. Ouchi N, Parker JL, Lugus JJ, et al. Adipokines in inflammation and metabolic disease. Nat Rev Immunol. 2011;11(2):85–97. doi:10.1038/nri2921.
12. Giovannucci E, Harlan DM, Archer MC, et al. Diabetes and cancer: a consensus report. Diabetes Care. 2010;33(7):1674–85. doi:10.2337/dc10-0666.
13. Kaplan MA, Pekkolay Z, Kucukoner M, et al. Type 2 diabetes mellitus and prognosis in early stage breast cancer women. Med Oncol. 2012;29(3):1576–80. doi:10.1007/s12032-011-0109-4.
14. Liu X, Ji J, Sundquist K, et al. The impact of type 2 diabetes mellitus on cancer-specific survival: a follow-up study in Sweden. Cancer. 2012;118(5):1353–61. doi:10.1002/cncr.26420.
15. Campbell PT, Newton CC, Patel AV, et al. Diabetes and cause-specific mortality in a prospective cohort of one million U.S. adults. Diabetes Care. 2012;35(9):1835–44. doi:10.2337/dc12-0002.
16. Calle EE, Rodriguez C, Walker-Thurmond K, et al. Overweight, obesity, and mortality from cancer in a prospectively studied cohort of U.S. adults. N Engl J Med. 2003;348(17):1625–38. doi:10.1056/NEJMoa021423.
17. DeFronzo RA. Pathogenesis of type 2 diabetes mellitus. Med Clin North Am. 2004;88(4):787–835, ix. doi:10.1016/j.mcna.2004.04.013.
18. Kahn CR, Vicent D, Doria A. Genetics of non-insulin-dependent (type-II) diabetes mellitus. Annu Rev Med. 1996;47:509–31. doi:10.1146/annurev.med.47.1.509.
19. Lillioja S, Mott DM, Spraul M, et al. Insulin resistance and insulin secretory dysfunction as precursors of non-insulin-dependent diabetes mellitus. Prospective studies of Pima Indians. N Engl J Med. 1993;329(27):1988–92. doi:10.1056/NEJM199312303292703.
20. LeRoith D, Gavrilova O. Mouse models created to study the pathophysiology of type 2 diabetes. Int J Biochem Cell Biol. 2006;38(5–6):904–12. doi:10.1016/j.biocel.2005.01.019.
21. Goodwin PJ, Ennis M, Pritchard KI, et al. Fasting insulin and outcome in early-stage breast cancer: results of a prospective cohort study. J Clin Oncol. 2002;20(1):42–51.

22. Lipscombe LL, Goodwin PJ, Zinman B, et al. The impact of diabetes on survival following breast cancer. Breast Cancer Res Treat. 2008;109(2):389–95. doi:10.1007/s10549-007-9654-0.
23. Mauro L, Bartucci M, Morelli C, et al. IGF-I receptor-induced cell-cell adhesion of MCF-7 breast cancer cells requires the expression of junction protein ZO-1. J Biol Chem. 2001;276(43):39892–7. doi:10.1074/jbc.M106673200.
24. Papa V, Pezzino V, Costantino A, et al. Elevated insulin receptor content in human breast cancer. J Clin Invest. 1990;86(5):1503–10. doi:10.1172/JCI114868.
25. Surmacz E. Function of the IGF-I receptor in breast cancer. J Mammary Gland Biol Neoplasia. 2000;5(1):95–105.
26. Law JH, Habibi G, Hu K, et al. Phosphorylated insulin-like growth factor-i/insulin receptor is present in all breast cancer subtypes and is related to poor survival. Cancer Res. 2008;68(24):10238–46. doi:10.1158/0008-5472.CAN-08-2755.
27. Belfiore A, Frasca F, Pandini G, et al. Insulin receptor isoforms and insulin receptor/insulin-like growth factor receptor hybrids in physiology and disease. Endocr Rev. 2009;30(6):586–623. doi:10.1210/er.2008-0047.
28. Frasca F, Pandini G, Scalia P, et al. Insulin receptor isoform A, a newly recognized, high-affinity insulin-like growth factor II receptor in fetal and cancer cells. Mol Cell Biol. 1999;19(5):3278–88.
29. Belfiore A, Malaguarnera R. Insulin receptor and cancer. Endocr Relat Cancer. 2011;18(4):R125–47. doi:10.1530/ERC-11-0074.
30. Sridhar SS, Goodwin PJ. Insulin-insulin-like growth factor axis and colon cancer. J Clin Oncol. 2009;27(2):165–7. doi:10.1200/JCO.2008.19.8937.
31. Baxter RC. IGF binding proteins in cancer: mechanistic and clinical insights. Nat Rev Cancer. 2014;14(5):329–41. doi:10.1038/nrc3720.
32. Becker S, Dossus L, Kaaks R. Obesity related hyperinsulinaemia and hyperglycaemia and cancer development. Arch Physiol Biochem. 2009;115(2):86–96. doi:10.1080/13813450902878054.
33. Thissen JP, Ketelslegers JM, Underwood LE. Nutritional regulation of the insulin-like growth factors. Endocr Rev. 1994;15(1):80–101. doi:10.1210/edrv-15-1-80.
34. D'Ercole AJ, Applewhite GT, Underwood LE. Evidence that somatomedin is synthesized by multiple tissues in the fetus. Dev Biol. 1980;75(2):315–28.
35. Yu H, Rohan T. Role of the insulin-like growth factor family in cancer development and progression. J Natl Cancer Inst. 2000;92(18):1472–89.
36. Vijayakumar A, Buffin NJ, Gallagher EJ, et al. Deletion of growth hormone receptors in post-natal skeletal muscle of male mice does not alter muscle mass and response to pathological injury. Endocrinology. 2013;154(10):3776–83. doi:10.1210/en.2013–1209.
37. Hankinson SE, Willett WC, Colditz GA, et al. Circulating concentrations of insulin-like growth factor-I and risk of breast cancer. Lancet. 1998;351(9113):1393–6. doi:10.1016/S0140-6736(97)10384-1.
38. Ikushima H, Miyazono K. TGFbeta signalling: a complex web in cancer progression. Nat Rev Cancer. 2010;10(6):415–24. doi:10.1038/nrc2853.
39. Kodaki T, Woscholski R, Hallberg, B et al. The activation of phosphatidylinositol 3-kinase by Ras. Curr Biol. 1994;4(9):798–806.
40. Rodriguez-Viciana P, Warne PH, Vanhaesebroeck B, et al. Activation of phosphoinositide 3-kinase by interaction with Ras and by point mutation. EMBO J. 1996;15(10):2442–51.
41. Samani AA, Yakar S, LeRoith D, et al. The role of the IGF system in cancer growth and metastasis: overview and recent insights. Endocr Rev. 2007;28(1):20–47. doi:10.1210/er.2006-0001.
42. Turner BC, Haffty BG, Narayanan L, et al. Insulin-like growth factor-I receptor overexpression mediates cellular radioresistance and local breast cancer recurrence after lumpectomy and radiation. Cancer Res. 1997;57(15):3079–83.
43. Belfiore A, Frittitta L, Costantino A, et al. Insulin receptors in breast cancer. Ann N Y Acad Sci. 1996;784:173–88.

44. Larue L, Bellacosa A. Epithelial-mesenchymal transition in development and cancer: role of phosphatidylinositol 3' kinase/AKT pathways. Oncogene. 2005;24(50):7443–54. doi:10.1038/sj.onc.1209091.
45. Testa JR, Bellacosa A. AKT plays a central role in tumorigenesis. Proc Natl Acad Sci U S A. 2001;98(20):10983–5. doi:10.1073/pnas.211430998.
46. Grille SJ, Bellacosa A, Upson J, et al. The protein kinase Akt induces epithelial mesenchymal transition and promotes enhanced motility and invasiveness of squamous cell carcinoma lines. Cancer Res. 2003;63(9):2172–8.
47. Kalluri R, Weinberg RA. The basics of epithelial-mesenchymal transition. J Clin Invest. 2009;119(6):1420–8. doi:10.1172/JCI39104.
48. Walsh LA, Damjanovski S. IGF-1 increases invasive potential of MCF 7 breast cancer cells and induces activation of latent TGF-beta1 resulting in epithelial to mesenchymal transition. Cell Commun Signal. 2011;9(1):10. doi:10.1186/1478-811X-9-10.
49. Kim HJ, Litzenburger BC, Cui X, et al. Constitutively active type I insulin-like growth factor receptor causes transformation and xenograft growth of immortalized mammary epithelial cells and is accompanied by an epithelial-to-mesenchymal transition mediated by NF-kappaB and snail. Mol Cell Biol. 2007;27(8):3165–75. doi:10.1128/MCB.01315-06.
50. Accili D, Drago J, Lee EJ, et al. Early neonatal death in mice homozygous for a null allele of the insulin receptor gene. Nat Genet. 1996;12(1):106–9. doi:10.1038/ng0196-106.
51. Joshi RL, Lamothe B, Cordonnier N, et al. Targeted disruption of the insulin receptor gene in the mouse results in neonatal lethality. EMBO J. 1996;15(7):1542–7.
52. Nagy A. Cre recombinase: the universal reagent for genome tailoring. Genesis. 2000;26(2):99–109.
53. Kulkarni RN, Bruning JC, Winnay JN, et al. Tissue-specific knockout of the insulin receptor in pancreatic beta cells creates an insulin secretory defect similar to that in type 2 diabetes. Cell. 1999;96(3):329–39.
54. Ulanet DB, Ludwig DL, Kahn CR, et al. Insulin receptor functionally enhances multistage tumor progression and conveys intrinsic resistance to IGF-1R targeted therapy. Proc Natl Acad Sci U S A. 2010;107(24):10791–8. doi:10.1073/pnas.0914076107.
55. Tamemoto H, Kadowaki T, Tobe K, et al. Insulin resistance and growth retardation in mice lacking insulin receptor substrate-1. Nature. 1994;372(6502):182–6. doi:10.1038/372182a0.
56. Bruning JC, Winnay J, Cheatham B, et al. Differential signaling by insulin receptor substrate 1 (IRS-1) and IRS-2 in IRS-1-deficient cells. Mol Cell Biol. 1997;17(3):1513–21.
57. Jenkins AB, Storlien LH. Insulin resistance and hyperinsulinaemia in insulin receptor substrate-1 knockout mice. Diabetologia. 1997;40(9):1113–4.
58. Zolzaya K, Nakamura A, Tajima K, et al. Role of insulin receptor substrate-1 for diethylnitrosamine plus high-fat diet-induced hepatic tumorigenesis in mice. J Diabetes Invest. 2014;5(1):27–30. doi:10.1111/jdi.12128.
59. Sayed-Ahmed MM, Aleisa AM, Al-Rejaie SS, et al. Thymoquinone attenuates diethylnitrosamine induction of hepatic carcinogenesis through antioxidant signaling. Oxid Med Cell Longev. 2010;3(4):254–61.
60. Luo M, Yang F, Huang SX, et al. Two-stage model of chemically induced hepatocellular carcinoma in mouse. Oncol Res. 2013;20(11):517–28. doi:10.3727/096504013X13747716581336.
61. Gibson SL, Ma Z, Shaw LM. Divergent roles for IRS-1 and IRS-2 in breast cancer metastasis. Cell Cycle. 2007;6(6):631–7.
62. Liang L, Zhou T, Jiang J, et al. Insulin receptor substrate-1 enhances growth hormone-induced proliferation. Endocrinology. 1999;140(5):1972–83. doi:10.1210/endo.140.5.6724.
63. Porter HA, Perry A, Kingsley C, et al. IRS1 is highly expressed in localized breast tumors and regulates the sensitivity of breast cancer cells to chemotherapy, while IRS2 is highly expressed in invasive breast tumors. Cancer Lett. 2013;338(2):239–48. doi:10.1016/j.canlet.2013.03.030.
64. Day E, Poulogiannis G, McCaughan F, et al. IRS2 is a candidate driver oncogene on 13q34 in colorectal cancer. Int J Exp Pathol. 2013;94(3):203–11. doi:10.1111/iep.12021.

65. Withers DJ, Gutierrez JS, Towery H, et al. Disruption of IRS-2 causes type 2 diabetes in mice. Nature. 1998;391(6670):900–4. doi:10.1038/36116.
66. Hennige AM, Burks DJ, Ozcan U, et al. Upregulation of insulin receptor substrate-2 in pancreatic beta cells prevents diabetes. J Clin Invest. 2003;112(10):1521–32. doi:10.1172/JCI18581.
67. Nagle JA, Ma Z, Byrne MA, et al. Involvement of insulin receptor substrate 2 in mammary tumor metastasis. Mol Cell Biol. 2004;24(22):9726–35. doi:10.1128/MCB.24.22.9726-9735.2004.
68. Peng XD, Xu PZ, Chen ML, et al. Dwarfism, impaired skin development, skeletal muscle atrophy, delayed bone development, and impeded adipogenesis in mice lacking Akt1 and Akt2. Genes Dev. 2003;17(11):1352–65. doi:10.1101/gad.1089403.
69. Garofalo RS, Orena SJ, Rafidi K, et al. Severe diabetes, age-dependent loss of adipose tissue, and mild growth deficiency in mice lacking Akt2/PKB beta. J Clin Invest. 2003;112 (2):197–208. doi:10.1172/JCI16885.
70. Watson KL, Moorehead RA. Loss of Akt1 or Akt2 delays mammary tumor onset and suppresses tumor growth rate in MTB-IGFIR transgenic mice. BMC Cancer. 2013;13:375. doi:10.1186/1471-2407-13-375.
71. LeRoith D, Werner H, Beitner-Johnson D, et al. Molecular and cellular aspects of the insulin-like growth factor I receptor. Endocr Rev. 1995;16(2):143–63.
72. Fernandez AM, Kim JK, Yakar S, et al. Functional inactivation of the IGF-I and insulin receptors in skeletal muscle causes type 2 diabetes. Genes Dev. 2001;15(15):1926–34. doi:10.1101/gad.908001.
73. Draznin B. Molecular mechanisms of insulin resistance: serine phosphorylation of insulin receptor substrate-1 and increased expression of p85alpha: the two sides of a coin. Diabetes. 2006;55(8):2392–7. doi:10.2337/db06-0391.
74. Myers MG Jr, White MF. Insulin signal transduction and the IRS proteins. Annu Rev Pharmacol Toxicol. 1996;36:615–58. doi:10.1146/annurev.pa.36.040196.003151.
75. Novosyadlyy R, Lann DE, Vijayakumar A, et al. Insulin-mediated acceleration of breast cancer development and progression in a nonobese model of type 2 diabetes. Cancer Res. 2010;70(2):741–51. doi:10.1158/0008-5472.CAN-09-2141.
76. Ferguson RD, Novosyadlyy R, Fierz Y, et al. Hyperinsulinemia enhances c-Myc-mediated mammary tumor development and advances metastatic progression to the lung in a mouse model of type 2 diabetes. Breast Cancer Res. 2012;14(1):R8. doi:10.1186/bcr3089.
77. Fierz Y, Novosyadlyy R, Vijayakumar A, et al. Insulin-sensitizing therapy attenuates type 2 diabetes-mediated mammary tumor progression. Diabetes. 2010;59(3):686–93. doi:10.2337/db09-1291.
78. Fierz Y, Novosyadlyy R, Vijayakumar A, et al. Mammalian target of rapamycin inhibition abrogates insulin-mediated mammary tumor progression in type 2 diabetes. Endocr Relat Cancer. 2010;17(4):941–51. doi:10.1677/ERC-10-0091.
79. Gallagher EJ, Fierz Y, Vijayakumar A, et al. Inhibiting PI3K reduces mammary tumor growth and induces hyperglycemia in a mouse model of insulin resistance and hyperinsulinemia. Oncogene. 2012;31(27):3213–22. doi:10.1038/onc.2011.495.
80. de Ostrovich KK, Lambertz I, Colby JK, et al. Paracrine overexpression of insulin-like growth factor-1 enhances mammary tumorigenesis in vivo. Am J Pathol. 2008;173(3):824–34. doi:10.2353/ajpath.2008.071005.
81. DiGiovanni J, Kiguchi K, Frijhoff A, et al. Deregulated expression of insulin-like growth factor 1 in prostate epithelium leads to neoplasia in transgenic mice. Proc Natl Acad Sci U S A. 2000;97(7):3455–60.
82. Abate-Shen C, Shen MM. Molecular genetics of prostate cancer. Genes Dev. 2000;14(19):2410–34.
83. Hadsell DL, Greenberg NM, Fligger JM, et al. Targeted expression of des(1-3) human insulin-like growth factor I in transgenic mice influences mammary gland development and IGF-binding protein expression. Endocrinology. 1996;137(1):321–30. doi:10.1210/endo.137.1.8536631.

84. Vargo-Gogola T, Rosen JM. Modelling breast cancer: one size does not fit all. Nat Rev Cancer. 2007;7(9):659–72. doi:10.1038/nrc2193.
85. Hadsell DL, Murphy KL, Bonnette SG, et al. Cooperative interaction between mutant p53 and des(1-3)IGF-I accelerates mammary tumorigenesis. Oncogene. 2000;19(7):889–98. doi:10.1038/sj.onc.1203386.
86. Carboni JM, Lee AV, Hadsell DL, et al. Tumor development by transgenic expression of a constitutively active insulin-like growth factor I receptor. Cancer Res. 2005;65(9):3781–7. doi:10.1158/0008-5472.CAN-04-4602.
87. Jones RA, Campbell CI, Gunther EJ, et al. Transgenic overexpression of IGF-IR disrupts mammary ductal morphogenesis and induces tumor formation. Oncogene. 2007;26(11):1636–44. doi:10.1038/sj.onc.1209955.
88. Jones RA, Campbell CI, Wood GA, et al. Reversibility and recurrence of IGF-IR-induced mammary tumors. Oncogene. 2009;28(21):2152–62. doi:10.1038/onc.2009.79.
89. Drelon C, Berthon A, Ragazzon B, et al. Analysis of the role of Igf2 in adrenal tumour development in transgenic mouse models. Plos One. 2012;7(8):e44171. doi:10.1371/journal.pone.0044171.
90. Hale LJ, Welsh GI, Perks CM, et al. Insulin-like growth factor-II is produced by, signals to and is an important survival factor for the mature podocyte in man and mouse. J Pathol. 2013;230(1):95–106. doi:10.1002/path.4165.
91. Rogler CE, Yang D, Rossetti L, et al. Altered body composition and increased frequency of diverse malignancies in insulin-like growth factor-II transgenic mice. J Biol Chem. 1994;269(19):13779–84.
92. Pravtcheva DD, Wise TL. Metastasizing mammary carcinomas in H19 enhancers-Igf2 transgenic mice. J Exp Zool. 1998;281(1):43–57.
93. Bates P, Fisher R, Ward A, et al. Mammary cancer in transgenic mice expressing insulin-like growth factor II (IGF-II). Br J Cancer. 1995;72(5):1189–93.
94. Moorehead RA, Hojilla CV, De Belle I, et al. Insulin-like growth factor-II regulates PTEN expression in the mammary gland. J Biol Chem. 2003;278(50):50422–7. doi:10.1074/jbc.M306894200.
95. Breuhahn K, Schirmacher P. Reactivation of the insulin-like growth factor-II signaling pathway in human hepatocellular carcinoma. World J Gastroenterol. 2008;14(11):1690–8.
96. Wang CF, Zhang G, Zhao LJ, et al. Effects of insulin, insulin-like growth factor-I and -II on proliferation and intracellular signaling in endometrial carcinoma cells with different expression levels of insulin receptor isoform A. Chinese Med J. 2013;126(8):1560–6.
97. Kalla Singh S, Tan QW, Brito C, et al. Differential insulin-like growth factor II (IGF-II) expression: a potential role for breast cancer survival disparity. Growth Horm IGF Res. 2010;20(2):162–70. doi:10.1016/j.ghir.2009.12.002.
98. Kalla Singh S, Brito C, Tan QW, et al. Differential expression and signaling activation of insulin receptor isoforms A and B: a link between breast cancer and diabetes. Growth Factors. 2011;29(6):278–89. doi:10.3109/08977194.2011.616200.
99. Moorehead RA, Fata JE, Johnson MB, et al. Inhibition of mammary epithelial apoptosis and sustained phosphorylation of Akt/PKB in MMTV-IGF-II transgenic mice. Cell Death Differ. 2001;8(1):16–29. doi:10.1038/sj.cdd.4400762.
100. Moorehead RA, Sanchez OH, Baldwin RM, et al. Transgenic overexpression of IGF-II induces spontaneous lung tumors: a model for human lung adenocarcinoma. Oncogene. 2003;22(6):853–7. doi:10.1038/sj.onc.1206188.
101. Iwamoto M, Ahnen DJ, Franklin WA, et al. Expression of beta-catenin and full-length APC protein in normal and neoplastic colonic tissues. Carcinogenesis. 2000;21(11):1935–40.
102. Diehl D, Oesterle D, Elmlinger MW, et al. IGF-II transgenic mice display increased aberrant colon crypt multiplicity and tumor volume after 1,2-dimethylhydrazine treatment. J Carcinog. 2006;5:24. doi:10.1186/1477-3163-5-24.
103. Wolf E, Kramer R, Blum WF, et al. Consequences of postnatally elevated insulin-like growth factor-II in transgenic mice: endocrine changes and effects on body and organ growth. Endocrinology. 1994;135(5):1877–86. doi:10.1210/endo.135.5.7525257.

104. Christofori G, Naik P, Hanahan D. A second signal supplied by insulin-like growth factor II in oncogene-induced tumorigenesis. Nature. 1994;369(6479):414–8. doi:10.1038/369414a0.
105. Olson P, Lu J, Zhang H, et al. MicroRNA dynamics in the stages of tumorigenesis correlate with hallmark capabilities of cancer. Genes Dev. 2009;23(18):2152–65. doi:10.1101/gad.1820109.
106. Wise TL, Pravtcheva DD. Delayed onset of Igf2-induced mammary tumors in Igf2r transgenic mice. Cancer Res. 2006;66(3):1327–36. doi:10.1158/0008-5472.CAN-05-3107.
107. Kornfeld S. Structure and function of the mannose 6-phosphate/insulinlike growth factor II receptors. Annu Rev Biochem. 1992;61:307–30. doi:10.1146/annurev.bi.61.070192.001515.
108. Yakar S, Liu JL, Stannard B, et al. Normal growth and development in the absence of hepatic insulin-like growth factor I. Proc Natl Acad Sci U S A. 1999;96(13):7324–9.
109. Yu R, Yakar S, Liu YL, et al. Liver-specific IGF-I gene deficient mice exhibit accelerated diabetes in response to streptozotocin, associated with early onset of insulin resistance. Mol Cell Endocrinol. 2003;204(1–2):31–42.
110. Wu Y, Yakar S, Zhao L, et al. Circulating insulin-like growth factor-I levels regulate colon cancer growth and metastasis. Cancer Res. 2002;62(4):1030–5.
111. Wu Y, Cui K, Miyoshi K, et al. Reduced circulating insulin-like growth factor I levels delay the onset of chemically and genetically induced mammary tumors. Cancer Res. 2003;63(15):4384–8.
112. Green JE, Shibata MA, Yoshidome K, et al. The C3(1)/SV40 T-antigen transgenic mouse model of mammary cancer: ductal epithelial cell targeting with multistage progression to carcinoma. Oncogene. 2000;19(8):1020–7. doi:10.1038/sj.onc.1203280.
113. Hong SH, Briggs J, Newman R, et al. Murine osteosarcoma primary tumour growth and metastatic progression is maintained after marked suppression of serum insulin-like growth factor I. Int J Cancer. 2009;124(9):2042–9. doi:10.1002/ijc.24169.
114. Konijeti R, Koyama S, Gray A, et al. Effect of a low-fat diet combined with IGF-1 receptor blockade on 22Rv1 prostate cancer xenografts. Mol Cancer Ther. 2012;11(7):1539–46. doi:10.1158/1535-7163.MCT-11-1003.
115. Reidy DL, Vakiani E, Fakih MG, et al. Randomized, phase II study of the insulin-like growth factor-1 receptor inhibitor IMC-A12, with or without cetuximab, in patients with cetuximab- or panitumumab-refractory metastatic colorectal cancer. J Clin Oncol;28(27):4240–6. doi:10.1200/JCO.2010.30.4154.
116. Schmitz S, Kaminsky-Forrett MC, Henry S, et al. Phase II study of figitumumab in patients with recurrent and/or metastatic squamous cell carcinoma of the head and neck: clinical activity and molecular response (GORTEC 2008-02). Ann Oncol. 2012;23(8):2153–61. doi:10.1093/annonc/mdr574.
117. Weickhardt A, Doebele R, Oton A, et al. A phase I/II study of erlotinib in combination with the anti-insulin-like growth factor-1 receptor monoclonal antibody IMC-A12 (cixutumumab) in patients with advanced non-small cell lung cancer. J Thorac Oncol. 2012;7(2):419–26. doi:10.1097/JTO.0b013e31823c5b11.
118. Yee D. Insulin-like growth factor receptor inhibitors: baby or the bathwater? J Natl Cancer Inst. 2012;104(13):975–81. doi:10.1093/jnci/djs258.

Chapter 9
Impact of Energy Balance on Chemically Induced Mammary Carcinogenesis in a Rat

Henry J. Thompson

Abstract Energy balance quantifies the amount of energy available to an organism, i.e., its energy exposure over any timeframe of interest. This chapter examines how energy availability impacts the carcinogenic response in a well-defined animal model for breast cancer with demonstrated relevance to the human disease. When animals are in positive energy balance, there is a direct association between the carcinogenic response and energy availability across a broad range of energy inputs. Both host systemic and cell autonomous processes are implicated in accounting for these effects. At the cellular level, the interplay between cell proliferation and apoptosis appears to dictate the rate of tumor mass accumulation with the integration of internal and external inputs regarding energy availability via the signaling network of which the mammalian target of rapamycin is a component. Gaps in the existing knowledge and research opportunities are identified, and new paradigms are presented with the goal of stimulating debate and discussion that will advance the field.

Keywords Energy availability · Energy balance · Mammary carcinogenesis

Introduction

Interest in the impact of energy balance on the development of cancer has expanded across cancer sites over several decades [1, 2]. Of the cancer sites investigated using preclinical models, an extensive amount of work has been done in experimentally induced breast cancer in the rat, particularly with regard to elucidating mechanisms. For that reason, this chapter will focus on the effects of energy balance on rat mam-

H. J. Thompson (✉)
Cancer Prevention Laboratory, Colorado State University,
1173 Campus Delivery, Fort Collins, CO 80523, USA
e-mail: henry.thompson@colostate.edu

N. A. Berger (ed.), *Murine Models, Energy Balance, and Cancer*,
Energy Balance and Cancer, DOI 10.1007/978-3-319-16733-6_9

mary carcinogenesis. Many investigators have conducted experiments on this topic over a period spanning at least four decades and Thompson's laboratory has been engaged in this field throughout that timeframe. The goal of this effort is not to simply summarize what has been reported. That has been done on a number of occasions [3–7]. Rather, the primary purpose of writing this chapter is to highlight knowledge gaps and unresolved questions and to identify areas of opportunity in order to provide a framework for advancing the field. As much of what is presented is the author's synthesis of developments in the field over an extended period of time and is not chronologically based, the first person, plural voice (we) is used to distinguish the viewpoints of the author while acknowledging that many people in his laboratory have contributed to the work on which those viewpoints are based.

Energy Balance Defined

Traditionally, energy balance is defined as the difference between energy intake and energy expenditure of an organism. Acutely, energy balance is usually computed for a period of 24 h [1]. From a chronic perspective, the timeframe can range from weeks to years and arguably is directed to the accumulation, loss, or maintenance of body weight relative to height, generally quantified as body mass index (body weight (kg)/height in meter^2) and to changes in body fat mass with particular interest in whether those changes occur in peripheral or central storage depots and/or ectopically, in tissues where fat is not supposed to accumulate, e.g., the liver (nonalcoholic fatty liver disease) [8–10]. Thus, energy balance establishes whether an individual is in positive or negative balance or in energy equilibrium over the reference timeframe of interest.

A Brief Summary of What Has Been Learned from the Rat About the Effects of Energy Balance on the Carcinogenic Response per se

Most studies of the effects of energy balance on experimental mammary carcinogenesis in the rat have been conducted using chemically induced models for this disease process. The most commonly used chemicals for initiation of the carcinogenic process in the mammary gland are either 7,12-dimethyl benz[α]anthracene or 1-methyl-1-nitrosourea (MNU) [11–13]. What is typically forgotten, given the current research focuses on genetically modified mouse models of mammary carcinogenesis, is that in both rat models between 70 and 80 % of the carcinomas that occur in ovary-intact animals are sex steroid hormone dependent and the remaining carcinomas are sex steroid hormone independent [14, 15]. When the ovaries are removed, greater than 80 % of the induced carcinomas are sex steroid hormone independent [11, 14]. Mammary carcinomas that are sex steroid hormone independent are illustrative of poor prognosis molecular subtypes of breast cancer and Thompson's laboratory has recently determined that within these poor prognosis carcinomas approximately 50% have focal overexpression of Her2/Neu (Unpublished observation, HJT). Thus, there is a strong basis for using these models to study how energy balance affects the natural pathogenic history of a spectrum of molecular subtypes of breast cancer that occur in women. However, little work has been done to exploit this opportunity.

The definition of energy balance provided above indicates that three states of balance can exist: weight gain, weight loss, and weight maintenance. Since rodents given sufficient dietary energy experience skeletal growth, albeit slowly, throughout their lives, weight maintenance, as it is understood in humans, is not studied in rodent models. Rather, the majority of work has focused on differential rates of weight gain and interpretation of those studies has been complicated both by differences among laboratories in technical approaches to controlled feeding and diet formulation, and by misunderstanding of what the research paradigms that have been used actually model [16]. A number of these issues have been previously discussed [3–5, 7, 17], but those that we consider to be pivotal are restated below.

Dietary Energy Restriction

The majority of work on energy balance in rat models of mammary carcinogenesis has focused on the effects of dietary energy restriction, which has often been referred to as caloric restriction or dietary restriction. However, these terms are not synonymous nor are the technical approaches used to implement them standardized. Major differences among studies include: (1) whether nutrient density per kcal of dietary energy is adjusted in diet formulations so that the only difference among treatment groups is the amount of energy ingested versus global restriction of all nutrients including energy (this distinguishes between studies of dietary energy restriction or caloric restriction versus those termed dietary restriction); (2) whether food is provided once a day or in multiple meals during a day (this highlights a critical metabolic distinction between effects of daily fasting-refeeding versus distributed meals); (3) whether the level of energy restriction is computed based on the intake of a control group fed ad libitum using some type of paired feeding algorithm or based on the weight of the animal being restricted (this affects the actual magnitude of restriction and its constancy throughout an experiment). It should be noted that in rat experiments, most work is done in the absence of any weight loss relative to fully fed controls [16]. Rather, what differs among groups given restricted amounts of dietary energy is their rate of growth. Thus, it is important to underscore that studies of dietary energy restriction in the rat are studies of the effects of positive energy balance on the carcinogenic response, i.e., the effects of different planes of energy nutrition.

Despite technical differences among laboratories in implementation, dietary energy has consistently been reported to inhibit one or more aspects of the carcinogenic process depending on the magnitude of the restriction imposed and the timeframe and frequency of dietary energy restriction relative to the stage of carcinogenesis during which restriction occurred. Specifically, and as illustrated in [18], dietary energy restriction during the post-initiation phase of mammary carcinogenesis results in energy dose-dependent reductions in cancer incidence, cancer multiplicity (number of carcinomas per rat), tumor burden (cancer mass per rat), and prolongation of cancer latency. The greatest degree of protection in the rat is afforded when dietary energy restriction is imposed chronically during the initiation and post-initiation stages of the disease process [19, 20].

Excessive Weight Gain

The number of studies of the effects of energy balance conducted in rat models of obesity is small. Nonetheless, there are already issues of terminology and technique that could affect data interpretation and its translation to human populations and thus hamper developments in the field.

With regard to terminology, we cite the work of Pariza et al. [21–24] for two reasons. It provides an excellent transition from the discussion of dietary energy restriction to excessive weight gain, and it emphasizes the importance of perspective in terms of what excessive weight gain can mean. A compelling case was made by that team of investigators that ad libitum-fed rats given either chow diet or recommended purified formulations, such as AIN76-A, are overweight and that animals fed restricted amounts of the same diet are actually an appropriate weight for their length and should serve as the referent control population. Data from those investigators and others clearly show that ad libitum-fed rats have more body mass and greater body fat stores than restricted fed rats, but we do not judge that this paradigm qualifies for a study of obesity. Rather, it sets the stage for considering positive energy balance as an energy continuum (underweight, normal weight, and excessive weight for length). Thus, Pariza et al.'s work recasts the studies of dietary energy restriction in this light.

From the perspective of energy continuum and the view that ad libitum-fed rats are overweight relative to dietary energy restricted "controls," the logical progression is to ask whether some populations of ad libitum-fed rats eat more than others. The answer in the rat, as it is in mice and humans, is yes. The experiments reported can be grouped into three categories: genetic models, dietary-induced models, and artificially selected models. Since we are focusing on chemically induced mammary carcinogenesis in the rat, in all cases it should be recognized that risk for cancer (carcinogen injection) is set at an early point in life and that the development of cancer and the accumulation of excess body weight are occurring concurrently. Thus, effects may be different than in experiments initiated after the "state of obesity" has been attained. We judge that this point is easily lost in a reading of the published literature.

Genetic Models

The effect of excessive weight gain on chemically induced mammary carcinogenesis has been studied in two models. In the Zucker rat model (fa/fa) in which leptin signaling is defective and in the LA/Ncp rat model of a recessive corpulence gene (cp/cp), the carcinogenic response was enhanced in the animals that developed an obese phenotype in comparison to their genetically lean counterparts [25–27]. In both models, it was also concluded that the enhanced carcinogenic response was not due to body fat per se. Since leptin has been implicated as playing a role in mammary carcinogenesis [28, 29], the interpretation of experiments using the Zucker rat may be confounded by alterations in leptin metabolism. The mechanism that accounts for excessive weight gain in the LA/Ncp model appears to be distinct from the Zucker model and thus points to the need to consider factors other than body fat accumulation per se in understanding how positive energy balance affects the carcinogenic process.

Dietary-Induced Obesity Models (DIO)

Some strains of rat have been reported to be obesity prone, meaning that they overeat relative to caloric needs when given a high fat diet to consume [30–32]. This situation is consistent with what is observed in mice where some mouse strains are considered sensitive (e.g., C57 B6) whereas others are resistant (e.g., AJ mice) to excessive weight gain when fed a high fat diet [33, 34]. However, what can be problematic about these models, an issue particularly prevalent in mouse models, is that some investigators use the same animal strain that is sensitive to diet-induced obesity and feed a low-fat diet (frequently an undefined chow diet) versus a high-fat-purified diet to study effects of excessive weight gain on the carcinogenic process not fully recognizing that the differences in dietary composition potentially confound the interpretation of results about the effects of excessive weight gain on the disease process.

In the rat, the animals in a DIO paradigm are generally referred to as either obesity prone or obesity resistant. In these models, the level of body fat accumulation does not reach that observed in genetically obese models, but is higher than observed in rat strains that are fed ad libitum and considered obesity resistant. While limited in number [35, 36], the studies reported have shown the carcinogenic response to be greater in obesity prone versus obesity resistant rats.

Artificially Selected

A different approach was used by B. E. Levin in order to develop a model of dietary-induced obesity. While Sprague Dawley rats are generally regarded as DIO resistant, there is a normal Gaussian distribution in any population of outbred Sprague Dawley rats, male and female, in terms of feed efficiency ratio when the same diet formulation is fed. Levin took advantage of this fact to artificially select two populations of Sprague Dawley rats that are resistant or sensitive to dietary-induced obesity when rats consume a purified diet containing 32 % of the dietary calories from fat. The artificial selection, over 20 generations, was done in a manner such that both lines are considered outbred, and they are commercially available from either Taconic Farms or Charles River. This model has been extensively characterized by Levin's group [37–39]. DIO-sensitive (DS) rats rapidly gain excess weight and have expanded peripheral and visceral fat depots by 3 months of age [38], display hyperlipidemia (total cholesterol and triglycerides) by 2 months, hyperleptinemia by 3 months, and pronounced fat infiltration of the liver by 6 months of age [38]. DS rats display prediabetic measures of glucose homeostasis including hyperinsulinemia by 2 months, insulin resistance by 3 months, worsened oral glucose tolerance by 2 months, and eventual reduced pancreatic insulin secretion by 9 months of age [38].

We took advantage of the existence of these rat strains and have shown profound enhancement of all aspects of the carcinogenic process in the mammary gland [40]. Relative to the dietary resistant strain, the dietary sensitive DS rats had higher incidence (26 % increase), multiplicity (2.5-fold increase), and burden (5.4-fold increase) of mammary carcinomas with a concomitant reduction in cancer latency (16 % earlier detection) compared to DIO-resistant (DR) rats ($P<.001$ for all analyses). In an as yet unpublished experiment, increases in the carcinogenic response were also observed in ovariectomized rats and the carcinomas induced were found to be predominantly poor prognosis molecular subtypes. The specific mechanisms that account for these effects are under investigation.

Weight Loss

A significant gap in our understanding of energy balance and breast cancer is the virtual absence of studies of the effects of weight loss when imposed on rats that are obese and at high risk of developing cancer. This is important since in other fields, it has been suggested that obesity creates a state of metabolic memory [41, 42]; whether weight loss results in complete reversion of phenotypes associated with excess accumulation of body fat, e.g., its effect on mammary carcinogenesis, is unknown.

The Clock Paradigm

As noted above, in the rat, the effects of energy balance on mammary carcinogenesis have been primarily studied in chemically induced models. Carcinogen needs to be administered only once, for example, using MNU, at a dose that induces no mortality. The use of a clock paradigm is adopted here to focus attention on questions not generally considered in energetics and cancer. For a moment, consider that the effect of injecting MNU is like placing a specific number of clocks (transformed cells with the potential to develop into cancers) that are not yet running in the mammary glands of a population of rats and this confers increased risk for cancer. Many of these clocks will remain in the tissue in a silent (dormant) state. However, other clocks will begin to tick at some point and the rate at which each clock runs ultimately determines whether cancer is observed in the specified period of time. In order not to overwork this analogy, it is used to ask two questions: does the state of energy balance of an organism affect when the clocks start ticking in a population of carcinogen injected rats, and once they start ticking, do the clocks run at the same apparent rate? These questions are asked in the context of dietary energy restriction.

As shown in Fig. 9.1, it is clear that limiting dietary energy results in fewer clocks that start to tick at every time point at which rats are monitored for detectable tumors by palpation. This analysis reveals a gap of knowledge that is in part technical (ability to detect single transformed cells in a tissue and follow their behavior), and in part theoretical (what causes transformed cells' clocks to start ticking and how does limiting energy suppress this process). To our knowledge, this question

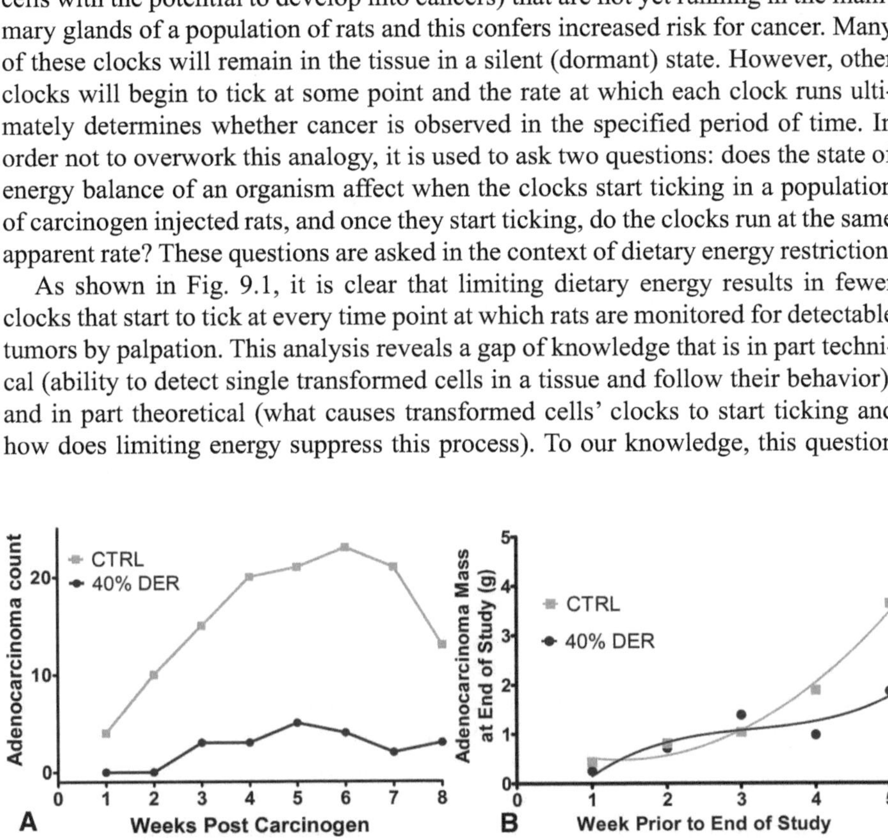

Fig 9.1 Panel A: In the rat model, animals are palpated for detectable mammary tumors weekly; tumors can be detected when they are the size of a pin head. At necropsy, tumors are removed and histopathologically classified. Thus, it is possible to determine when new clocks (when a tumor is initially detected) are activated (turned-on) throughout an experiment. This graph shows the number of new tumors being detected at each week of palpation. What is concluded from this analysis is that fewer clocks turn on in energy restricted rats (*DER*) versus ad libitum-fed *control* animals at every time point that rats were palpated. Panel B: In the same rats reported in Panel A, we show the tumor mass of each tumor measured at the end of the study based on when the tumor was first detected. As expected, the greater the length of time from detection to end of study, the larger the tumor mass. At every point the tumor mass of *DER* rats was smaller than the mass observed in *control* rats

has never been addressed but has potentially important translational implications in human populations with defined cancer risk, since the ultimate goal is to stop their "clocks" from ever undergoing activation.

The second question about the rate at which the ticking clocks run is more amenable to investigation. In Fig. 9.1, Panel B shows that the apparent rate at which the clocks run is slower in restricted than ad libiutum fed rats. Interestingly, in the obesity model that we have published [40], the clocks in the obese rats actually run at a faster rate than in the control group (data not shown; publication under review, HJT). The cellular mechanisms at play in the apparent rate at which the clocks run are discussed in subsequent sections.

The Ticking Clock

This section considers available evidence as to why tumors grow at a slower rate over the same period of time in proportion to energy restriction. To do this we ask: (1) what cellular process(es) is likely to underlie the effect on tumor mass, e.g., cell proliferation and/or apoptotic cell death rates, (2) what cellular machinery is required for the effects on cell proliferation and/or death to be exerted, (3) what signaling pathway(s) and/or network(s) regulate the altered components of cellular machinery, and (4) how are that signaling pathway(s)/ network(s) regulated by extracellular and intracellular mediators and their respective cell surface receptors and intracellular detection sensors?

Cellular Processes

From the clock paradigm presented above, it is clear that the clocks run at different rates depending on the amount of energy made available to the organism and that the apparent rate is positively associated to the magnitude of energy available, at least at the organismal level. What this means is that tumor mass over a defined period of observation is greater with increasing energy availability. While the explanation that might immediately come to mind is that energy availability drives tumor cell proliferation, that view is tempered by the hallmarks of cancer [43, 44], in particular by the hallmark that relates to resistance to apoptosis. As summarized extensively in Steel's classic work on tumor growth kinetics [45–47], and as applied to the prevention context by us [48], the accumulation of tumor cell mass is the net result of the following factors: the growth fraction of the cells within the tumor, the rate of their transit through the cell cycle, and the rate of cell loss, with necrosis, apoptosis, and autophagy being important potential contributors to cell loss. We judge that the effects of energy availability on this entire macro process have not been cohesively investigated and may never be for a number of reasons. They include: (1) limitations in methodology particularly in the face of tumor heterogeneity which changes over time, and (2) the perception that while an information gap exists, it does not constitute a hurdle to progress. The counter argument is that without an understanding of dominant cellular process(es) affected by energy availability and the context within which it (they) operates, the opportunity to exploit the targets of energy availability for cancer prevention and control is being limited.

Our laboratory has made progress on how tumor mass accumulates when energy availability is limited, but not in the comprehensive manner described above. What has been determined is that BRDU incorporation (rate of DNA synthesis) is reduced in a manner inversely proportional to energy availability [49]. In addition, the rate of apoptosis is increased dependent on the dose in mammary carcinomas from energy-restricted animals [49]. To our knowledge, effects on amounts of necrosis or autophagy in response to energy availability have never been quantified. Simply based on the observation that the process of cell loss is quantitatively more powerful than cell proliferation within any 24 h period of time (theoretically by a factor of 8), we have judged that effects of energy availability on cell loss are likely to be the dominating factor determining the rate at which tumor mass accumulates in response to different levels of energy availability.

The Cellular Machinery Underlying Accumulation of Cell Mass

The regulation of cell cycle machinery is complex and the effects of energy availability have been incompletely investigated. Areas of neglect include the regulation of entry and exit from the cell cycle, which may relate to how energy availability affects clocks being turned on in the clock paradigm (Question 1), and the regulation of G2/M. Where most work has been directed is on the factors regulating the G1/S transition since that appears to be a target of energy availability. Limiting energy availability has been reported to block cell-cycle progression at G1/S. Observed effects include significant reduction in levels of phosphorylated Rb and E2F-1; reductions in CDK2 (82 %) and CDK4 (77 %) kinase activity were likely to account in part for the observed effects of limiting energy availability on Rb and E2F-1; both Cipl/p21 and Kipl/p27 and levels of these proteins complexed with CDK2 were significantly elevated in when dietary energy was limited; and levels of cyclin E were reduced. The observed decrease in CDK4 kinase activity was likely attributable to effects on cyclin D1 as well as the increased binding of P16 and P19 to CDK4 [50, 51].

With regard to apoptosis, limiting energy availability has been reported to induce a proapoptotic state via the coordinated regulation of pro- and anti-apoptotic factors involved in the mitochondrial pathway of caspase activation [49, 52]. Specifically, in addition to the dose-dependent effects observed on the rate of apoptosis determined morphologically, cleaved products of poly(ADP-ribose) polymerase 1 were elevated by limiting energy availability, providing biochemical evidence of apoptosis induction. cDNA microarray analysis identified the *Bcl-2*, *CARD*, and *IAP* functional gene groupings as being involved in apoptosis induction. Consistent with the microarray data, the activities of caspases 9 and 3 were observed to be two-fold higher in carcinomas from energy-restricted rats, whereas, caspase 8 activity was similar in carcinomas from ad libitum-fed and energy-restricted animals. Collectively, this evidence indicated that energy restriction induced apoptosis mediated by the mitochondrial pathway. This conclusion was further supported by the finding that levels of Bcl-2, Bcl-xl, and XIAP protein were significantly lower, and levels of Bax and Apaf-1 were elevated in carcinomas from energy-restricted versus ad libitum-fed rats. Nonetheless, effects on the extrinsic pathway of apoptosis induction have not been intensively investigated and should not be dismissed as being completely uninvolved until studied in other animal models and disease contexts.

Implicated Signaling Pathways and Their Regulation

There are undoubtedly multiple signaling pathways that are regulated by energy availability. We provide evidence for one network and its regulation, which we refer to as the *mTOR network hypothesis:* limiting energy availability inhibits tumor growth by suppressing the activation of the mTOR signaling network in mammary carcinomas. Suppression is mediated through effects of energy availability on concentrations of the circulating growth factors and hormones and of the substrates used in intermediary metabolism to synthesize high-energy phosphates and reducing equivalents. As a consequence, the drive for cell proliferation is reduced [50, 51], a proapoptotic environment is maintained [49, 52], and the stimulus for new blood vessel formation is suppressed in mammary carcinomas [53]. The translational relevance of this hypothesis is that one or more elements of the mTOR network are deregulated in the majority of human breast cancers [54]. Various components of this network are discussed in subsequent sections.

1. *The mammalian target of rapamycin (mTOR)* is an intracellular protein that plays a key role in integrating information received from the extracellular environment via the binding of growth factors and hormones with their cognate receptor tyrosine kinases (e.g., IGF-1: IGF1-R) with signals from metabolic checkpoints within the cell in a manner that affects cell growth, cell division, and cell survival (or death) [55]. mTOR is an evolutionarily conserved serine–threonine kinase that is a key regulator of protein translation and synthesis. mTOR is centrally involved in cell growth, i.e., increase in cell size and cell mass, and these processes are tightly coupled to cell division (reviewed in [56]). The regulation of mTOR is multifaceted, and still being investigated. It has two biochemically and functionally distinct complexes: mTOR complex 1 (TORC1) and mTOR complex 2 (TORC2) [57]. TORC1 is comprised of mTOR, regulatory-associated protein of mTOR (Raptor) and G protein beta subunit-like (Gβ1/also known as mLST8) and its activity is nutrient/energy sensitive, whereas TORC2 is comprised of mTOR, rapamycin-insensitive companion of mTOR (Rictor), stress-activated protein-kinase-interacting protein 1 (SIN1) and Gβ1) and plays a role in regulating the signaling pathway of which Akt is a component. A primary

locus for control of mTOR is via the tuberous sclerosis protein complex (TSC), specifically TSC2, which is phosphorylated on different sites by either activated AMPK (Thr1227 and Ser1345 residues that correspond to Thr1271 and Ser1387 respectively in human TSC2) or activated Akt (Thr1462 and Ser939 residues). A second locus for control of mTOR via AMPK and Akt is via the phosphorylation of TORC1 complex components raptor and PRAD, respectively [55]. mTOR mediates its effects on downstream targets via site specific phosphorylation. Relative to its effects on cell growth and cell division, two principal targets of mTOR are 70-kDa ribosomal protein S6 kinase (p70S6K) and 4E binding protein 1 (4E-BP1). Activated mTOR phosphorylates p70S6K and this leads to increased ribosomal biogenesis [58, 59]. 4E-BP1 is a repressor of translation initiation [60–62]. Activated mTOR phosphorylates 4E-BP1 which inactivates the protein. When it is hypophosphorylated, 4E-BP1 binds to and inhibits the rate-limiting translation initiation factor eIF4E (eukaryotic translation initiation factor 4E). Upon phosphorylation, eIF4E is released from 4E-BP1, allowing eIF4E to assemble with other translation initiation factors to initiate cap-dependent translation [61]. We have reported that limiting energy availability alters multiple regulatory nodes of this network in a manner consistent with reduced mTOR activity [63, 64] decreases the levels of both phosphorylated S6K and 4E-BP1.

2. *AMP-activated protein kinase (AMPK)* serves as a metabolic checkpoint, down regulating cell growth and cell division in the absence of an adequate supply of biosynthetic and energy substrates [55]. In this respect, AMPK has been likened in its function to p53 that serves as a guardian of genome integrity. AMPK has been shown to be an exquisitely sensitive detector of small changes in the intracellular ratio of AMP to ATP, and some investigators have even proposed that AMPK plays a central role in homeostatic regulation of whole body energy metabolism [65]. We have reported that limiting energy availability results in AMPK activation [66]. This suggests that energy availability is either altering substrate availability (the fuel mixture presented to tissues throughout the body) or that activation is being induced via a mechanism independent of the AMP to ATP ratio. In this regard, it is becoming clear that additional factors control the activation of AMPK, including various cytokines such as adiponectin [67].
3. *Protein kinase B (Akt)*: Limiting energy availability has been reported to decrease circulating levels of IGF-1 [68], an observation confirmed in our work under some conditions of energy availability [66, 69–71] Lower levels of IGF-1 would be expected to down-regulate signaling via the pathway of which IGF-1 receptor, phosphoinositide kinase-3 (PI3K), and Akt are components. Of these proteins, activated Akt, a serine/threonine kinase, is the critical affecter molecule. Akt is activated by its phosphorylation on Ser473. Phospho-Akt serves important roles in cell proliferation, cell survival and new blood vessel formation that are associated with tumor development [72]. While it is clear that reduced levels of activated Akt are likely to affect proliferation, apoptosis and angiogenesis by mechanisms independent of mTOR (reviewed in [73]), the finding that limiting energy availability induced both AMPK activation and down regulation of

growth factor signaling via Akt, points to mTOR as a downstream target mediating the effects of energy availability.

4. *Insulin like growth factor-1 (IGF-1)*: The insulin-like growth factor (IGF) signaling system plays a critical role in the growth and development of many tissues and regulates overall growth ([74] and references therein). The IGF system has also been implicated in various pathophysiological conditions, and is thought to play a particularly prominent role in tumorigenesis. The IGF system is comprised of the IGF ligands (IGF-I and IGF-II), cell-surface receptors that mediate the biological effects of the IGFs, including the IGF-I receptor (IGF-IR), the IGF-II receptor (IGF-IIR), and the insulin receptor (IR), as well as a family of IGF-binding proteins (IGFBPs). IGFBPs affect the half-lives and bioavailability of the IGFs in the circulation, in extracellular fluids, and may exert IGF-independent effects under certain conditions. Most, if not all, of the effects of IGF-I result from its activation of the IGF-IR and lead to activation of the mitogen-activated protein kinase (MAP kinase) and PI3 kinase cascades. The ultimate targets of the MAP kinase and PI3 kinase cascades include members of mTOR network and forkhead transcription factor families. Regulation of transcription factors provides a mechanism by which IGF action at the cell surface can elicit changes in gene expression that eventually mediate the proliferative, differentiating, and apoptotic effects of IGF-1. In 1998, Hawkinson et al. reported that pre-menopausal, but not post-menopausal women in the highest tertile of serum IGF-I levels had a significantly increased risk of developing breast cancer [75]. We have found that the level of plasma IGF-1 was reduced by limiting energy availability [69]. However, we also have reported that infusion of IGF-1 does not fully reverse the effects of limiting energy availability on the carcinogenic process [68].

The Broader Framework: Toward an Integrated View

Given that energy availability has the capacity, at least theoretically, to: (1) affect all tissues in the body, (2) three of the tissues known to be affected by energy availability, liver, muscle, and adipose, are among the largest of the endocrine tissues, and (3) the mammary gland is highly vascularized, it is expected that energy availability impacts the environmental exposures to which mammary epithelial cells are subjected. Current evidence indicates that this includes the plasma metabolome, e.g., monosaccharides such as glucose, fatty acids, and amino acids, and intermediates in core metabolism that have been reported to influence cell behavior [76–79]. At the level of the secretome, a rich body of literature indicates that biomarkers (growth factors, cytokines, and hormone) associated with a number of metabolic processes, (glucose homeostasis, chronic inflammation and potentially its relationship to the gut microbiome, cellular oxidation, adipokine metabolism, and sex steroid metabolism) are affected by energy availability and many of these processes are also linked to the development of breast cancer [5, 7, 80, 81]. As noted above, only a limited

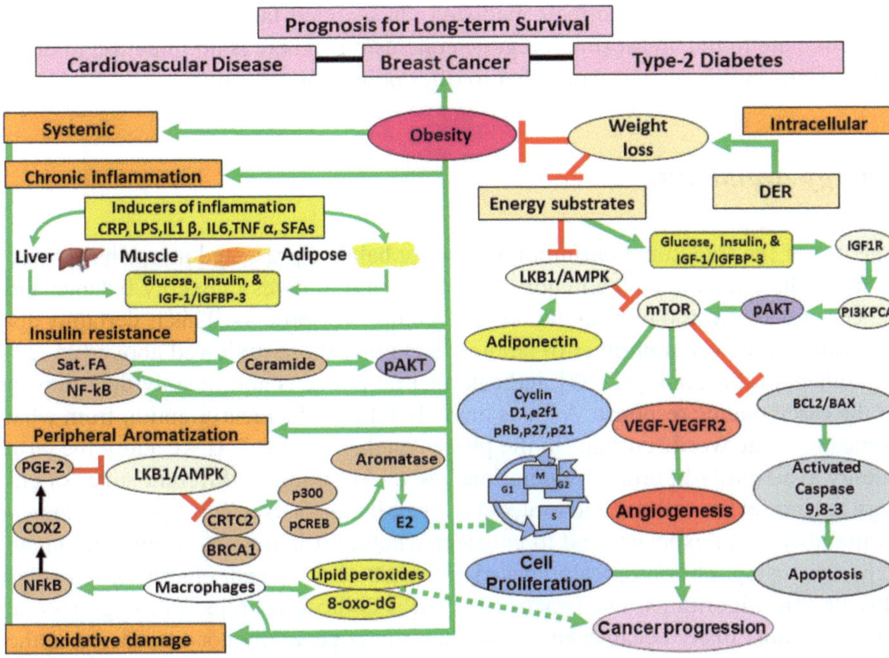

Fig. 9.2 Many mechanisms are likely to be involved in determining how the amount of energy available to a rat (or human) impacts not only the occurrence of breast cancer but also prognosis for long term survival following treatment for breast cancer. Survival prognosis is affected not only by recurrence and metastasis but also the common comorbidities of breast cancer survivors primary amongst which are cardiovascular disease and type-2 diabetes. Both *Systemic* and *Intracellular* processes likely to be impacted by dietary energy availability are shown. Systemic processes include *chronic inflammation*, glucose homeostasis which involves *insulin resistance*, *peripheral aromatization*, and cellular oxidation. Within the cell, the availability of energy substrates and systemic factors ultimately determine the behavior of transformed clones of cells relative to cell proliferation, cell death (apoptosis, autophagy, and necrosis), and angiogenesis. One of candidate integrators of extracellular and intracellular information that signals to the proliferative, death and vascularization machinery within the cell is the mammalian target of rapamycin (mTOR)

amount of work has been done in the rat on the effects of excessive weight gain and there is a void of knowledge about the effects of weight loss. Hypothesized interrelationships among candidate mediator processes that require investigation are diagrammed in Fig. 9.2.

Deeper Reflections

We would like to address, what for us has been a nagging concern and that for others has been more or less a foregone conclusion, the effects of energy balance on breast cancer mediated by direct effects in mammary epithelial cells (our nagging concern), or the observed effects mediated via secreted factors from other tissues that circulate to the breast (the foregone conclusion). There is no intention to draw conclusions in this section, but rather to offer points for reflection.

Energy Availability

The use of the traditional definition of energy balance presented above has served to focus the attention of the majority of research initiatives in energy balance and cancer on the effect(s) of body fat on the carcinogenic process. This is certainly appropriate. However, a number of studies, including those reviewed above [25–27] and others in this book [82, 83], indicate that fat per se is not the only determinant of the carcinogenic response. A focus on body fat also distracts attention from what is meant by the word "energy" in the phase "energy balance". If we adopt the term energy availability in order to operationalize the meaning of different states of energy balance at the cellular and molecular level, it allows us to conceptualize energy availability as an exposure and to consider whether it is possible to model human exposure using physiologically based pharmacokinetic and pharmacodynamics (PB-PK/PB) algorithms. The lack of PB-PK/PD approach applied to energetics and cancer is most likely because energy availability is generally not considered in the context of being an exposure, and that the hormone, growth factors, and small molecules affected by energy availability have not been regarded as chemical surrogates for energy exposure. A PB-PK/PD approach, if applied to Fig. 9.2, might identify energy availability drivers versus bystanders.

Energy Homeostasis

Perhaps because the number of disciplines engaged in research on energetics and cancer has become very broad and also because the role of changes in core elements of intermediary metabolism are being shown to play a central role in the carcinogenic process [44], we judge that the understanding of the term energy, as in energy balance, and the understanding of energy homeostasis, as occurs within the cell throughout a 24 h time cycle, are exceptionally varied and that this limits progress in the field. While an effort to address a complex issue such as this is beyond the scope of this chapter, a number of observations and questions are provided in order to foster thinking and discussion.

One of the most traditional ways to assess the energy status of cells within a tissue is referred to as energy charge, the ratio of the concentrations of ATP:ADP:AMP:P_i. To illustrate the point made in the preceding paragraph, we ask this question: "How is the energy charge within a cell, e.g., in a clone of transformed epithelial cells in breast, affected when an individual's 24-h energy balance is positive or negative?" Our experience has been that there is a widely held notion that the answer is, "there is no effect," because the energy charge of cells, particularly the concentration of ATP, is held at a constant level. However, scientifically, it is recognized that measured energy charge is an average, and that variation exists among cells in their concentration of ATP at any snapshot in time. Emerging evidence indicates the existence of very wide temperature gradients within cells, suggesting that there are also likely to be parallel variations in high-energy phosphate concentrations across subcellular compartments [84].

Recognizing that homeostatic regulatory mechanisms are always operative within cells in an effort to balance demand for and supply of high-energy phosphates and reducing equivalent and cells can remain viable over a considerable range in intracellular concentrations of these molecules, we question whether it would be beneficial to focus attention on how cells attempt to satisfy their needs for high-energy phosphates, as well as reducing equivalents, within the constraints of their genetics and their environment and how this affects individual transformed cells and clonally expanding cell populations. To this end, we note that it has been shown that there is considerable heterogeneity among tumors in their glucose concentration [85, 86]. Moreover, a recent report indicates that proliferative activity of different cell lines varies markedly in response to low glucose concentrations, an effect mediated in part by whether cells generate ATP primarily via aerobic glycolysis or oxidative phosphorylation [87]. These findings suggest that energy substrates, i.e., carbohydrates, amino acids, and fatty acids, whose availability to cells is defined by energy exposure, are the moment-to-moment drivers of cell behavior. This is consistent with the highly conserved energy sensing systems that exist within cells and signal directly to pathways that regulated cell proliferation, apoptosis, and autophagy [7]. Possibly, paracrine and endocrine factors serve the purpose of a second tier of regulation that not only optimizes cellular responses to environmental stressors, but also coordinate the behaviors of cells within tissues and across organ systems.

Energy Hormesis

As an extension of the discussion in the preceding section, which argues that there is an optimal level of energy availability for protection against breast cancer, we leave the rat model for a moment and return to the traditional definition of energy balance that has as an index of chronic energy balance, body mass index (BMI). In reviewing studies that presented breast cancer mortality relative to BMI, with a BMI of 22–23 being considered as the referent group, breast cancer deaths are higher at BMIs on either side of the referent value, e.g., [88]. This U or J-shaped response is identified with hormetic responses that are typical of how many organisms adapt to stress. We introduce this concept for three reasons: (1) if the disease endpoint shows a hormetic response to energy balance quantified as BMI, then evaluating the large number of biomarkers that energy availability has been reported to affect (Fig. 9.2) for hometic response profiles might help separate driver changes from those that are bystanders; (2) in the metabolic reprogramming literature, availability of energy is frequently viewed as a challenge or stressor to which the cell adapts and responds, yet the concept that limiting energy availability (balance) is a healthy physiological stressor is rarely discussed; (3) the impact of stress on the development of breast cancer at all stages has been inadequately studied and represents a rich opportunity for new and translatable discoveries.

Energy Sensitive Cancers

It appears to us, in a simplistic view, that energy availability acts like a rheostat that regulates tissue size to either expand or contract, and this can function only when the circuitry that the rheostat controls remains subject to regulation [89]. This gives rise to two questions: (1) how do cells solve the problem of getting what they need to proliferate in the energy restricted state, and (2) what is that cell's fate when the problem cannot be solved in a particular snapshot in time? We propose that relative to availability of energy substrates/reducing equivalents, the metabolic flexibility of the "target cell" defines whether it is energy sensitive or insensitive. Sensitive cells lack ability to make adaptive changes in core metabolism to accommodate their current replicative potential; whereas, insensitive cells are metabolically flexible and use alterative pathways to satisfy their requirements for proliferation. If correct, this provides a basis for intervention development through the lens of imposing constraints on metabolic flexibility in order to prevent or control the fate of transformed cells.

Limitations

There is obvious danger in and limitations to using one animal model to understand any human disease process. Given the complexity of the heterogeneous group of diseases referred to as breast cancer, it is clear that many models are needed to maximize our understanding of the effects of energy availability on this disease process [15]. Nonetheless, the rat model has been incredibly valuable in developing concepts that have translated well to the clinic, particularly, the protective roles of ovariectomy and of pregnancy and the develop of selective estrogen receptor modulators like tamoxifen and raloxifene [90].

This chapter identifies many unaddressed questions that the rat model can be used to answer expediently. While the availability of transgenic and knockout models in the mouse has facilitated rapid progress, similar approaches are becoming available in the rat. Moreover, given the many limitations of current, widely used mouse models for breast cancer, particularly with regard to organ specific expression/knockout of genes, the rat is likely to continue to play an important role in this field.

Translational Implications

Understanding how energy availability regulates all aspects of the carcinogenic process is likely to identify individuals most and least likely to benefit from interventions that target energy sensitive processes linked to the development of cancer. Opportunities range from strategies to suppress clonal expansion of transformed cells to the clearing of tissues of premalignant pathologies to affect early cure, perhaps through lifestyle interventions in combination with acutely administered drugs. Energy exposures are both unavoidable and modifiable, offering an opportunity to prevent and control cancer that is limited only by the imagination and creativity of those who engaged in this field of investigation.

Concluding Comments

Limiting energy availability to organisms such as the rat has been repeatedly shown to be one of the most powerful, physiological, nontoxic approaches to protect against the development of breast cancer. We are now on the verge of understanding how excessive energy intake in the same species drives the carcinogenic process. It is clear that multiple mechanisms are involved and that the intracellular demands for high-energy phosphates and reducing equivalents plays an important role in regulating cellular behavior. What is being learned has direct translational relevance across the lifespan and at all stages in the development and progression of breast cancer. Despite this, the gaps in knowledge are considerable and argue for the intensification of efforts to unravel the complex mechanisms underlying protection so that simple and reliable guidance can be given to the public about reducing risks and improving survival benefits in the battle against cancer.

Conflicts of Interest The author discloses no potential conflicts of interest.

Funding Supported by PHS grant R01-CA52626 from the National Cancer Institute.

References

1. IARC. Weight control and physical activity. Vol. 6. Lyon: IARC; 2002. pp. 1–355 (IARC Handbook of cancer prevention).
2. WCRF/AICR. Food, nutrition, physical activity, and the prevention of cancer: a global per spective. Washington, DC: AICR; 2007 (5-14–2008).
3. Thompson HJ, Jiang W, Zhu Z. Mechanisms by which energy restriction inhibits carcinogenesis. Adv Exp Med Biol. 1999;470:77–84.
4. Thompson HJ, Zhu Z, Jiang W. Dietary energy restriction in breast cancer prevention. J Mammary Gland Biol Neoplasia. 2003;8:133–42.

5. Thompson HJ, Jiang W, Zhu Z. Mechanisms associating obesity with cancer Incidence: animal Models. In: McTiernan A, editor. Cancer prevention and management through exercise and weight control. Boco Raton: CRC Press; 2006. pp. 329–40.
6. Thompson HJ, Jiang W, Zhu Z. Obesity as a cancer risk factor: potential mechanisms of action. In: Awad AB, Bradford PG, editors. Nutrition and cancer prevention. Boco Raton: CRC Press; 2006. pp. 565–78.
7. Thompson HJ, Jiang W, Zhu Z. Energetics and cancer: exploring a road less traveled. In: McTiernan A, editor. Physical activity, dietary calorie restriction, and cancer. New York: Springer; 2011. pp. 55–67.
8. Fair AM, Montgomery K. Energy balance, physical activity, and cancer risk. Methods Mol Biol. 2009;472:57–88.
9. Demark-Wahnefried W, Platz EA, Ligibel JA, Blair CK, Courneya KS, Meyerhardt JA, Ganz PA, Rock CL, Schmitz KH, Wadden T, Philip EJ, Wolfe B, Gapstur SM, Ballard-Barbash R, McTiernan A, Minasian L, Nebeling L, Goodwin PJ. The role of obesity in cancer survival and recurrence. Cancer Epidemiol Biomark Prev. 2012;21:1244–59.
10. Arsenault BJ, Beaumont EP, Despres JP, Larose E. Mapping body fat distribution: a key step towards the identification of the vulnerable patient? Ann Med. 2012;44:758–72.
11. Thompson HJ, Sporn MB. Mammary cancer in rats. In: Teichler B, editor. Tumor models in cancer research. New York: Humana Press; 2002. pp 173–81.
12. Thompson HJ. Mammary cancer in rats. In: Teichler B, editor. Tumor models in cancer research. New York: Humana Press; 2011. pp. 245–55.
13. Thompson HJ. Methods for the induction of mammary carcinogenesis in the rat using either 7, 12-Dimethylbenz [+ ¦] anthracene or MNU. Methods in mammary gland biology and breast cancer research. New York: Springer US; 2000, pp 19–29.
14. Medina D, Thompson HJ. A comparison of the salient features of mouse, rat, and human mammary tumorigenesis. Methods in mammary gland biology and breast cancer research. New York: Springer US; 2000, pp 31–6.
15. Medina D. Chemical carcinogenesis of rat and mouse mammary glands. Breast Dis. 2007;28:63–8.
16. Thompson HJ, Zhu Z, Jiang W. Protection against cancer by energy restriction: all experimental approaches are not equal. J Nutr. 2002;132:1047–9.
17. Thompson HJ, Zhu Z, Jiang W. Dietary energy restriction in breast cancer prevention. J Mammary Gland Biol Neoplasia. 2003;8:133–42.
18. Zhu Z, Haegele AD, Thompson HJ. Effect of caloric restriction on pre-malignant and malignant stages of mammary carcinogenesis. Carcinogenesis. 1997;18:1007–12.
19. Sylvester PW, Aylsworth CF, Meites J. Relationship of hormones to inhibition of mammary tumor development by underfeeding during the "critical period" after carcinogen administration. Cancer Res. 1981;41:1384–8.
20. Sylvester PW, Aylsworth CF, Van Vugt DA, Meites J. Influence of underfeeding during the "critical period" or thereafter on carcinogen-induced mammary tumors in rats. Cancer Res. 1982;42:4943–7.
21. Pariza MW. Fat, calories, and mammary carcinogenesis: net energy effects. Am J Clin Nutr. 1987;45:261–3.
22. Pariza MW. Dietary fat, calorie restriction, ad libitum feeding, and cancer risk. Nutr Rev. 1987;45:1–7.
23. Pariza MW. Calorie restriction, ad libitum feeding, and cancer. Proc Soc Exp Biol Med. 1986;183:293–8.
24. Pariza MW. Calories and energy expenditure in carcinogenesis. Bol Asoc Med PR. 1986;78:456–8.
25. Hakkak R, MacLeod S, Shaaf S, Holley AW, Simpson P, Fuchs G, Jo CH, Kieber-Emmons T, Korourian S. Obesity increases the incidence of 7,12-dimethylbenz(a)anthracene-induced mammary tumors in an ovariectomized Zucker rat model. IntJ Oncol. 2007;30:557–63.

26. Hakkak R, Holley AW, Macleod SL, Simpson PM, Fuchs GJ, Jo CH, Kieber-Emmons T, Korourian S. Obesity promotes 7,12-dimethylbenz(a)anthracene-induced mammary tumor development in female zucker rats. Breast Cancer Res. 2005;7:R627–R33.
27. Klurfeld DM, Lloyd LM, Welch CB, Davis MJ, Tulp OL, Kritchevsky D. Reduction of enhanced mammary carcinogenesis in LA/N-cp (corpulent) rats by energy restriction. Proc Soc Exp Biol Med. 1991;196:381–4.
28. Ando S, Catalano S. The multifactorial role of leptin in driving the breast cancer microenvironment. Nat Rev Endocrinol. 2012;8:263–75.
29. Barone I, Catalano S, Gelsomino L, Marsico S, Giordano C, Panza S, Bonofiglio D, Bossi G, Covington KR, Fuqua SA, Ando S. Leptin mediates tumor-stromal interactions that promote the invasive growth of breast cancer cells. Cancer Res. 2012;72:1416–27.
30. Rosini TC, Silva AS, Moraes C. Diet-induced obesity: rodent model for the study of obesity-related disorders. Rev Assoc Med Bras. 2012;58:383–7.
31. Hariri N, Thibault L. High-fat diet-induced obesity in animal models. Nutr Res Rev. 2010;23:270–99.
32. Tschop M, Heiman ML. Rodent obesity models: an overview. Exp Clin Endocrinol Diabetes. 2001;109:307–19.
33. Spiezio SH, Amon LM, McMillen TS, Vick CM, Houston BA, Caldwel M, Ogimoto K, Morton GJ, Kirk EA, Schwartz MW, Nadeau JH, LeBoeuf RC. Genetic determinants of atherosclerosis, obesity, and energy balance in consomic mice. Mamm Genome. 2014; Dec;25(11–12):549–63. doi:10.1007/s00335-014-9530-2. Epub 2014 Jul 8.
34. Cleary MP. Impact of obesity on development and progression of mammary tumors in preclinical models of breast cancer. J Mammary Gland Biol Neoplasia. 2013;18:333–43.
35. Giles ED, Jackman MR, Johnson GC, Schedin PJ, Houser JL, MacLean PS. Effect of the estrous cycle and surgical ovariectomy on energy balance, fuel utilization, and physical activity in lean and obese female rats. Am J Physiol Regul Integr Comp Physiol. 2010;299:R1634–R42.
36. MacLean PS, Giles ED, Johnson GC, McDaniel SM, Fleming-Elder BK, Gilman KA, Andrianakos AG, Jackman MR, Shroyer KR, Schedin PJ. A surprising link between the energetics of ovariectomy-induced weight gain and mammary tumor progression in obese rats. Obesity (SilverSpring). 2010;18:696–703.
37. Levin BE, Dunn-Meynel AA, Balkan B, Keesey RE. Selective breeding for diet-induced obesity and resistance in Sprague-Dawley rats. AmJ Physiol. 1997;273:R725–R30.
38. Madsen AN, Hansen G, Paulsen SJ, Lykkegaard K, Tang-Christensen M, Hansen HS, Levin BE, Larsen PJ, Knudsen LB, Fosgerau K, Vrang N. Long-term characterization of the diet-induced obese and diet-resistant rat model: a polygenetic rat model mimicking the human obesity syndrome. J Endocrinol. 2010;206:287–96.
39. Paulsen SJ, Jelsing J, Madsen AN, Hansen G, Lykkegaard K, Larsen LK, Larsen PJ, Levin BE, Vrang N. Characterization of beta-cell mass and insulin resistance in diet-induced obese and diet-resistant rats. Obesity. 2010;18:266–73.
40. Matthews SB, Zhu Z, Jiang W, McGinley JN, Neil ES, Thompson HJ. Excess weight gain accelerates 1-methyl-1-nitrosourea-induced mammary carcinogenesis in a rat model of premenopausal breast cancer. Cancer Prev Res(Phila). 2014;7:310–8.
41. Standl E. Dysglycemia and abdominal obesity. Curr Vasc Pharmacol. 2012;10:678–9.
42. de Mello VD, Pulkkinen L, Lalli M, Kolehmainen M, Pihlajamaki J, Uusitupa M. DNA methylation in obesity and type 2 diabetes. Ann Med. 2014;46:103–13.
43. Hanahan D, Weinberg RA. The hallmarks of cancer. Cell. 2000;100:57–70.
44. Hanahan D, Weinberg RA. Hallmarks of cancer: the next generation. Cell. 2011;144:646–74.
45. Fu KK, Steel GG. Growth-kinetics of a rat mammary-tumor transplanted into immune-suppressed mice. Cell Tissue Kinet. 1979;12:493–9.
46. Steel GG. Some holes in tumor-growth kinetics. Cell Tissue Kinet. 1978;11:689.
47. Tannock IF, Steel GG. Tumor growth and cell kinetics in chronically hypoxic animals. J Natl Cancer Inst. 1970;45:123–33.

48. Thompson HJ, Strange R, Schedin PJ. Apoptosis in the genesis and prevention of cancer. Cancer Epidemiol Biomark Prev. 1992;1:597–602.
49. Zhu Z, Jiang W, Thompson HJ. Effect of energy restriction on tissue size regulation during chemically induced mammary carcinogenesis. Carcinogenesis. 1999;20:1721–6.
50. Jiang W, Zhu Z, Thompson HJ. Effect of energy restriction on cell cycle machinery in 1-methyl-1-nitrosourea-induced mammary carcinomas in rats. Cancer Res. 2003;63:1228–34.
51. Zhu Z, Jiang W, Thompson HJ. Effect of energy restriction on the expression of cyclin D1 and p27 during premalignant and malignant stages of chemically induced mammary carcinogenesis. Mol Carcinog. 1999;24:241–5.
52. Thompson HJ, Zhu Z, Jiang W. Identification of the apoptosis activation cascade induced in mammary carcinomas by energy restriction. Cancer Res. 2004;64:1541–5.
53. Thompson HJ, McGinley JN, Spoelstra NS, Jiang W, Zhu Z, Wolfe P. Effect of dietary energy restriction on vascular density during mammary carcinogenesis. Cancer Res. 2004;64:5643–50.
54. Wood LD, Parsons DW, Jones S, Lin J, Sjoblom T, Leary RJ, Shen D, Boca SM, Barber T, Ptak J, Silliman N, Szabo S, Dezso Z, Ustyanksky V, Nikolskaya T, Nikolsky Y, Karchin R, Wilson PA, Kaminker JS, Zhang Z, Croshaw R, Willis J, Dawson D, Shipitsin M, Willson JKV, Sukumar S, Polyak K, Park BH, Pethiyagoda CL, Pant PVK, Ballinger DG, Sparks AB, Hartigan J, Smith DR, Suh E, Papadopoulos N, Buckhaults P, Markowitz SD, Parmigiani G, Kinzler KW, Velculescu VE, Vogelstein B. The genomic landscapes of human breast and colorectal cancers. Science. 2007;318:1108–13.
55. Gwin DM, Shackelford DB, Egan DF, Mihaylova MM, Mery A, Vasquez DS, Turk BE, Shaw RJ. AMPK phosphorylation of raptor mediates a metabolic checkpoint. Mol Cell. 2008;30:214–26.
56. Motoshima H, Goldstein BJ, Igata M, Araki E. AMPK and cell proliferation–AMPK as a therapeutic target for atherosclerosis and cancer. J Physiol. 2006;574:63–71.
57. Sabatini DM. mTOR and cancer: insights into a complex relationship. Nat Rev Cancer. 2006;6:729–34.
58. Le CC, Huang CC, Wu MY, Hsu KS. Insulin stimulates postsynaptic density-95 protein translation via the phosphoinositide 3-kinase-Akt-mammalian target of rapamycin signaling pathway. J Biol Chem. 2005;280:18543–50.
59. Martin KA, Rzucidlo EM, Merenick BL, Fingar DC, Brown DJ, Wagner RJ, Powel RJ. The mTOR/p70 S6K1 pathway regulates vascular smooth muscle cell differentiation. Am J Physiol Cell Physiol. 2004;286:C507–C17.
60. Gingras AC, Raught B, Sonenberg N. Regulation of translation initiation by FRAP/mTOR. Genes Dev. 2001;15:807–26.
61. Sonenberg N, Gingras AC. The mRNA 5′ cap-binding protein eIF4E and control of cell growth. Curr Opin Cell Biol. 1998;10:268–75.
62. Martin KA, Blenis J. Coordinate regulation of translation by the PI 3-kinase and mTOR pathways. Adv Cancer Res. 2002;86:1–39.
63. Jiang W, Zhu Z, Thompson HJ. Effects of limiting energy availability via diet and physical activity on mammalian target of rapamycin-related signaling in rat mammary carcinomas. Carcinogenesis. 2013;34:378–87.
64. Jiang W, Zhu Z, Thompson HJ. Dietary energy restriction modulates the activity of AMP-activated protein kinase, Akt, and mammalian target of rapamycin in mammary carcinomas, mammary gland, and liver. Cancer Res. 2008;68:5492–9.
65. Hardie DG. The AMP-activated protein kinase pathway–new players upstream and downstream. J Cell Sci. 2004;117:5479–87.
66. Zhu Z, Jiang W, Sells JL, Neil ES, McGinley JN, Thompson HJ. Effect of nonmotorized wheel running on mammary carcinogenesis: circulating biomarkers, cellular processes, and molecular mechanisms in rats. Cancer Epidemiol Biomark Prev. 2008;17:1920–9.
67. Kahn BB, Alquier T, Carling D, Hardie DG. AMP-activated protein kinase: ancient energy gauge provides clues to modern understanding of metabolism. Cell Metab. 2005;1:15–25.

68. Zhu Z, Jiang W, McGinley J, Wolfe P, Thompson HJ. Effects of dietary energy repletion and IGF-1 infusion on the inhibition of mammary carcinogenesis by dietary energy restriction. Mol Carcinog. 2005;42:170–6.
69. Zhu Z, Jiang W, McGinley J, Thompson HJ. Energetics and mammary carcinogenesis: effects of moderate intensity running and energy intake on cellular processes and miolecular mechanisms in rats. J Appl Physiol. In Press 2008.
70. Jiang W, Zhu Z, Thompson HJ. Effects of physical activity and restricted energy intake on chemically induced mammary carcinogenesis. Cancer Prev Res. In Press 2008.
71. Thompson HJ, Neil ES, Sells JL, McGinley J, Jiang W, Zhu Z. Motorized wheel running and mammary carcinogenesis in rats: effects of physical activity duration. Nutrition and Cancer. In Press 2008.
72. Hursting SD, Lavigne JA, Berrigan D, Perkins SN, Barret JC. Calorie restriction, aging, and cancer prevention: mechanisms of action and applicability to humans. Annu Rev Med. 2003;54:131–52.
73. Younes H, Leleu X, Hatjiharissi E, Moreau AS, Hideshima T, Richardson P, Anderson KC, Ghobrial IM. Targeting the phosphatidylinositol 3-kinase pathway in multiple myeloma. Clin Cancer Res. 2007;13:3771–5.
74. LeRoith D, Roberts CT. The insulin-like growth factor system and cancer. Cancer Lett. 2003;195:127–37.
75. Hankinson SE, Willet WC, Colditz GA, Hunter DJ, Michaud DS, Dero B, Rosner B, Speizer FE, Pollak M. Circulating concentrations of insulin-like growth factor I and risk of breast cancer. Lancet. 1998;351:1393–6.
76. Jove M, Naudi A, Ramirez-Nunez O, Portero-Otin M, Selman C, Withers DJ, Pamplona R. Caloric restriction reveals a metabolomic and lipidomic signature in liver of male mice. Aging Cell. 2014; Oct;13(5):828–37. doi:10.1111/acel.12241. Epub 2014 Jul 23.
77. De Guzman JM, Ku G, Fahey R, Youm YH, Kas I, Ingram DK, Dixit VD, Kheterpal I. Chronic caloric restriction partially protects against age-related alteration in serum metabolome. Age (Dordr). 2013;35:1091–104.
78. Mellert W, Kap M, Straus V, Wiemer J, Kamp H, Walk T, Looser R, Prokoudine A, Fabian E, Krennrich G, Herold M, van RB. Nutritional impact on the plasma metabolome of rats. Toxicol Lett. 2011;207:173–81.
79. Kristal BS, Shurubor YI, Kaddurah-Daouk R, Matson WR. Metabolomics in the study of aging and caloric restriction. Methods Mol Biol. 2007;371:393–409.
80. Ford NA, Devlin KL, Lashinger LM, Hursting SD. Deconvoluting the obesity and breast cancer link: secretome, soil and seed interactions. J Mammary Gland Biol Neoplasia. 2013;18:267–75.
81. Hursting SD. Obesity, energy balance, and cancer: a mechanistic perspective. Cancer Treat Res. 2014;159:21–33.
82. Hursting SD, Nunez NP, Varticovski L, Vinson C. The obesity-cancer link: lessons learned from a fatless mouse. Cancer Res. 2007;67:2391–3.
83. Nunez NP, Oh WJ, Rozenberg J, Perella C, Anver M, Barret JC, Perkins SN, Berrigan D, Moitra J, Varticovski L, Hursting SD, Vinson C. Accelerated tumor formation in a fatless mouse with type 2 diabetes and inflammation. Cancer Res. 2006;66:5469–76.
84. Okabe K, Inada N, Gota C, Harada Y, Funatsu T, Uchiyama S. Intracellular temperature mapping with a fluorescent polymeric thermometer and fluorescence lifetime imaging microscopy. Nat Commun. 2012;3:705.
85. Hirayama A, Kami K, Sugimoto M, Sugawara M, Toki N, Onozuka H, Kinoshita T, Saito N, Ochiai A, Tomita M, Esumi H, Soga T. Quantitative metabolome profiling of colon and stomach cancer microenvironment by capillary electrophoresis time-of-flight mass spectrometry. Cancer Res. 2009;69:4918–25.
86. Urasaki Y, Heath L, Xu CW. Coupling of glucose deprivation with impaired histone H2B monoubiquitination in tumors. PLoS One. 2012;7:e36775.

87. Birsoy K, Possemato R, Lorbeer FK, Bayraktar EC, Thiru P, Yucel B, Wang T, Chen WW, Clish CB, Sabatini DM. Metabolic determinants of cancer cell sensitivity to glucose limitation and biguanides. Nature. 2014;508:108–12.
88. Moore SC, Mayne ST, Graubard BI, Schatzkin A, Albanes D, Schairer C, Hoover RN, Leitzman MF. Past body mass index and risk of mortality among women. Int J Obes (Lond). 2008;32:730–9.
89. Kalaany NY, Sabatini DM. Tumours with PI3K activation are resistant to dietary restriction. Nature. 2009;458:725–31.
90. Welsch CW. Host factors affecting the growth of carcinogen-induced rat mammary carcinomas: a review and tribute to Charles Brenton Huggins. Cancer Res. 1985;45:3415–43.

Chapter 10
Models and Mechanisms of High-Fat Diet (HFD) Promotion of Pancreatic Cancer

Hui-Hua Chang, Guido Eibl and Enrique Rozengurt

Abstract There is epidemiologic evidence that obesity increases the risk of cancers. Several underlying mechanisms, including inflammation and insulin resistance, are proposed. However, the driving mechanisms in pancreatic cancer are poorly understood. In this chapter, we discuss models and mechanisms of diet-induced obesity and pancreatic cancer development. The focus is on a state-of-the-art mouse model, the conditional $Kras^{G12D}$ mouse model. High-fat, high-calorie diet-fed animals showed early pancreatic neoplasia and important clinical features of human obesity, including weight gain and metabolic disturbances such as hyperinsulinemia, hyperglycemia, hyperleptinemia, and elevated levels of insulin-like growth factor (IGF-1). Consequently, the signal transduction pathways initiated by insulin/IGF-induced in pancreatic cancer cells, including the PI3K/Akt/mTORC1, are discussed. The high-fat, high-calorie diet-fed conditional $Kras^{G12D}$ mouse model will provide the

E. Rozengurt (✉)
Division of Digestive Diseases, Department of Medicine, David Geffen School of Medicine at UCLA, 900 Veterans Avenue, Warren Hall Room 11-124, 90095 Los Angeles, CA, USA
CURE: Digestive Diseases Research Center, David Geffen School of Medicine at UCLA, Los Angeles, CA, USA
e-mail: ERozengurt@mednet.ucla.edu

H.-H. Chang
Department of Surgery, David Geffen School of Medicine at UCLA, 675 Charles E. Young Dr. South, MRL 2535, Los Angeles, CA 90095, USA
e-mail: huihua@ucla.edu

G. Eibl
Department of Surgery, David Geffen School of Medicine at UCLA, 10833 LeConte Avenue, 72-236 CHS, Los Angeles, CA 90095, USA
CURE: Digestive Diseases Research Center, David Geffen School of Medicine at UCLA, Los Angeles, CA, USA
e-mail: GEibl@mednet.ucla.edu

N. A. Berger (ed.), *Murine Models, Energy Balance, and Cancer*,
Energy Balance and Cancer, DOI 10.1007/978-3-319-16733-6_10

basis for more robust studies attempting to unravel the mechanisms underlying the cancer-promoting properties of obesity, as well as to evaluate dietary- and chemopreventive strategies targeting obesity-associated pancreatic cancer development.

Keywords Obesity · Inflammation · Pancreatic intraepithelial neoplasia · Tumor promotion · *Kras* · Insulin · IGF-1 · mTORC1 · Signaling cross-talk

Obesity and Pancreatic Cancer

Pancreatic ductal adenocarcinoma (PDAC) is one of the most lethal human diseases with overall 5-year survival rate of only about 5 % and a median survival period of 4–6 months. The incidence of this disease in the USA has increased to 46,420 new cases in 2014 and is now the fourth leading cause of cancer mortality in both men and women [1]. Despite advances in understanding the molecular mechanisms of PDAC development, molecularly targeted therapy has not been translated into reduced mortality or improved survival in this deadly disease [2]. Indeed, total deaths due to PDAC are projected to increase dramatically to become the second leading cause of cancer-related deaths before 2030 [3]. Consequently, the focus of research, which was placed mostly on development of therapeutic compounds, has shifted gradually toward its prevention. Novel targets and agents for chemoprevention are urgently needed and will most likely arise from a more detailed understanding of the signaling mechanisms that stimulate the promotion and progression of sub-malignant (initiated) cells into pancreatic cancer cells and from the identification of modifiable risk factors for PDAC.

One of the known modifiable risk factors for pancreatic cancer is obesity [4], which is partially attributable to the consumption of a Western-style diet rich in fats and calories, and is associated with chronic inflammation and insulin resistance. As shown recently in a large pooled analysis from the National Cancer Institute (NCI) Pancreatic Cancer Cohort Consortium (PanScan), there is strong support for a positive association between obesity and pancreatic cancer risk [5]. Considering that almost 70 % of adults in the USA are either overweight or obese [6], studying the complex relationship between obesity and pancreatic cancer will help identify targets for preventive or therapeutic strategies for this deadly disease. In addition to the epidemiologic evidence, tumor-promoting effects of high-fat diets (HFDs) or positive energy balance have also been demonstrated in a number of animal models of pancreatic cancer. These studies are of importance to elucidate the biological mechanisms of obesity-induced pancreatic cancer promotion. In this review, we first present an overview of common animal models of pancreatic cancer followed by a discussion of murine models used specifically to study obesity-associated pancreatic cancer.

Modeling Pancreatic Cancer In Vivo

Among various malignancies in the pancreas, infiltrating PDAC is the predominant histopathological form, which accounts for over 95 % of all pancreatic tumors. Analogous to many other epithelial cancers, PDAC is generally believed to originate from precursor lesions (pancreatic intra-epithelial neoplasias, PanINs) through a stepwise process with increasing degrees of morphologic atypia and accumulating genetic alterations [7]. Remarkably, oncogenic *Kras* mutation, found in early PanIN lesions and essentially all invasive pancreatic tumors, is thought to drive tumor initiation as well as to sustain tumor progression [8, 9]. Another hallmark of PDAC is the dense desmoplasia characterized by an extensive fibrosis and inflammatory reaction, which is thought to play an important role in tumor progression and chemoresistance [10]. Our recent understanding of pancreatic cancer has enabled the refinement of animal models aiming at recapitulating many key aspects of the human disease. As summarized below, there are three major types of pancreatic cancer animal models that are being used in the research community.

Xenograft Mouse Models

Xenograft tumor models can be generated by implanting cultured human pancreatic cancer cells (e.g., MIA PaCa-2, BxPC-3, PANC1, AsPC-1, Capan-1, and Capan-2) or resected human tumor blocks, either subcutaneously (ectopic) or orthotopically (into the pancreas) [11, 12]. Despite being convenient, relatively inexpensive, and useful for testing tumorigenicity and drug responses, there are a number of major limitations of these models. For example, xenografts do not allow the detailed study of immunological and inflammatory tumor–host interactions, as immunocompromised mice are generally used to avoid transplant rejection. Besides, transplanted models often fail to faithfully recapitulate certain important features of pancreatic cancer, such as extensive desmoplasia. Furthermore, early steps of tumorigenesis and disease progression naturally cannot be investigated using xenografts.

Carcinogen-Induced Models

Chemically induced models are among the first and widely used pancreatic cancer models, where carcinogens are administered to generate pancreatic lesions and adenocarcinoma in rodents. Examples include the use of azaserine in rats [13] and 7,12-dimethylbenz(a)anthracene (DMBA) in rats or mice [14, 15]. In azaserine-injected rats, the predominant histological type of pancreatic malignancy is acinar carcinoma rather than tumors with ductal features, which remains to be the most prevalent type of human pancreatic cancer [16]. In contrast, implantation of crystalline DMBA into the head of the pancreas results in the development of ductal adenocarcinoma and metastases similar to that observed in human PDAC [14, 15]. Another well-known model of pancreatic carcinogenesis is the *N*-nitrosobis(2-oxopropyl)amine (BOP)-treated hamster model, which develops PDAC with histology and genetic lesions resembling that in human cases, including *Kras*-activating and $p16^{INK4A}$-inactivating mutations [17–21]. These chemically induced models display an evolutionary spectrum of pancreatic carcinogenesis in a relatively short time period. However, the carcinogenic agents can also induce tumors or tissue damages in other organs, as they are often delivered systemically, e.g., intraperitoneally. These "off-target effects" may be limited by direct implantation of carcinogens into the pancreas, as exemplified by the aforementioned DMBA model [22], although dangers posed by carcinogen manipulation are still of concern. In addition, mice are generally more resistant to carcinogen-induced pancreatic cancer, which jointly with other drawbacks has precluded these models from extensive applications.

Genetically Engineered Mouse (GEM) Models

Over the past decade, as progress was made in technologies for mouse genomic manipulation and characterization of the genetic basis in pancreatic cancer, GEM models soon became the gold standard of modeling human PDAC, as they are able to overcome many of the limitations of other model systems. In 2001, an important breakthrough was the generation of the mouse harboring a latent $LSL\text{-}Kras^{G12D}$ allele, in which the activating G12D point mutation was introduced to exon 1 of the mouse *Kras* gene, preceded by a loxP-STOP-loxP cassette that prohibits the expression of mutant *Kras* until removed by Cre recombination [23, 24]. This major advance allowed tissue-restricted expression of oncogenic mutant $Kras^{G12D}$ gene at its endogenous locus to maintain physiological expression levels (Fig. 10.1a). Since oncogenic *Kras* mutation is the most frequent and one of the earliest genetic alterations in human PDAC, the expression of $Kras^{G12D}$ was targeted to the pancreas with Cre recombinase, whose expression is driven by the Pdx-1 or Ptf1a-p48 (p48) promoters; both are transcriptional factors labeling multipotent progenitors in the embryonic pancreas [25, 26]. In the resulting mice, which are usually referred to as KC mice for "Kras/Cre," essentially all pancreatic cells express mutant *Kras* beginning from embryological development. This conditional Kras^{G12D} model recapitulates the full spectrum of progressive development of precursor lesions (PanIN 1-3) as seen in humans [25], and is currently the major GEM model in the field. Noteworthy is that, without additional genetic alterations or environmental stimuli, the development of invasive pancreatic cancer in this model usually occurs very late (> 12 months) and at low frequency (5–10 %) [25, 27, 28], making it a suitable model for studying early tumor promotion as well as for testing preventive strategies.

The KC mouse model, complemented with a variety of other genetic modifications, has been widely utilized to investigate the roles of selected genes in PDAC development, yielding several other useful models of advanced pancreatic cancer [29]. For example, invasive PDAC in the KC model is markedly accelerated by in-

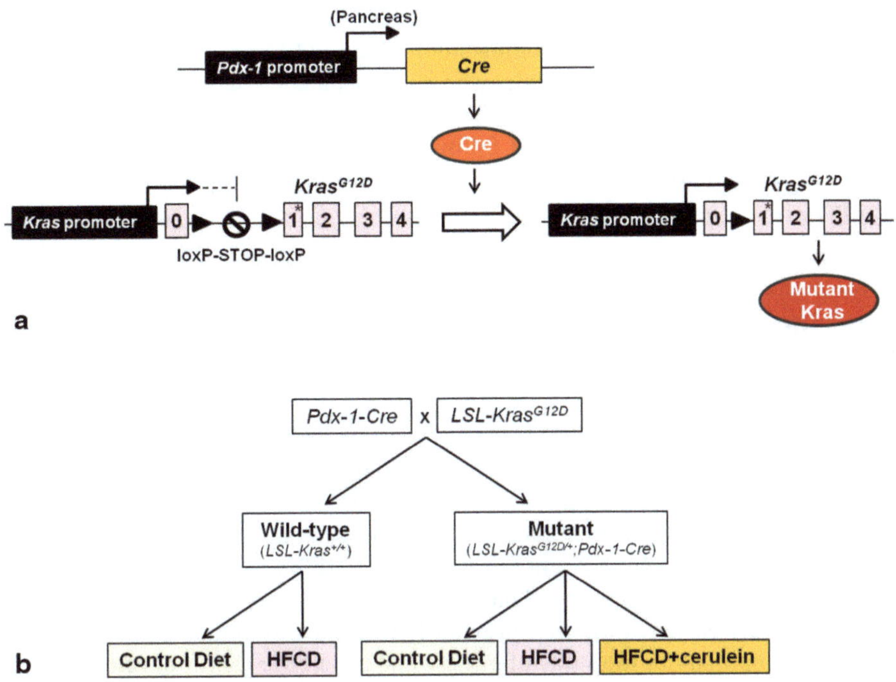

Fig. 10.1 A GEM model of diet-promoted pancreatic neoplasm. **a** In a conditional $Kras^{G12D}$ model, pancreas-restricted expression of oncogenic $Kras^{G12D}$ mutant protein is dependent on the *Pdx-1* promoter-driven Cre recombinase, which can excise the stop cassette inserted upstream of the genetically modified *Kras* coding exon 1. **b** Scheme of the experimental design for studying obesity-associated pancreatic cancer. After weaning, offspring of *LSL-KRas*^{G12D} × *Pdx-1-Cre* intercrosses were randomly assigned to either AIN-76A-based control diet or a high-fat, high-calorie diet (HFCD). In addition, some HFCD-fed mutant mice were simultaneously given intraperitoneal (*i.p.*) injections of cerulein, a cholecystokinin analogue that stimulates pancreatic inflammation

troducing additional mutations in tumor suppressor genes such as *Tp53* (resulting in KPC mice) [28], *p16*^{INK4A} *and p19*^{ARF} (Ink4a/Arf locus) [26], and *Smad4* [30] consistent with genetic signatures frequently observed in human PDAC. Furthermore, by genetically manipulating the KC mice, numerous groups have begun to study the significance of stromal interactions and immune responses in the context of oncogenic *Kras*-induced tumorigenesis. Overall, the conditional $Kras^{G12D}$ mouse model, with or without additional crosses into a variety of other genetic backgrounds, is regarded extremely valuable for studying the development of autochthonous pancreatic cancer.

The conditional expression of oncogenic *Kras* can also be targeted to different types of pancreatic cells by various promoter constructs (e.g., elastase promoter for mature acinar cells), which has provided new insights into the cell of origin for PDAC, although acinar-targeted expression of *Kras* tend to result in lesions with mixed acinar/ductal histotype [31, 32]. Further, expression of mutant *Kras* in adult mice may be achieved by introducing inducible alleles of Cre recombinase such as *Cre*^{ERT} activated by tamoxifen administration and *tetO-Cre* regulated by doxycycline. Interestingly, in a mouse model where an oncogenic *LSL-Kras*^{G12V} allele was activated in the adult pancreas (using the *Elastase-tTA/tetO-Cre* system), formation of PanINs or PDAC only occurred following cerulein-induced pancreatitis [27], underscoring the importance of injury and inflammatory response in PDAC development.

Experimental Models of Obesity-Associated Pancreatic Cancer

There is great interest in elucidating the biological mechanisms by which diet-driven obesity increases the risk of cancer development. Links between obesity, energy balance, and pancreatic cancer have been verified and investigated by utilizing some of the animal models mentioned above. For example, in mice with orthotopically implanted human MIA PaCa-2 cancer cells, a HFD was shown to increase tumor burden, as well as change lipid metabolism and cell turnover of the tumor [33]. Because of the weakness described previously for xenograft tumors in general, this model is not optimal for the study of obesity and pancreatic cancer development.

Carcinogen-induced models have also been employed in early studies validating the role of obesity or diets in modulating pancreatic carcinogenesis. In DMBA-implanted Sprague–Dawley rats fed on a high-fat/high-protein diet for 9 months, incidence of pancreatic displasia and ductal adenocarcinoma is significantly greater than that in the regular chow-fed rats, with a proportional increase of *Kras* mutation rate [34]. In addition, the tumor-promoting effects of dietary fat on pancreatic carcinogenesis have been examined in a number of other studies using BOP-treated hamsters [35–38]. Unfortunately, these models are not particularly satisfactory for further mechanistic studies, due to possible irrelevant mutations, tissue damages, and off-pancreas tumorigenesis induced by the carcinogens.

In a recent study, a HFD was shown to accelerate PanIN development in a well-established KC model (*LSL*-Kras$^{G12D/+}$; p48-*Cre*) [39]. The authors showed that the promotional effect of the HFD may be associated with inflammation mediated by TNF-α signaling, as PanIN development was abrogated on a TNF-α receptor-defective background (TNFR1$^{-/-}$). Despite the enhancement of inflammation and advanced PanINs, however, the animals on the HFD did not have greater weight gain than control mice, retained insulin sensitivity, and experienced aggravated pancreatic exocrine insufficiency with a dramatic alteration in energy metabolism, which collectively led to improved glucose tolerance [39]. These observations indicate that these animals were refractory to diet-induced obesity and insulin resistance, and therefore may not serve as an optimal model recapitulating obesity-associated pancreatic cancer in human.

More recently, we developed a model of diet-induced obesity in *Kras*-driven pancreatic neoplasia [40]. Remarkably, this model has a greater resemblance to the human condition of diet-induced obesity with ensuing hyperinsulinemia and hyperglycemia. Specifically, a high-fat, high-calorie diet (HFCD) was given to conditional *LSL-Kras*$^{G12D/+}$*;Pdx-1-Cre* mice (Fig. 10.1b). Compared with the control diet, the experimental diet is defined by increased calorie content stemming from corn oil-based fat (40 vs. 12% of total caloric intake). Mice fed with HFCD for 3 months had greater weight gain accompanied by hyperglycemia, and elevated levels of insulin, IGF-1 and leptin in serum, as compared with control animals. In addition, there were no signs of pancreatic exocrine insufficiency (weight loss and steatorrhea). Importantly, the pancreata of these HFCD-fed mutant animals displayed higher percentage of advanced PanIN lesions as well as robust signs of inflammation characterized by acinar cell loss, inflammatory cell infiltrates, and stromal fibrosis [40]. In particular, infiltrating inflammatory cells observed in the pancreas and surrounding fat tissue are mainly macrophages and T cells. This is consistent with other studies demonstrating marked infiltration of leukocytes with immunosuppressive properties (e.g., myeloid-derived suppressor cells, tumor-associated macrophages, and regulatory T cells) during early development of pancreatic cancer in GEM models [41, 42]. Additionally, in HFCD-fed mice we observed substantial changes in cytokine/chemokine levels, including IL-6 and TNF-α in the pancreas as well as in visceral white adipose tissue, which deserve further investigation. In wild type mice fed either with control diet or the HFCD (Fig. 10.1b), there were no PanIN lesions observed. However, the HFCD led to a higher chronic pancreatitis index reflecting minor loss of acinar cells, a moderate infiltration of inflammatory cells, and weak stromal fibrosis in wild type mice [40].

In a separate cohort of the conditional *Kras*^{G12D} (KC) mice fed with HFCD, additional intraperitoneal (*i.p.*) injections of cerulein (5 μg/mouse) were given five times a week to enhance chronic pancreatic inflammation. Consequently, we observed marked desmoplasia and acinar cell loss in the severely inflamed pancreas. Importantly, these mice displayed significantly less normal pancreatic ducts and more advanced PanIN lesions compared with animals fed with control diet (Fig. 10.2). Also, repeated administration of cerulein further accelerated PanIN progression in HFCD-fed mice. These results further confirm the key role of inflammation in pancreatic tumorigenesis. Noteworthy is that, in cerulein-treated KC mice fed with control diet, there was less severe chronic inflammation and progressively less advanced PanINs as compared with cerulein-injected, HFCD-fed animals. These findings suggest that HFCD may contribute to promoting PanIN development and progression by creating a pro-inflammatory microenvironment and possibly other mechanisms associated with the obese state.

Together, our findings not only reinforce the significance of obesity-associated chronic inflammation but also provide a highly relevant model for defining the role of diet-induced obesity during PDAC development as seen in humans. The observed discrepancy between the two GEM models might be attributed to different background strains (mixed C57BL/6;129 vs. pure C57BL/6 background). In fact, the influence of strain background on the responses to diets is highlighted by another report, where the promotional effects of high omega-6 fat diet on inflammation and PanIN formation were examined in elastase-driven *Kras*^{G12D} mice with differ-

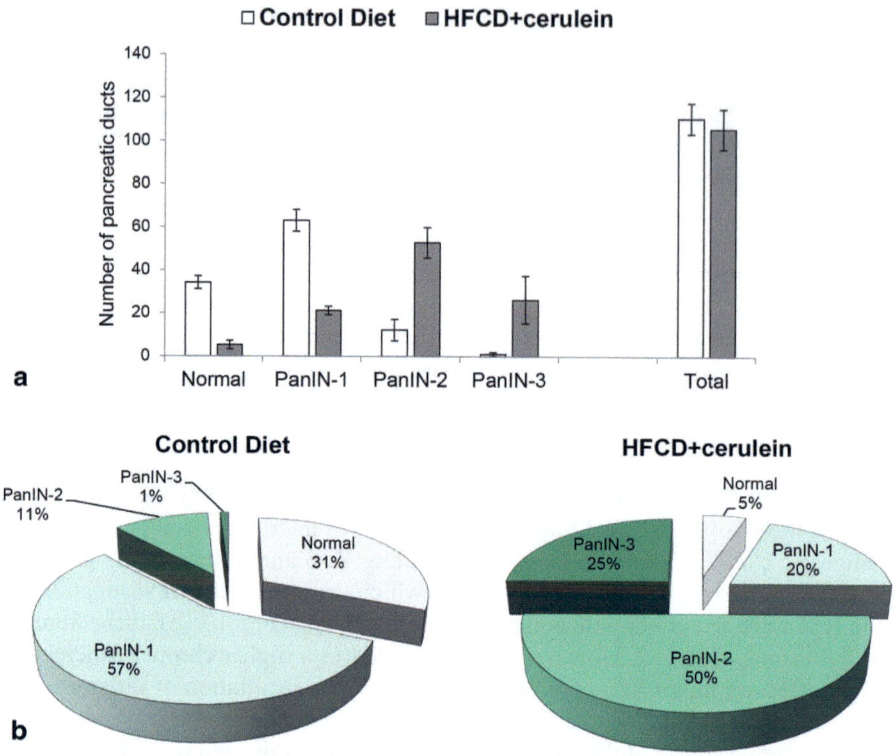

Fig. 10.2 Conditional *Kras*^{G12D} mice fed with HFCD plus repetitive cerulein injections showed substantially more advanced PanIN lesions compared to control animals (mutant mice on a control diet). **a** Quantitative analyses of PanIN lesions (graded from stage 1 to 3) in the pancreata of KC mice fed with control diet or HFCD with additional cerulein injection. **b** Pie charts depicting different distributions (in %) of normal pancreatic ducts and mouse PanIN-1, -2, and -3 lesions in the pancreata of KC mice fed with control diet or HFCD with additional cerulein injection

ent backgrounds [43]. Besides, there may be dynamic changes regarding metabolic disturbances at different stages of disease progression, which is an area currently under active investigation.

Another group also investigated the impact of a HFD (isocaloric to control diet, with 60 vs. 10% energy from fats) in a conditional Pdx-1-Cre;Kras^{G12D} model, which confirmed our results demonstrating enhanced adiposity, PanIN lesions, pancreatic inflammation, and fibrosis in HFD-fed KC mice [44]. In the same report, the HFD was also tested in elastase-Cre^{ERT} LSL-Kras^{G12D} mice, an adult acinar-specific model which displayed similar results, but with fewer spontaneous PanINs as compared with the developmental model. Interestingly, the authors described a model, where mutant Kras can be further activated by an environmental stimulus such as COX-2-mediated inflammation induced by a HFD, thereby promoting PanIN progression and PDAC development [44].

Collectively, the well-established KC mouse fed with HFCD serves as a faithful model for diet-driven obesity and early pancreatic cancer promotion, providing a scientific platform to study the mechanistic links between diet-induced obesity and pancreatic cancer. The effect of modulating energy balance on advanced PDAC development, besides precancerous lesions, has also been evaluated using a conditional Kras/Ink4a$^{+/-}$ model with both a *Kras*^{G12D} mutation and an Ink4a/Arf deficiency common to human pancreatic tumors [45]. An obesity-inducing HFCD (~30 % more total energy), in contrast to calorie restriction (CR) diet (30 % less total energy), accelerated the progression of PanINs to PDAC, enhanced pancreatic desmoplasia and liver and lung metastases, and shortened pancreatic tumor-free survival in *LSL-Kras*^{G12D}*;Pdx-1-Cre;Ink4a/Arf*$^{lox/+}$ mice. The cancer promoting effects of obesity-inducing diet and the anti-cancer effects of CR were further shown to be largely mediated by circulating IGF-1 and downstream Akt/mTOR signaling [45].

Insulin/IGF Signaling in the Mediation of Obesity-Induced Pancreatic Cancer

Diet-related metabolic disturbances (obesity, metabolic syndrome, and early stages of T2DM), which are reaching alarming rates in the Western world [46–49], are multifaceted but characterized by peripheral insulin resistance, compensatory overproduction of insulin by the β cells of the islet and increased bioavailability of IGF-1 [50]. Given the complexity of the pancreatic microcirculation [51–56], characterized by a portal system that conveys islet hormone (e.g., insulin) enriched blood to the acinar-ductal parenchyma [52–56], locally overproduced insulin is thought to act directly on insulin receptors expressed by pancreatic cells. The highly related IGF-1 receptor (IGF-1R) can also be activated by insulin [57], in particular at the high concentrations of intra-pancreatic insulin. Furthermore, IGF-1, produced by pancreatic stromal cells in response to factors released by the cancer cells, including hedgehog ligands [58, 59], can also interact with homo- or heterodimers of insulin/IGF-1 receptors in the cancer cells. Accordingly, PDAC cells express insulin and IGF-1 receptors and over-express insulin receptor substrate (IRS)-1 and IRS-2 [60–63] and PDAC (but not normal) tissue expresses activated IGF-1R [63]. Furthermore, IGF-1 is dramatically upregulated in highly metastatic human PDAC cells [64]. Recently, individual gene variations in the IGF-1 signaling system have been associated to worse survival in patients with PDAC [65]. Collectively, these studies underscore the significance of the insulin/IGF-1 signaling pathway in PDAC promotion and progression [66].

Many studies support the notion that crosstalk between the insulin/IGF-1 receptor and G protein-coupled receptor (GPCR) signaling systems plays a key role in the regulation of normal and abnormal functions, including the pathogenesis of cardiovascular and renal pathologies in obesity and T2DM [67–74]. GPCRs and their cognate agonists also mediate autocrine/paracrine growth stimulation [75–82] and

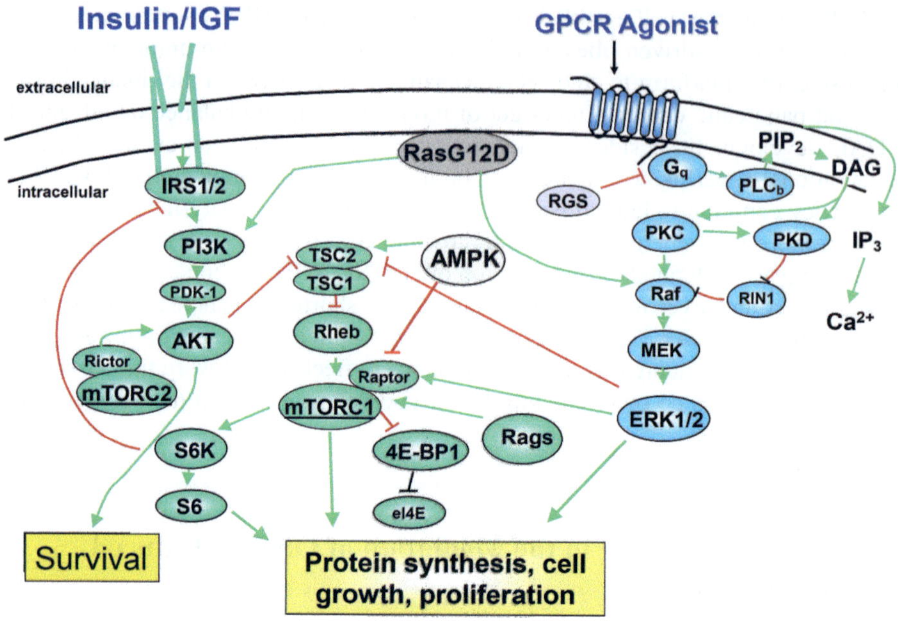

Fig. 10.3 Signal transduction pathways and crosstalk activated by insulin/IGF-1 receptor and GPCR systems. The binding of an agonist to its cognate GPCR induces Gq/PLC activation, hydrolysis of PIP_2, generation of Ins(1,4,5)P_3, and Ca^{2+} mobilization. DAG, the other product of PLC, activates novel PKCs (δ, ε, θ, η,) and, in synergy with Ca^{2+}, conventional PKCs (α, β1, β2, γ). PKD (PKD1, PKD2, and PKD3) operate downstream of DAG and PKCs and lead via inactivation of the Ras/Raf inhibitor RIN1 to ERK pathway activation, potentiating signaling via mutated *Kras*. The binding of insulin/IGF to its tetrameric receptor results in receptor autophosphorylation and activation of the receptor tyrosine kinase, followed by tyrosine phosphorylation of insulin receptor substrates (IRS). Insulin/IGF induces the PI3-kinase/Akt/TSC/Rheb/mTORC1 pathway. Negative feedback from S6K and mTORC1 on IRS is also indicated. AMPK opposes mTORC1 by phophorylating TSC2 and raptor, as indicated. See text for further details

dramatically synergize with insulin/IGF-1 in inducing mitogenic signaling [83–89]. A recent characterization of cancer genomes demonstrated frequent mutations in GPCRs and G proteins [90]. Consequently, crosstalk between insulin/IGF-1 receptor and GPCR signaling systems is also a mechanism for enhancing the development of pancreatic cancer [66] and a target for therapeutic intervention [91]. As depicted in Fig. 10.3, many GPCRs activate G proteins of the Gq family, promoting its dissociation into Gαq and Gβγ and the exchange of GDP bound to Gαq for GTP [92]. The resulting GTP-Gαq complex activates the β isoforms of phospholipase C (PLC), identified as one of the "core" signaling pathways that undergo somatic alterations in nearly all pancreatic cancers [93].

Cellular responses to insulin have been extensively characterized in diverse normal tissues, including liver, muscle, and fat. Similarly, the mechanisms that mediate obesity-induced peripheral insulin resistance in these tissues via inflammatory mediators are increasingly clarified [94] but the precise pathways and cross-talks of insulin/IGF-1 receptor signaling in PDAC cells remain incompletely understood [66]. In most cells, binding of insulin to its tetrameric receptor results in receptor autophosphorylation and activation of the receptor tyrosine kinase, followed by tyrosine phosphorylation of insulin receptor substrates (IRS 1-4) which propagate downstream signals [57, 95–97]. A key insulin/IGF1R-induced pathway via IRS is phosphatidylinositol 3-kinase (PI3K)/Akt leading to mammalian target of rapamycin (mTOR) activation [57]. This signaling module plays a pivotal role in stimulating proliferation of PDAC cells [98], is activated in PDAC tissues [99, 100], and limits catabolic processes, including autophagy [101]. These findings, together with the intriguing possibility that mTOR suppression may be associated with the antitumor actions of caloric restriction [102], suggest that mTOR signaling may play an important role in obesity-induced pancreatic cancer and a potential target for chemoprevention.

mTOR, a master regulator of cell metabolism, growth, and proliferation functions as a catalytic subunit in two structurally distinct multi-protein complexes, mTORC1 and mTORC2 [103, 104]. mTORC1, a complex of mTOR, the substrate binding subunit Raptor, GβL/mLST8, and PRAS40, phosphorylates and controls at least two regulators of protein synthesis, the 40S ribosomal protein subunit S6 kinase (S6K) and the inhibitor of protein synthesis 4E-binding protein 1, referred as 4EBP1 [96, 105–107]. mTORC1 is acutely inhibited by rapamycin whereas mTORC2, which consists of mTOR, Rictor, Protor-1, GβL/mLST8, and Sin1 is not inhibited by this agent [108, 109]. The heterodimer of the tumor suppressor TSC2 (tuberin) and TSC1 (hamartin) represses mTORC1 signaling [110, 111] by acting as the GTPase-activator protein for the small G protein Rheb (Ras homolog enriched in brain), a potent activator of mTORC1 signaling in its GTP-bound state [112, 113]. Phosphorylation of TSC2 by Akt and/or ERK/p90RSK (at different sites) suppresses its GTPase activating activity toward Rheb, leading to mTORC1 activation (reviewed in [103]). The Rag GTPases interact with mTORC1 via Raptor, and activate it in response to amino acids, by promoting mTORC1 translocation to a compartment (e.g., lysosomes) that contains Rheb-GTP [114, 115]. Inactivation of p53, as seen during the progression of 50–70 % of PDAC, is recognized to potently upregulate the insulin/IGF-1/mTORC1 pathway [116, 117]. AMPK, a conserved sensor of cellular energy activated when adenosine triphosphate (ATP) decreases and 5′-AMP (adenosine monophosphate) increases [118], is a potent inhibitor of mTORC1 via AMPK-mediated phosphorylation of TSC2 and Raptor [119].

It is pertinent to mention that the PI3K/Akt pathway is subject to potent negative feedback regulation. In addition to becoming phosphorylated at multiple Tyr residues that promote downstream signaling, the IRS family is also phosphorylated at multiple serine and threonine residues that attenuate signaling and promote degradation. In this context, it is important that activation of the mTORC1/S6K cascade inhibits IRS-1 function, including PI3K/Akt activation, following its phosphorylation at multiple residues, including $Ser^{636/639}$ by mTORC1 and $Ser^{270/307/636/1001}$ by S6K [120]. Consequently, a prominent consequence of mTORC1/S6K inhibition by rapamycin and its analogs (known as rapalogs) in cells, preclinical cancer models and clinical trials has been a striking increase in Akt phosphorylation at Thr^{308} by

PDK1 and at Ser^{473} by mTORC2 [121–123]. Suppression of these feedback loops by inhibitors of mTORC1/S6K causes compensatory over-activation of upstream signaling nodes, including PI3K and Akt that potentially oppose the anti-proliferative effects of the inhibitors and lead to drug resistance.

The preceding discussion emphasizes that mTORC1 may play a pivotal role in obesity-induced pancreatic cancer promotion. It is plausible that mTORC1 signaling acts cooperatively with other pathways elicited by inflammatory mediators as part of a complex signaling network that links obesity to pancreatic cancer. The precise impact of obesity on pancreatic cancer cells is not fully understood and a fertile ground for additional studies.

Concluding Remarks

The use of murine models for the study of energy balance and PDAC development has significantly improved our knowledge of the tumor-promoting effects of obesity in pancreatic cancer. Further refinement of murine models provides us opportunities to uncover the complex biological basis of the obesity–cancer association. Based on epidemiological and experimental studies, the mechanisms linking obesity with pancreatic cancer are multifaceted. Among the factors that are currently implicated, chronic inflammation as well as peripheral insulin resistance, compensatory overproduction of insulin by the cells of the islet and increased bioavailability of IGF-1 occupy a prominent position. In this chapter, we discussed animal models that recapitulate these cardinal elements of obesity-induced pancreatic carcinogenesis. These models not only validate the clinical observations but also serve as a valuable tool to elucidate the underlying molecular mechanisms and to evaluate potential preventive or therapeutic strategies, with an ultimate goal of reducing PDAC mortality in the increasingly obese community.

Grant Support The studies in the laboratories of Drs. Eibl and Rozengurt are supported by the National Institute of Health (P01 CA163200; P30DK41301), Department of Veterans Affairs (ER), and the Hirshberg Foundation for Pancreatic Cancer Research (GE and ER).

References

1. Siegel R, Ma J, Zou Z, Jemal A. Cancer statistics, 2014. CA Cancer J Clin. 2014;64(1):9–29. doi:10.3322/caac.21208.
2. Morgan MA, Parsels LA, Maybaum J, Lawrence TS. Improving gemcitabine-mediated radiosensitization using molecularly targeted therapy: a review. Clin Cancer Res. 2008;14(21):6744–50. doi:10.1158/1078-0432.ccr-08-1032.
3. Rahib L, Smith BD, Aizenberg R, Rosenzweig AB, Fleshman JM, Matrisian LM. Projecting cancer incidence and deaths to 2030: the unexpected burden of thyroid, liver, and pancreas cancers in the United States. Cancer Res. 2014;74(11):2913–21. doi:10.1158/0008-5472.can-14-0155.

4. Bracci PM. Obesity and pancreatic cancer: overview of epidemiologic evidence and biologic mechanisms. Mol Carcinog. 2012;51(1):53–63. doi:10.1002/mc.20778.
5. Arslan AA, Helzlsouer KJ, Kooperberg C, Shu XO, Steplowski E, Bueno-de-Mesquita HB, Fuchs CS, Gross MD, Jacobs EJ, Lacroix AZ, Petersen GM, Stolzenberg-Solomon RZ, Zheng W, Albanes D, Amundadottir L, Bamlet WR, Barricarte A, Bingham SA, Boeing H, Boutron-Ruault MC, Buring JE, Chanock SJ, Clipp S, Gaziano JM, Giovannucci EL, Hankinson SE, Hartge P, Hoover RN, Hunter DJ, Hutchinson A, Jacobs KB, Kraft P, Lynch SM, Manjer J, Manson JE, McTiernan A, McWilliams RR, Mendelsohn JB, Michaud DS, Palli D, Rohan TE, Slimani N, Thomas G, Tjonneland A, Tobias GS, Trichopoulos D, Virtamo J, Wolpin BM, Yu K, Zeleniuch-Jacquotte A, Patel AV. Anthropometric measures, body mass index, and pancreatic cancer: a pooled analysis from the pancreatic cancer cohort consortium (PanScan). Arch Intern Med. 2010;170(9):791–802. doi:10.1001/archinternmed.2010.63 170/9/791 [pii].
6. Flegal KM, Carroll MD, Ogden CL, Curtin LR. Prevalence and trends in obesity among US adults, 1999–2008. JAMA. 2010;303(3):235–41. doi:10.1001/jama.2009.2014 2009.2014 [pii].
7. Hezel AF, Kimmelman AC, Stanger BZ, Bardeesy N, Depinho RA. Genetics and biology of pancreatic ductal adenocarcinoma. Genes Dev. 2006;20(10):1218–49.
8. Deramaudt T, Rustgi AK. Mutant KRAS in the initiation of pancreatic cancer. Biochim Biophys Acta. 2005;1756(2):97–101. doi:S0304-419X(05)00044-2 [pii] 10.1016/j.bbcan.2005.08.003.
9. Ji B, Tsou L, Wang H, Gaiser S, Chang DZ, Daniluk J, Bi Y, Grote T, Longnecker DS, Logsdon CD. Ras activity levels control the development of pancreatic diseases. Gastroenterology. 2009;137(3):1072–82, 1082.e1071–1076. doi:10.1053/j.gastro.2009.05.052 S0016-5085(09)00900-7 [pii].
10. Neesse A, Michl P, Frese KK, Feig C, Cook N, Jacobetz MA, Lolkema MP, Buchholz M, Olive KP, Gress TM, Tuveson DA. Stromal biology and therapy in pancreatic cancer. Gut. 2011;60(6):861–8. doi:10.1136/gut.2010.226092 gut2010226092 [pii].
11. Fu X, Guadagni F, Hoffman RM. A metastatic nude-mouse model of human pancreatic cancer constructed orthotopically with histologically intact patient specimens. Proc Natl Acad Sci U S A. 1992;89(12):5645–9.
12. Ding Y, Cravero JD, Adrian K, Grippo P. Modeling pancreatic cancer in vivo: from xenograft and carcinogen-induced systems to genetically engineered mice. Pancreas. 2010;39(3):283–92. doi:10.1097/MPA.0b013e3181c15619 00006676-201004000-00001 [pii].
13. Longnecker DS, Curphey TJ. Adenocarcinoma of the pancreas in azaserine-treated rats. Cancer Res. 1975;35(8):2249–58.
14. Dissin J, Mills LR, Mains DL, Black O Jr, Webster PD 3rd. Experimental induction of pancreatic adenocarcinoma in rats. J Natl Cancer Inst. 1975;55(4):857–64.
15. Osvaldt AB, Wendt LR, Bersch VP, Backes AN, de Cassia ASR, Edelweiss MI, Rohde L. Pancreatic intraepithelial neoplasia and ductal adenocarcinoma induced by DMBA in mice. Surgery. 2006;140(5):803–9. doi:S0039-6060(06)00160-7 [pii] 10.1016/j.surg.2006.02.012.
16. Longnecker DS. Lesions induced in rodent pancreas by azaserine and other pancreatic carcinogens. Environ Health Perspect. 1984;56:245–51.
17. Pour P, Althoff J, Kruger FW, Mohr U. A potent pancreatic carcinogen in Syrian hamsters: N-nitrosobis(2-oxopropyl)amine. J Natl Cancer Inst. 1977;58(5):1449–53.
18. Fujii H, Egami H, Chaney W, Pour P, Pelling J. Pancreatic ductal adenocarcinomas induced in Syrian hamsters by N-nitrosobis(2-oxopropyl)amine contain a c-Ki-ras oncogene with a point-mutated codon 12. Mol Carcinog. 1990;3(5):296–301.
19. Cerny WL, Mangold KA, Scarpelli DG. K-ras mutation is an early event in pancreatic duct carcinogenesis in the Syrian golden hamster. Cancer Res. 1992;52(16):4507–13.
20. Chang KW, Laconi S, Mangold KA, Hubchak S, Scarpelli DG. Multiple genetic alterations in hamster pancreatic ductal adenocarcinomas. Cancer Res. 1995;55(12):2560–8.
21. Li J, Weghorst CM, Tsutsumi M, Poi MJ, Knobloch TJ, Casto BC, Melvin WS, Tsai MD, Muscarella P. Frequent p16INK4A/CDKN2A alterations in chemically induced Syrian

golden hamster pancreatic tumors. Carcinogenesis. 2004;25(2):263–8. doi:10.1093/carcin/bgh007 bgh007 [pii].
22. Rivera JA, Graeme-Cook F, Werner J, Z'Graggen K, Rustgi AK, Rattner DW, Warshaw AL, Fernandez-del Castillo C. A rat model of pancreatic ductal adenocarcinoma: targeting chemical carcinogens. Surgery. 1997;122(1):82–90.
23. Jackson EL, Willis N, Mercer K, Bronson RT, Crowley D, Montoya R, Jacks T, Tuveson DA. Analysis of lung tumor initiation and progression using conditional expression of oncogenic K-ras. Genes Dev. 2001;15(24):3243–8. doi:10.1101/gad.943001.
24. Johnson L, Mercer K, Greenbaum D, Bronson RT, Crowley D, Tuveson DA, Jacks T. Somatic activation of the K-ras oncogene causes early onset lung cancer in mice. Nature. 2001;410(6832):1111–6. doi:10.1038/35074129 35074129 [pii].
25. Hingorani SR, Petricoin EF, Maitra A, Rajapakse V, King C, Jacobetz MA, Ross S, Conrads TP, Veenstra TD, Hitt BA, Kawaguchi Y, Johann D, Liotta LA, Crawford HC, Putt ME, Jacks T, Wright CV, Hruban RH, Lowy AM, Tuveson DA. Preinvasive and invasive ductal pancreatic cancer and its early detection in the mouse. Cancer Cell. 2003;4(6):437–50. doi:S153561080300309X [pii].
26. Aguirre AJ, Bardeesy N, Sinha M, Lopez L, Tuveson DA, Horner J, Redston MS, DePinho RA. Activated Kras and Ink4a/Arf deficiency cooperate to produce metastatic pancreatic ductal adenocarcinoma. Genes Dev. 2003;17(24):3112–26. doi:10.1101/gad.1158703 1158703 [pii].
27. Guerra C, Schuhmacher AJ, Canamero M, Grippo PJ, Verdaguer L, Perez-Gallego L, Dubus P, Sandgren EP, Barbacid M. Chronic pancreatitis is essential for induction of pancreatic ductal adenocarcinoma by K-Ras oncogenes in adult mice. Cancer Cell. 2007;11(3):291–302. doi:S1535-6108(07)00027-X [pii] 10.1016/j.ccr.2007.01.012.
28. Hingorani SR, Wang L, Multani AS, Combs C, Deramaudt TB, Hruban RH, Rustgi AK, Chang S, Tuveson DA. Trp53R172H and KrasG12D cooperate to promote chromosomal instability and widely metastatic pancreatic ductal adenocarcinoma in mice. Cancer Cell. 2005;7(5):469–83. doi:S1535-6108(05)00128-5 [pii] 10.1016/j.ccr.2005.04.023.
29. Mazur PK, Siveke JT. Genetically engineered mouse models of pancreatic cancer: unravelling tumour biology and progressing translational oncology. Gut. 2012;61(10):1488–500. doi:10.1136/gutjnl-2011-300756 gutjnl-2011-300756 [pii].
30. Bardeesy N, Cheng KH, Berger JH, Chu GC, Pahler J, Olson P, Hezel AF, Horner J, Lauwers GY, Hanahan D, DePinho RA. Smad4 is dispensable for normal pancreas development yet critical in progression and tumor biology of pancreas cancer. Genes Dev. 2006;20(22):3130–46. doi:20/22/3130 [pii] 10.1101/gad.1478706.
31. Grippo PJ, Nowlin PS, Demeure MJ, Longnecker DS, Sandgren EP. Preinvasive pancreatic neoplasia of ductal phenotype induced by acinar cell targeting of mutant Kras in transgenic mice. Cancer Res. 2003;63(9):2016–9.
32. Tuveson DA, Zhu L, Gopinathan A, Willis NA, Kachatrian L, Grochow R, Pin CL, Mitin NY, Taparowsky EJ, Gimotty PA, Hruban RH, Jacks T, Konieczny SF.Mist1-KrasG12D knock-in mice develop mixed differentiation metastatic exocrine pancreatic carcinoma and hepatocellular carcinoma. Cancer Res. 2006;66(1):242–7. doi:66/1/242 [pii] 10.1158/0008-5472.CAN-05-2305
33. Wang F, Kumagai-Braesch M, Herrington MK, Larsson J, Permert J.Increased lipid metabolism and cell turnover of MiaPaCa2 cells induced by high-fat diet in an orthotopic system. Metabolism 2009;58(8):1131–6. doi:10.1016/j.metabol.2009.03.027 S0026-0495(09)00119-X [pii]
34. Z'Graggen K, Warshaw AL, Werner J, Graeme-Cook F, Jimenez RE, Fernandez-Del Castillo C. Promoting effect of a high-fat/high-protein diet in DMBA-induced ductal pancreatic cancer in rats. Ann Surg. 2001;233(5):688–95.
35. Birt DF, Salmasi S, Pour PM. Enhancement of experimental pancreatic cancer in Syrian golden hamsters by dietary fat. J Natl Cancer Inst. 1981;67(6):1327–32.
36. Birt DF, Julius AD, White LT, Pour PM. Enhancement of pancreatic carcinogenesis in hamsters fed a high-fat diet ad libitum and at a controlled calorie intake. Cancer Res. 1989;49(21):5848–51.

37. Schneider MB, Matsuzaki H, Haorah J, Ulrich A, Standop J, Ding XZ, Adrian TE, Pour PM. Prevention of pancreatic cancer induction in hamsters by metformin. Gastroenterology. 2001;120(5):1263–70. doi:10.1053/gast.2001.23258 S0016508501564938 [pii].
38. Hori M, Kitahashi T, Imai T, Ishigamori R, Takasu S, Mutoh M, Sugimura T, Wakabayashi K, Takahashi M. Enhancement of carcinogenesis and fatty infiltration in the pancreas in N-nitrosobis(2-oxopropyl)amine-treated hamsters by high-fat diet. Pancreas. 2011;40(8):1234–40. doi:10.1097/MPA.0b013e318220e742.
39. Khasawneh J, Schulz MD, Walch A, Rozman J, Hrabe de Angelis M, Klingenspor M, Buck A, Schwaiger M, Saur D, Schmid RM, Kloppel G, Sipos B, Greten FR, Arkan MC. Inflammation and mitochondrial fatty acid beta-oxidation link obesity to early tumor promotion. Proc Natl Acad Sci U S A. 2009;106(9):3354–9. doi:10.1073/pnas.0802864106 0802864106 [pii].
40. Dawson DW, Hertzer K, Moro A, Donald G, Chang HH, Go VL, Pandol SJ, Lugea A, Gukovskaya AS, Li G, Hines OJ, Rozengurt E, Eibl G. High-fat, high-calorie diet promotes early pancreatic neoplasia in the conditional KrasG12D mouse model. Cancer Prev Res (Phila). 2013;6(10):1064–73. doi:10.1158/1940-6207.CAPR-13-0065 1940-6207CAPR-13-0065 [pii].
41. Clark CE, Hingorani SR, Mick R, Combs C, Tuveson DA, Vonderheide RH. Dynamics of the immune reaction to pancreatic cancer from inception to invasion. Cancer Res. 2007;67(19):9518–27. doi:67/19/9518 [pii] 10.1158/0008-5472.CAN-07-0175.
42. Clark CE, Beatty GL, Vonderheide RH.Immunosurveillance of pancreatic adenocarcinoma: insights from genetically engineered mouse models of cancer. Cancer Lett. 2009;279(1):1–7. doi:10.1016/j.canlet.2008.09.037 S0304-3835(08)00799-4 [pii].
43. Cheon EC, Strouch MJ, Barron MR, Ding Y, Melstrom LG, Krantz SB, Mullapudi B, Adrian K, Rao S, Adrian TE, Bentrem DJ, Grippo PJ. Alteration of strain background and a high omega-6 fat diet induces earlier onset of pancreatic neoplasia in EL-Kras transgenic mice. Int J Cancer. 2011;128(12):2783–92. doi:10.1002/ijc.25622.
44. Philip B, Roland CL, Daniluk J, Liu Y, Chatterjee D, Gomez SB, Ji B, Huang H, Wang H, Fleming JB, Logsdon CD, Cruz-Monserrate Z. A high-fat diet activates oncogenic Kras and COX2 to induce development of pancreatic ductal adenocarcinoma in mice. Gastroenterology. 2013;145(6):1449–58. doi:10.1053/j.gastro.2013.08.018 S0016-5085(13)01191-8 [pii].
45. Lashinger LM, Harrison LM, Rasmussen AJ, Logsdon CD, Fischer SM, McArthur MJ, Hursting SD. Dietary energy balance modulation of Kras- and Ink4a/Arf+/-driven pancreatic cancer: the role of insulin-like growth factor-I. Cancer Prev Res (Phila). 2013;6(10):1046–55. doi:10.1158/1940-6207.CAPR-13-0185 1940-6207CAPR-13-0185 [pii].
46. Flier JS. Obesity wars: molecular progress confronts an expanding epidemic. Cell. 2004;116(2):337–50.
47. Calle EE, Kaaks R. Overweight, obesity and cancer: epidemiological evidence and proposed mechanisms. Nat Rev Cancer. 2004;4(8):579–91.
48. Calle EE, Thun MJ. Obesity and cancer. Oncogene. 2004;23(38):6365–78.
49. Muoio DM, Newgard CB. Obesity-related derangements in metabolic regulation. Annu Rev Biochem. 2006;75:367–401.
50. Tam CS, Viardot A, Clément K, Tordjman J, Tonks K, Greenfield JR, Campbell LV, Samocha-Bonet D, Heilbronn LK. Short-term overfeeding may induce peripheral insulin resistance without altering subcutaneous adipose tissue macrophages in humans. Diabetes. 2010;59(9):2164–70. doi:10.2337/db10-0162.
51. Ballian N, Brunicardi FC. Islet vasculature as a regulator of endocrine pancreas function. World J Surg. 2007;31(4):705–14.
52. Lee KY, Zhou L, Ren XS, Chang TM, Chey WY. An important role of endogenous insulin on exocrine pancreatic secretion in rats. Am J Physiol. 1990;258(2 Pt 1):G268–74.
53. Lee KY, Lee YL, Kim CD, Chang TM, Chey WY. Mechanism of action of insulin on pancreatic exocrine secretion in perfused rat pancreas. Am J Physiol. 1994;267(2 Pt 1):G207–12.

54. Murakami T, Hitomi S, Ohtsuka A, Taguchi T, Fujita T. Pancreatic insulo-acinar portal systems in humans, rats, and some other mammals: scanning electron microscopy of vascular casts. Microsc Res Tech. 1997;37(5/6):478–88.
55. Wayland H. Microcirculation in pancreatic function. Microsc Res Tech. 1997;37(5/6):418–33.
56. Bertelli E, Regoli M, Orazioli D, Bendayan M. Association between islets of Langerhans and pancreatic ductal system in adult rat. Where endocrine and exocrine meet together? Diabetologia. 2001;44(5):575–84.
57. Taniguchi CM, Emanuelli B, Kahn CR. Critical nodes in signalling pathways: insights into insulin action. Nat Rev Mol Cell Biol. 2006;7(2):85–96.
58. Nakamura K, Sasajima J, Mizukami Y, Sugiyama Y, Yamazaki M, Fujii R, Kawamoto T, Koizumi K, Sato K, Fujiya M, Sasaki K, Tanno S, Okumura T, Shimizu N, Kawabe J-i, Karasaki H, Kono T, Ii M, Bardeesy N, Chung DC, Kohgo Y. Hedgehog promotes neovascularization in pancreatic cancers by regulating Ang-1 and IGF-1 expression in bone-marrow derived proangiogenic cells. PLoS One. 2010;5(1):e8824.
59. Yauch RL, Gould SE, Scales SJ, Tang T, Tian H, Ahn CP, Marshall D, Fu L, Januario T, Kallop D, Nannini-Pepe M, Kotkow K, Marsters JC, Rubin LL, de Sauvage FJ. A paracrine requirement for hedgehog signalling in cancer. Nature. 2008;455(7211):406.
60. Kornmann M, Maruyama H, Bergmann U, Tangvoranuntakul P, Beger HG, White MF, Korc M. Enhanced expression of the insulin receptor substrate-2 docking protein in human pancreatic cancer. Cancer Res. 1998;58(19):4250–4.
61. Kolb S, Fritsch R, Saur D, Reichert M, Schmid RM, Schneider G. HMGA1 controls transcription of insulin receptor to regulate cyclin D1 translation in pancreatic cancer cells. Cancer Res. 2007;67(10):4679–86.
62. Kwon J, Stephan S, Mukhopadhyay A, Muders MH, Dutta SK, Lau JS, Mukhopadhyay D. Insulin receptor substrate-2 mediated insulin-like growth factor-I receptor overexpression in pancreatic adenocarcinoma through protein kinase Cdelta. Cancer Res. 2009;69(4):1350–7.
63. Stoeltzing O, Liu W, Reinmuth N, Fan F, Parikh AA, Bucana CD, Evans DB, Semenza GL, Ellis LM. Regulation of hypoxia-inducible factor-1alpha, vascular endothelial growth factor, and angiogenesis by an insulin-like growth factor-I receptor autocrine loop in human pancreatic cancer. Am J Pathol. 2003;163(3):1001–11.
64. Shi W-D, Meng Z-Q, Chen Z, Lin J-H, Zhou Z-H, Liu L-M. Identification of liver metastasis-related genes in a novel human pancreatic carcinoma cell model by microarray analysis. Cancer Lett. 2009;283(1):84.
65. Dong X, Javle M, Hess KR, Shroff R, Abbruzzese JL, Li D. Insulin-like growth factor axis gene polymorphisms and clinical outcomes in pancreatic cancer. Gastroenterology. 2010;139:464–73.
66. Rozengurt E, Sinnett-Smith J, Kisfalvi K. Crosstalk between insulin/insulin-like growth factor-1 receptors and G protein-coupled receptor signaling systems: a novel target for the antidiabetic drug metformin in pancreatic cancer. Clin Cancer Res. 2010;16:2505–11. doi:10.1158/1078-0432.ccr-09-2229.
67. Gavi S, Shumay E, Wang HY, Malbon CC. G-protein-coupled receptors and tyrosine kinases: crossroads in cell signaling and regulation. Trends Endocrinol Metab. 2006;17(2):48–54.
68. Velloso LA, Folli F, Perego L, Saad MJ. The multi-faceted cross-talk between the insulin and angiotensin II signaling systems. Diabetes Metab Res Rev. 2006;22(2):98–107.
69. Horiuchi M, Mogi M, Iwai M. Signaling crosstalk angiotensin II receptor subtypes and insulin. Endocr J. 2006;53(1):1–5.
70. Sowers JR. Insulin resistance and hypertension. Am J Physiol Heart Circ Physiol. 2004;286(5):H1597–1602.
71. Kakoki M, Kizer CM, Yi X, Takahashi N, Kim HS, Bagnell CR, Edgell CJ, Maeda N, Jennette JC, Smithies O. Senescence-associated phenotypes in Akita diabetic mice are enhanced by absence of bradykinin B2 receptors. J Clin Invest. 2006;116(5):1302–9.
72. Kakoki M, Takahashi N, Jennette JC, Smithies O. Diabetic nephropathy is markedly enhanced in mice lacking the bradykinin B2 receptor. Proc Natl Acad Sci U S A. 2004;101(36):13302–5.
73. Aguilar D, Solomon SD. ACE inhibitors and angiotensin receptor antagonists and the incidence of new-onset diabetes mellitus: an emerging theme. Drugs. 2006;66(9):1169–77.

74. Haffner SM. Risk constellations in patients with the metabolic syndrome: epidemiology, diagnosis, and treatment patterns. Am J Med. 2006;119(5 Suppl 1):S3–9.
75. Rozengurt E. Growth factors and cell proliferation. Curr Opin Cell Biol. 1992;4(2):161–5.
76. Rozengurt E. Signal transduction pathways in the mitogenic response to G protein-coupled neuropeptide receptor agonists. J Cell Physiol. 1998;177:507–17.
77. Gutkind JS. Cell growth control by G protein-coupled receptors: from signal transduction to signal integration. Oncogene. 1998;17:1331–42.
78. Rozengurt E. Autocrine loops, signal transduction, and cell cycle abnormalities in the molecular biology of lung cancer. Curr Opin Oncol. 1999;11:116–22.
79. Rozengurt E, Walsh JH. Gastrin, CCK, signaling, and cancer. Annu Rev Physiol. 2001;63:49–76.
80. Heasley LE. Autocrine and paracrine signaling through neuropeptide receptors in human cancer. Oncogene. 2001;20(13):1563–9.
81. Rozengurt E. Neuropeptides as growth factors for normal and cancer cells. Trends Endocrinol Metabol. 2002;13:128–34.
82. Arafat HA, Gong Q, Chipitsyna G, Rizvi A, Saa CT, Yeo CJ. Antihypertensives as novel antineoplastics: angiotensin-I-converting enzyme inhibitors and angiotensin II type 1 receptor blockers in pancreatic ductal adenocarcinoma. J Am Coll Surg. 2007;204(5):996–1005.
83. Rozengurt E, Legg A, Pettican P. Vasopressin stimulation of mouse 3T3 cell growth. Proc Natl Acad Sci U S A. 1979;76(3):1284–7.
84. Dicker P, Rozengurt E. Phorbol esters and vasopressin stimulate DNA synthesis by a common mechanism. Nature. 1980;287(5783):607–12.
85. Rozengurt E, Sinnett-Smith J. Bombesin stimulation of DNA synthesis and cell division in cultures of Swiss 3T3 cells. Proc Natl Acad Sci U S A. 1983;80(10):2936–40.
86. Rozengurt E. Early signals in the mitogenic response. Science. 1986;234(4773):161–6.
87. Seufferlein T, Withers DJ, Rozengurt E. Reduced requirement of mitogen-activated protein kinase (MAPK) activity for entry into the S phase of the cell cycle in Swiss 3T3 fibroblasts stimulated by bombesin and insulin. J Biol Chem. 1996;271(35):21471–7.
88. Withers DJ, Seufferlein T, Mann D, Garcia B, Jones N, Rozengurt E. Rapamycin dissociates p70(S6K) activation from DNA synthesis stimulated by bombesin and insulin in Swiss 3T3 cells. J Biol Chem. 1997;272(4):2509–14.
89. Mann DJ, Higgins T, Jones NC, Rozengurt E. Differential control of cyclins D1 and D3 and the cdk inhibitor p27Kip1 by diverse signalling pathways in Swiss 3T3 cells. Oncogene. 1997;14(15):1759–66.
90. Kan Z, Jaiswal BS, Stinson J, Janakiraman V, Bhatt D, Stern HM, Yue P, Haverty PM, Bourgon R, Zheng J, Moorhead M, Chaudhuri S, Tomsho LP, Peters BA, Pujara K, Cordes S, Davis DP, Carlton VE, Yuan W, Li L, Wang W, Eigenbrot C, Kaminker JS, Eberhard DA, Waring P, Schuster SC, Modrusan Z, Zhang Z, Stokoe D, de Sauvage FJ, Faham M, Seshagiri S. Diverse somatic mutation patterns and pathway alterations in human cancers. Nature. 2010;466:869–73.
91. Kisfalvi K, Eibl G, Sinnett-Smith J, Rozengurt E. Metformin disrupts crosstalk between G protein-coupled receptor and insulin receptor signaling systems and inhibits pancreatic cancer growth. Cancer Res. 2009;69(16):6539–45.
92. Rozengurt E. Mitogenic signaling pathways induced by G protein-coupled receptors. J Cell Physiol. 2007;213(3):589–602.
93. Olive KP, Jacobetz MA, Davidson CJ, Gopinathan A, McIntyre D, Honess D, Madhu B, Goldgraben MA, Caldwell ME, Allard D, Frese KK, DeNicola G, Feig C, Combs C, Winter SP, Ireland-Zecchini H, Reichelt S, Howat WJ, Chang A, Dhara M, Wang L, Ruckert F, Grutzmann R, Pilarsky C, Izeradjene K, Hingorani SR, Huang P, Davies SE, Plunkett W, Egorin M, Hruban RH, Whitebread N, McGovern K, Adams J, Iacobuzio-Donahue C, Griffiths J, Tuveson DA. Inhibition of hedgehog signaling enhances delivery of chemotherapy in a mouse model of pancreatic cancer. Science. 2009;324(5933):1457–61. doi:10.1126/science.1171362.
94. Pessin JE, Kwon H.Adipokines mediate inflammation and insulin resistance. Front Endocrinol. 2013;4:71. doi:10.3389/fendo.2013.00071.

95. Saltiel AR, Pessin JE. Insulin signaling pathways in time and space. Trends Cell Biol. 2002;12(2):65–71.
96. Um SH, D'Alessio D, Thomas G. Nutrient overload, insulin resistance, and ribosomal protein S6 kinase 1, S6K1. Cell Metab. 2006;3(6):393–402.
97. Metz HE, Houghton MA. Insulin receptor substrate regulation of phosphoinositide 3-kinase. Clin Cancer Res. 2011;17(2):206–11.
98. Asano T, Yao Y, Shin S, McCubrey J, Abbruzzese JL, Reddy SA. Insulin receptor substrate is a mediator of phosphoinositide 3-kinase activation in quiescent pancreatic cancer cells. Cancer Res. 2005;65(20):9164–8.
99. Asano T, Yao Y, Zhu J, Li D, Abbruzzese JL, Reddy SA. The rapamycin analog CCI-779 is a potent inhibitor of pancreatic cancer cell proliferation. Biochem Biophys Res Commun. 2005;331(1):295–302.
100. Pham NA, Schwock J, Iakovlev V, Pond G, Hedley DW, Tsao MS. Immunohistochemical analysis of changes in signaling pathway activation downstream of growth factor receptors in pancreatic duct cell carcinogenesis. BMC Cancer. 2008;8:43.
101. Lee JW, Park S, Takahashi Y, Wang H-G. The association of AMPK with ULK1 regulates autophagy. PLoS One. 2010;5(11):e15394.
102. Kopelovich L, Fay JR, Sigman CC, Crowell JA. The mammalian target of rapamycin pathway as a potential target for cancer chemoprevention. Cancer Epidemiol Biomarkers Prev. 2007;16(7):1330.
103. Foster KG, Fingar DC. Mammalian target of rapamycin (mTOR): conducting the cellular signaling symphony. J Biol Chem. 2010;285(19):14071–7. doi:10.1074/jbc.R109.094003.
104. Laplante M, Sabatini DM. mTOR signaling at a glance. J Cell Sci. 2009;122(20):3589–94.
105. Ali SM, Sabatini DM. Structure of S6 kinase 1 determines whether raptor-mTOR or rictor-mTOR phosphorylates its hydrophobic motif site. J Biol Chem. 2005;280:19445–8.
106. Long X, Muller F, Avruch J. TOR action in mammalian cells and in *Caenorhabditis elegans*. Curr Top Microbiol Immunol. 2004;279:115–38.
107. Armengol G, Rojo F, Castellvi J, Iglesias C, Cuatrecasas M, Pons B, Baselga J, Ramon y Cajal S. 4E-binding protein 1: a key molecular "funnel factor" in human cancer with clinical implications. Cancer Res. 2007;67(16):7551–5.
108. Wullschleger S, Loewith R, Hall MN. TOR signaling in growth and metabolism. Cell. 2006;124(3):471–84.
109. Inoki K, Guan KL. Complexity of the TOR signaling network. Trends Cell Biol. 2006;16(4):206–12.
110. Tee AR, Fingar DC, Manning BD, Kwiatkowski DJ, Cantley LC, Blenis J. Tuberous sclerosis complex-1 and -2 gene products function together to inhibit mammalian target of rapamycin (mTOR)-mediated downstream signaling. Proc Natl Acad Sci U S A. 2002;99(21):13571–6.
111. Manning BD, Tee AR, Logsdon MN, Blenis J, Cantley LC. Identification of the tuberous sclerosis complex-2 tumor suppressor gene product tuberin as a target of the phosphoinositide 3-kinase/akt pathway. Mol Cell. 2002;10(1):151–62.
112. Garami A, Zwartkruis FJ, Nobukuni T, Joaquin M, Roccio M, Stocker H, Kozma SC, Hafen E, Bos JL, Thomas G. Insulin activation of Rheb, a mediator of mTOR/S6K/4E-BP signaling, is inhibited by TSC1 and 2. Mol Cell. 2003;11(6):1457–66.
113. Zhang Y, Gao X, Saucedo LJ, Ru B, Edgar BA, Pan D. Rheb is a direct target of the tuberous sclerosis tumour suppressor proteins. Nat Cell Biol. 2003;5(6):578–81.
114. Sancak Y, Peterson TR, Shaul YD, Lindquist RA, Thoreen CC, Bar-Peled L, Sabatini DM. The rag GTPases bind raptor and mediate amino acid signaling to mTORC1. Science. 2008;320(5882):1496–501. doi:10.1126/science.1157535.
115. Sancak Y, Bar-Peled L, Zoncu R, Markhard AL, Nada S, Sabatini DM. Ragulator-Rag complex targets mTORC1 to the lysosomal surface and is necessary for its activation by amino acids. Cell. 2010;141(2):290.
116. Feng Z. p53 regulation of the IGF-1/AKT/mTOR pathways and the endosomal compartment. Cold Spring Harb Perspect Biol. 2010;2(2):a001057.

117. Feng Z, Levine AJ. The regulation of energy metabolism and the IGF-1/mTOR pathways by the p53 protein. Trends Cell Biol. 2010;20(7):427–34.
118. Kahn BB, Alquier T, Carling D, Hardie DG. AMP-activated protein kinase: ancient energy gauge provides clues to modern understanding of metabolism. Cell Metab. 2005;1(1):15–25.
119. Gwinn DM, Shackelford DB, Egan DF, Mihaylova MM, Mery A, Vasquez DS, Turk BE, Shaw RJ. AMPK phosphorylation of raptor mediates a metabolic checkpoint. Mol Cell. 2008;30(2):214–26.
120. Tanti J-F, Jager J. Cellular mechanisms of insulin resistance: role of stress-regulated serine kinases and insulin receptor substrates (IRS) serine phosphorylation. Curr Opin Pharmacol. 2009;9(6):753–62.
121. Shi Y, Yan H, Frost P, Gera J, Lichtenstein A. Mammalian target of rapamycin inhibitors activate the AKT kinase in multiple myeloma cells by up-regulating the insulin-like growth factor receptor/insulin receptor substrate-1/phosphatidylinositol 3-kinase cascade. Mol Cancer Ther. 2005;4(10):1533–40. doi:10.1158/1535-7163.mct-05-0068.
122. O'Reilly KE, Rojo F, She QB, Solit D, Mills GB, Smith D, Lane H, Hofmann F, Hicklin DJ, Ludwig DL, Baselga J, Rosen N. mTOR inhibition induces upstream receptor tyrosine kinase signaling and activates Akt. Cancer Res. 2006;66(3):1500–8.
123. Lane HA, Breuleux M. Optimal targeting of the mTORC1 kinase in human cancer. Curr Opin Cell Biol. 2009;21(2):219–29.

Chapter 11
Maternal Energetics and the Developmental Origins of Prostate Cancer in Offspring

Emily C. Benesh and Kelle H. Moley

Abstract In 2014, Prostate cancer was the most commonly diagnosed cancer in men in the Western world, but few modifiable risk factors exist for this disease. Many studies have examined the association between obesity and prostate cancer incidence, producing conflicting results. However, maternal energetics during periconception and gestation has barely been considered as a potential risk factor, despite the fact that prostate tissue patterning and early development occur in utero. Maternal diabetes and obesity dramatically affect health at the time of pregnancy, and rodent and human studies demonstrate that the effects manifest in a variety of noncommunicable conditions in offspring, including early initiation of mammary tumors. Importantly, maternal diabetes and obesity directly alter the health of the early embryo and the maternal germ cell, or oocyte. Obesity-exposed oocytes exhibit alterations in size, gene expression, chromosome structure, and metabolism, and adult offspring present with defects including hyperplastic prostate tissues. We propose here that maternal obesity alters epigenetic and metabolic functions in the oocyte, which are passed to offspring, sensitizing them to precancerous conditions. Such an in utero environment could provide a "first hit" for prostate cancer development and should be considered when making dietary recommendations for expectant mothers.

Keywords Maternal obesity · Maternal undernutrition · Prostate cancer · Oocyte · Epigenetics · Metabolism · Noncommunicable disease

K. H. Moley (✉)
James P. Crane Professor, Department of Obstetrics and Gynecology, Washington University in Saint Louis, School of Medicine, 425 South Euclid Ave., BJC-IH 10th Floor, Campus Box 8064, Saint Louis, MO 63110, USA
e-mail: moleyk@wudosis.wustl.edu

E. C. Benesh
Department of Obstetrics and Gynecology, Washington University School of Medicine, St. Louis, MO 63110,

N. A. Berger (ed.), *Murine Models, Energy Balance, and Cancer*,
Energy Balance and Cancer, DOI 10.1007/978-3-319-16733-6_11

Introduction

Cancer is a condition in which abnormal cells divide without control and eventually invade other tissues [1]. Cells become abnormal through alterations in genes and gene products (e.g., pro-proliferative mutations) that change cellular behaviors and increase the fitness of the affected cells relative to their neighbors. The etiology of cancer within a given organ system is highly complex, and this complexity is compounded by the more than 100 types of cancers that arise through distinct molecular mechanisms [1]. Few cancers derived from the same organ are identical, and the perturbations that lead to cancer vary in each individual. Cancer can be initiated by inherited mutations, spontaneously arising or carcinogen-induced mutations, autonomous gene expression changes occurring during normal division, or cancer-permissive perturbations in an organ system environment (e.g., chronic inflammation, metabolic syndrome) [1]. Additionally, these causal factors are not mutually exclusive. Thus, pinpointing the perturbations that initiate a particular cancer and identifying generalizable risk factors for cancer subtypes are extremely challenging propositions.

Prostate adenocarcinoma (PCa) is a major concern for the Western world. In 2014, PCa was expected to be the most commonly diagnosed cancer in the USA (233,000 new cases; 14 % of all cancers) [2]. Approximately 15.3 % of American men will be diagnosed with PCa over their lifetime. In 2010, PCa-associated costs were $ 11.9 billion in the USA, alone [3]. Preventing PCa is particularly challenging because the strongest risk factors for the disease (age, ethnicity, and genetic inheritance) are not modifiable [2]. Small positive associations, such as with vitamin E supplementation [4], have been identified in human population studies, but data establishing other potential risk factors have been largely unreliable. For example, dietary supplementation with 1 mg daily of folic acid, the synthetic form of the nutrient folate, positively associated with prostate cancer [2]. However, adequate consumption of folate, gained from whole foods rather than supplementation, was protective against PCa incidence [5, 6]. Similarly, a diet high in dairy foods and calcium has been shown to positively associate with prostate cancer risk, but the Swedish Apolipoprotein MOrtality RISk study demonstrated a weak negative association between dietary calcium and prostate cancer when correcting for serum albumin, a protein to which calcium binds [7–9]. These examples illustrate the conflicting data that are pervasive throughout the studies of prostate cancer dietary risk factors.

Inconsistent PCa risk factor data may be best exemplified in the body of epidemiologic and experimental literature seeking to define the link between PCa and obesity at the time of diagnosis. Such studies began as early as 1976 [10], yet the links have been inconsistent [11]. Recent meta-analyses propose that body mass index (BMI) at diagnosis has either no effect on PCa incidence or may be protective [12, 13]. In contrast, in a man diagnosed with prostate cancer, obesity has been consistently associated with increased risk of *advanced* PCa, cancer recurrence, and mortality [12–16]. Clearly, the role of obesity in the etiology of PCa is complex, and it is very difficult to pinpoint an individual nutrient that is a definitive culprit. There are at least three possible explanations for the inconsistencies surrounding obesity and PCa incidence. First, the results from population studies fail to account for the differences in detection, treatment strategies, and drug efficacy in obese versus lean patients. For example, Allott et al. propose that the association between obesity and advanced prostate cancer in American men is indirect; prostate cancer is detected later in obese men because of obesity-driven changes in prostrate-specific antigen (PSA) level and ineffective digital rectal exams [16]. Second, it is unclear whether exposure to individual dietary components consumed by the patient directly initiate signaling events and changes in gene expression, or whether the prostate phenotypes are secondary consequences of comorbidities associated with obesity, or both. Finally, the time window of exposure in which obesity most influences the health of the prostate is unknown. Together, these factors make standardization of observational data and experimental techniques unreliable, leading to the high level of variability.

In humans, prostate tissue undergoes normal growth and development at two points in the life course. As described by Humphrey (2003), the human prostate originates during embryonic development from the urogenital sinus region and is first visible at 9 weeks' gestation [17]. During this time, androgen hormone signals initiate prostatic ductal morphogenesis and differentiation. This ductal differentiation stimulates the urogenital sinus mesenchyme to develop into smooth muscle cells that surround the ducts. The signaling interplay between the differentiating prostatic epithelium and urogenital mesenchyme regulates a coordinated interplay of ductal budding, branching morphogenesis, proliferation, and secretory function. At 13 weeks' gestation, many primary ducts are present, and between 20 and 30 weeks, the bud stage occurs in which solid cellular buds cap the ends of ducts. From 31 to 36 weeks, both the cellular buds and prostate glands, or acini, appear and characterize the bud-tubule stage. From 36 weeks through the end of gestation, prostate acinotubular structures become organized into lobules. After birth, the immature prostate settles into a semi-quiescent state in which little growth occurs, but prostate ductal epithelia are generated and branching morphogenesis occurs [17]. At puberty, the prostate transitions into a second period of rapid proliferation driven by androgens, which continues until the end of puberty when the prostate has reached adult size. Immature acinotubular structures differentiate into the adult basal and luminal epithelial cells. A healthy prostate will then usually maintain its postpubescent size for 25–30 years with proliferation occurring to replenish functional prostatic epithelial cells that are sloughed off during the normal secretion of prostatic fluid. The adult human prostate is morphologically separated into the peripheral, transition, and central zones (Fig. 11.1a). Prostatic carcinoma arises in the peripheral zone, whereas after age 50, prostate gland size may increase in the transition zone due to benign prostatic hyperplasia [18].

Alternatively, in rodents, the tubule-ductal prostate is separated into ventral, anterior, dorsal, and lateral lobular structures that encircle the urethra directly posterior to the bladder, and do not spontaneously develop prostatic carcinoma (Fig. 11.1b). Rodent prostate buds first appear in late embryogenesis and undergo rapid growth and differentiation until 3 weeks of age; more than 5 weeks prior to sexual maturity

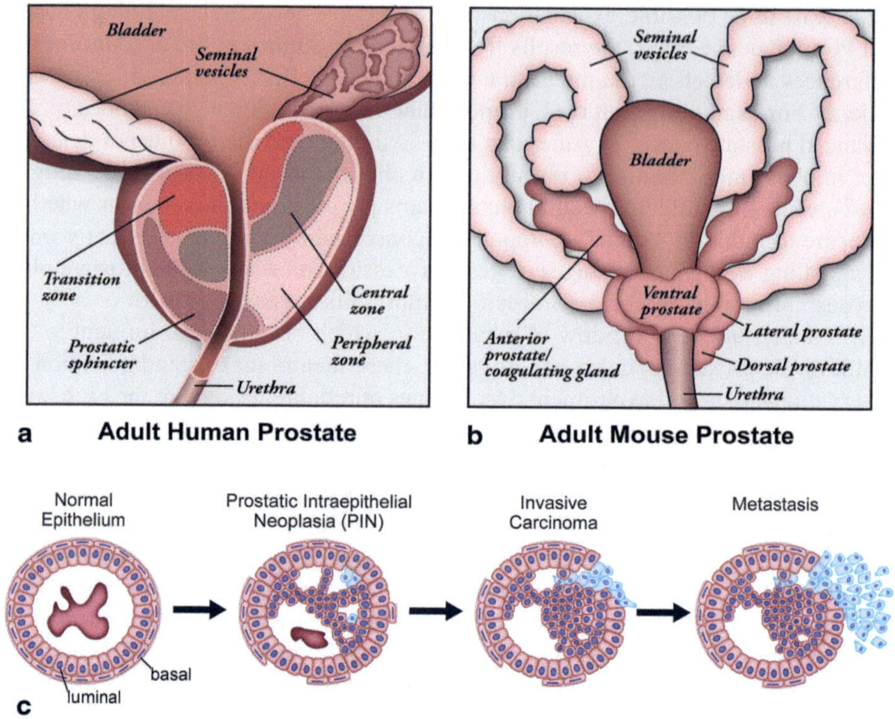

Fig. 11.1 An illustration of human (**a**) and mouse (**b**) prostate glands was adapted from [18]. Human prostate tissue is histologically and spatially organized into peripheral, central, and transition zones (**a**). Mouse prostate is similarly oriented around the bladder but exists in four pairs of tubule-ductal glandular structures: the anterior (or coagulating gland), ventral, lateral, and dorsal lobes (**b**). Individual mouse and human prostate glands share cellular anatomy in both normal tissue and cancer progression phenotypes (**c**). In normal prostate glands, a monolayer luminal epithelial secretory cells surround an open lumen into which prostate fluid is secreted, and are supported by a monolayer of basal and neuroendocrine cells, as well as a surrounding basal lamina. As prostate cancer progresses, some transformed cells (*blue cells*) proliferate, invade the surrounding stroma, and eventually metastasize to remote sites, such as bone

[18]. Importantly, when prostate cancer is initiated in rodent models, the progression of the disease in the dorsal and lateral lobes recapitulates human disease phenotypes (Fig. 11.1c). This makes rodent models powerful tools to investigate the contribution of genetic and environmental disturbances to prostate cancer outcomes.

Interestingly, a life-course approach assessing anthropometric measurements of body size in humans suggested that early-life body size was correlated to prostate cancer incidence [19]. Because the prostate develops during gestation and early life, some reviews have suggested that exposure to external factors, such as obesity, during early life (under 30 years of age) are relevant risk factors for prostate cancer [11, 20]. Furthermore, early time windows of exposure could explain the inconsistent associations between prostate cancer and obesity of the patient at diagnosis.

Despite these compelling motivations, few studies have experimentally investigated prenatal and prepubescent exposure to obesity and diabetes and prostate cancer outcomes.

In this chapter, we will explore the possibility that development of chronic diseases such as prostate cancer are affected by maternal energetics, particularly overnutrition due to an obesogenic diet, during gestation. The fetal origins of adult disease hypothesis states that exposures occurring during fetal life are critical to the developmental patterning of tissues in offspring [21]. This has been best exemplified by studies identifying cardiometabolic defects in offspring of mothers who were obese during gestation (extensively reviewed in [22–25]). However, diseases other than cardiometabolic conditions have also been linked to deregulated maternal energetics including asthma, atopic conditions, behavioral disorders such as attention deficit hyperactivity disorder and autism, mammary tumor development, and prostate hyperplasia. Interestingly, mechanistic studies suggest that generational transmission of all of these conditions may be attributed to alterations in epigenetic programming or metabolic defects. Because prostate tissue first develops in utero, a time window when both epigenetic marks and metabolic programming of the offspring are set, it is conceivable that prostate health is also patterned during gestation. To explore this possibility, this chapter includes: (1) a brief overview of epigenetic and metabolic mechanisms that can be altered during gestation, (2) a description of the phenotype of an obese pregnancy, (3) a discussion of the noncommunicable diseases (NCDs), including cancer, in which maternal nutrition is an influential regulator, and (4) presentation of a unifying proposal: inappropriate development of the oocyte is the origin of both epigenetic and metabolic perturbations that manifest as prostate hyperplasia and, potentially, other NCDs in offspring.

Regulatory Mechanisms That Can Be Altered During Gestation

Epigenetics

In 1953, C. H. Waddington coined the term "epigenetics", which he defined as the "study of mechanisms of temporal and spatial control of gene activity during the development of complex organisms" [26]. Since those early studies, the field of epigenetics has broadened and is now defined as regulation of gene expression that occurs without altering the DNA code. Underscoring their critical roles in normal organism health, epigenetic regulatory systems are frequently deregulated in chronic diseases [27]. Currently, epigenetic regulatory mechanisms are categorized in one of three general paradigms: (1) DNA methylation changes that act as on/off switches regulating DNA transcription, (2) histone modifications that alter the accessibility of DNA for transcription, and (3) noncoding microRNAs that can alter transcriptional properties of genes or posttranscriptional utilization of newly synthesized mRNAs [28].

DNA Methylation

DNA methylation, the transfer of a methyl group from *S*-adenosyl-methionine to the C-5 position of cytosine, is responsible for many changes in promoter activity and subsequent gene expression [29]. DNA methylation is carried out by DNA methyltransferase (Dnmt) enzymes such as Dnmt1, which performs maintenance methylation (Fig. 11.2), and Dnmt3a and Dnmt3b, which perform de novo methylation [27]. CpG dinucleotides are present throughout the DNA sequence but are enriched in short "CpG island" regions in the promoters of approximately 60 % of genes. The common pattern of methylation of an active gene is hypomethylation of promoter CpG islands and hypermethylation of CpG dinucleotides in the corresponding coding region [30, 31]. Methylation of a promoter silences gene expression through the methylation-dependent binding proteins, such as methyl-CpG binding proteins (MBPs) and CCCTC-binding factor [27, 32]. For example, the presence of methyl-

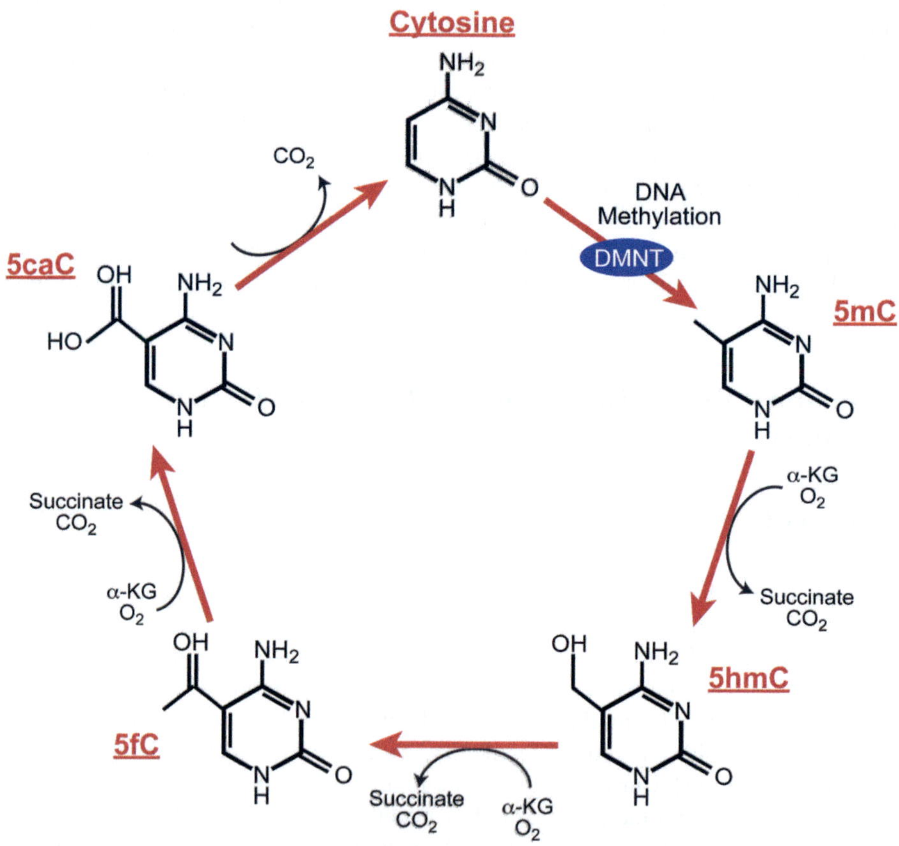

Fig. 11.2 An illustration of the methylation reaction is depicted in which Dnmt1 transfers a methyl group to cytosine. Through a series of intermediate steps, 5-methyl cytosine can be subsequently oxidized back to unmethylated cytosine

ated CpG dinucleotides recruits MBPs that repress gene expression both directly through transcriptional repression domains and indirectly by recruiting repressor proteins and influencing chromatin conformation [33]. Methylation marks are removed by proteins such as the ten-eleven translocation family of enzymes, which oxidize 5-methyl cytosine (5mC) to 5-hydroxymethylcytosine (5hmC), oxidize 5hmC to 5-formylcytosine (5fC) then to 5-carboxycytosine (5caC), and convert this back to cytosine, completing the methylation cycle (Fig. 11.2; [34]).

Histone Modification

Chromatin, or the nuclear complex of DNA and DNA-organizing proteins, is a dynamic structure that is amenable to external cues. The basic unit of chromatin, the nucleosome, comprises 146 DNA base pairs wrapped around a histone octamer made of two each of histones H2A, H2B, H3, and H4 [35]. Chromatin organization in the nucleus is dependent on the activity of the DNA and vice versa; DNA in euchromatin is accessible to transcriptional machinery, whereas DNA in heterochromatin is tightly associated with histones and is not transcribed. Although factors such as the stage of the cell cycle play an important role in the accessibility or compactness of chromatin, stimulated expression of specific genes requires local changes in the chromatin structures [36, 37]. Posttranslational modifications of histone proteins are epigenetic occurrences: they are regulatory factors that control the accessibility of a gene to transcription factors but do not change the DNA sequence. The pattern of histone modifications has been coined "the histone code" [38]. Modifications are posttranslationally added or removed from histone tails, small domains that protrude from the nucleosome [39], by specific enzymes. One such enzyme, histone methyltransferase, can add up to three methyl groups to histone H3 at lysine 27. Histones can also undergo other modifications including acetylation, phosphorylation, sumoylation, ubiquitination, ADP-ribosylation, deamination, and proline isomerization [40]. Many histone posttranslational modifications contribute to the formation of open or closed chromatin. For example, methylation of histone H3 at lysine 27 is generally considered a repressive chromatin mark that inhibits gene expression [41]. Given the variety of specific combinations of marks and the number of marks that can occur on a given histone residue, the number of potential combinations is vast. This variety is thought to provide epigenetic fine-tuning of chromatin accessibility and contribute to spatial and temporal patterning of gene expression [42].

MicroRNA and RNA Interference

MicroRNA (miRNA) molecules, first discovered in the 1990s in the nematode *Caenorhabditis elegans,* are approximately 22-base pair noncoding RNAs [43]. Multiple functions have been identified for the more than 1000 individual miRNAs that have been discovered; most are related to the control of gene posttranscriptionally.

Mature miRNAs are generated from longer transcripts and often function through association with the RNA-induced silencing complex [44]. One canonical function of miRNA is to inhibit the translation of select mRNA molecules. The sequence of a given miRNA is at least partially complementary to the specific mRNA target, thus allowing it to hybridize with the mRNA. This binding initiates a number of inhibitory functions, depending on the miRNA, such as degradation of mRNA or new polypeptides, ribosome drop-off, and inhibition of translation initiation or elongation [43, 45]. Interestingly, miRNAs also may initiate transcription and translation of some targets [44]. Altogether, miRNAs provide a host of potential mechanisms to further fine-tune gene expression in response to the environment.

Fetal Programming of Epigenetic Regulation

Each of these three epigenetic mechanisms functions to pattern the development of an individual organism [46], and the mechanisms are coordinated to regulate particular developmental programs [47]. Furthermore, epigenetic regulation can be programmed during early life by fetal exposures [48]. The best example of epigenetic regulation by maternal energetics during early life is embryonic methylation patterning. Semipermanent DNA methylation marks are patterned during early embryogenesis and modified throughout gestation (Fig. 11.3). The program in which methylation is set in utero was nicely summarized by Edwards and Peterson Myers [49]. During gametogenesis, all methylations, including parent-of-origin-specific, mono-allelically expressed, or "imprinted," methylation patterns are reset [50]. Most well-characterized imprinted genes regulate fetal growth and development. Shortly after fertilization, the embryo demethylates most of its DNA, except for the imprinted genes, and remethylation occurs between cleavage and gastrulation. In female embryos, X chromosome inactivation (via promoter methylation) occurs between implantation and organogenesis. Primordial germ cells, which develop at gastrulation, proliferate and undergo migration to the genital ridge where they then globally demethylate, including imprints and inactive X chromosomes. The germ cells remethylate later, during gametogenesis, to complete the cycle. In the somatic cells, methylation marks dynamically modify the genome throughout the life of the organism.

Because methylation is so dynamic in utero, it is unsurprising that offspring methylation patterns are sensitive to maternal exposures during specific gestational time windows. Maternal energetics has been demonstrated to deregulate the embryonic methylation process for specific genes. For example, maternal diets lacking nutrients critical to the DNA methylation machinery (e.g., folic acid) are causally linked to impaired methylation in offspring and subsequent neural tube defects [51]. Likewise, imprinting of the insulin-like growth factor 2 (*IGF2*) gene is deregulated in obesity-exposed large animal models [52]. As *IGF2* expression is linked to many NCDs, including prostate cancer and atherosclerosis [53, 54], this suggests that maternal obesity critically regulates epigenetic mechanisms in offspring that are linked to the later development of NCDs.

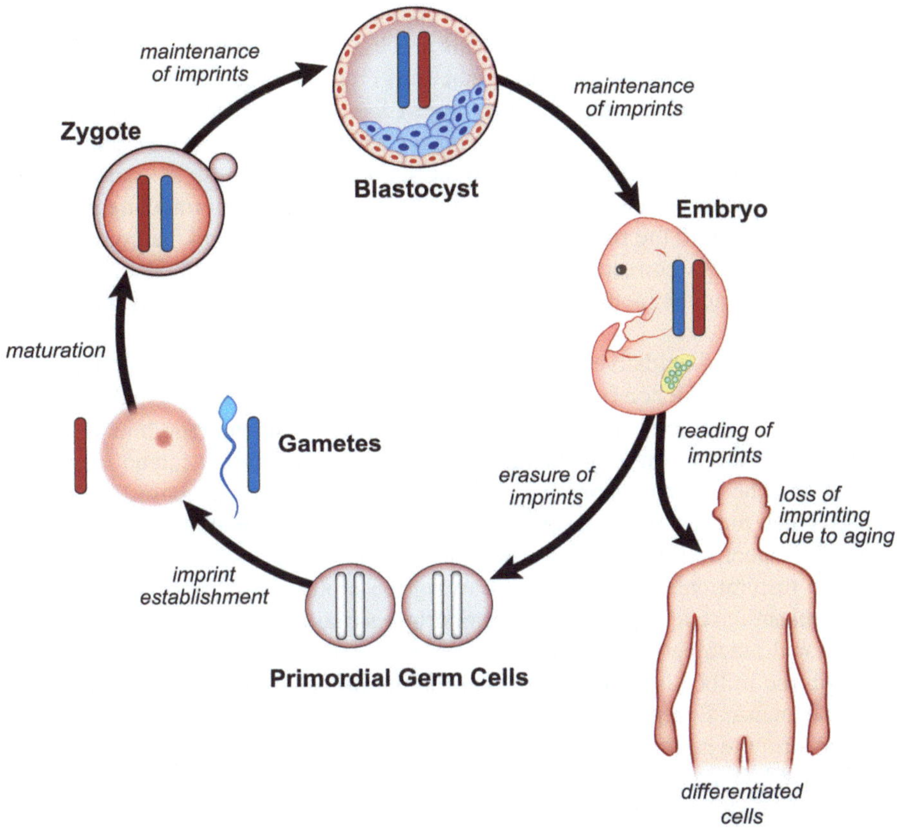

Fig. 11.3 The patterning process of methylation-based epigenetic gene regulation in humans. Primordial germ cells are devoid of all methylation marks, which are acquired during gametogenesis. With the exception of imprinted genes, demethylation occurs during fertilization and remethylation during early embryogenesis. In the offspring, primordial germ cells are demethylated until gametogenesis, while all other cell-type specific methylations are maintained and read to instruct gene expression. Methylation of genes including imprints can be gradually lost due to aging

Metabolism and Mitochondria

Metabolism is the cellular process by which energy is converted through catabolic respiration from ingested or stored nutrients to adenosine triphosphate (ATP), which is used to perform the functions of the organism. In addition to catabolic respiration, anabolic pathways convert macromolecules into other organic compounds needed by the body. In a healthy individual, the interplay of respiration and biosynthetic processes is regulated by supply and demand. The major substrate of catabolic metabolism is free glucose which is broken down through a series of enzymatic conversions resulting in capture of energy in ATP molecules. ATP production occurs through both the highly efficient oxidative phosphorylation process of the Krebs cycle coupled with the electron transport chain, which generates carbon dioxide, and the nonoxidative inefficient glycolysis process, which generates pyruvate. ATP is then used to catalyze the molecular functions required of the cellular machinery. Importantly, mammals do not regularly ingest free glucose. Rather, they consume foods that comprise ratios of macromolecules such as fats-, proteins-, and carbohydrates such as disaccharides and polysaccharides. All of these macromolecules are digested and either stored, or converted into glucose or intermediate products that are then fed into the respiration cycle [36]. The rate at which macromolecules are digested, stored, converted, and utilized differs in each individual and is influenced by many factors including genetics, epigenetics, exercise pattern, and dietary makeup.

ATP-generating oxidative phosphorylation and the electron transport chain both occur within the mitochondria; hence, this organelle is commonly referred to as "the powerhouse of the cell." The mitochondrion is a double-membrane bound organelle found in most eukaryotic cells that specializes in facilitating catabolic aerobic respiration. The inner membrane is infolded into cristae that increase the surface area in which mitochondrial proteins can perform the electron transport chain. Mitochondria-specific ribosomes translate the mitochondrial proteins participating in catabolic respiration in the cytosol, allowing the proteins to bypass processing in the endoplasmic reticulum. The mitochondrion contains DNA, which encodes the translational machinery for the mitochondrial proteins, and is enclosed inside the cristae in the mitochondrial matrix. Importantly, mitochondria replicate by binary fission; in mammals replication is driven by the respiratory needs of a given cell and mitochondria are randomly dispersed to daughter cells during cytokinesis and cell division [36]. In a zygote, the mother is the sole contributor of mitochondria and mitochondrial DNA to offspring [55]. Thus, the efficiency of mothers' mitochondria may have germ-line effects on metabolic activity in her offspring.

Fetal Programming of Metabolism

Although proper metabolic function is absolutely essential to health, the overall function and efficiency of metabolic processes vary between individual organisms. Strong evidence suggests that this variation is regulated by in utero exposures. For example, maternal energetics can influence macromolecular substrate utilization in the mother and downstream metabolic function and patterning in offspring. In murine models, ingestion of a high-fat diet alters the mother's lipid storage and metabolism rates by changing serum fatty acid levels and the expression of enzymes involved in lipid biosynthesis and metabolism in the liver [56]. The male offspring of mice fed a high-fat diet exhibit increased lipid storage (adiposity) and glucose intolerance, which is indicative of metabolic inefficiency. Humans exposed to maternal obesity also exhibit glucose intolerance, increased adiposity, and resistance to insulin, the circulating hormone that stimulates glucose uptake and metabolism or storage by peripheral tissues [57]. The risk of metabolic defects is also substantially increased in animals exposed to maternal undernutrition (e.g., from famine) [58].

Thus, the evidence strongly supports the importance of a balanced in utero environment for the transmission of healthy metabolic function to offspring.

There are two possible, and likely interdependent, mechanisms by which maternal energetics can influence the metabolism of offspring. First, the epigenetic regulation of metabolic enzymes is altered in obesity-exposed offspring [59, 60]. For instance, in a murine maternal obesity model, exposed offspring upregulated expression of the rate-limiting fatty acid synthesis enzyme stearoyl-CoA desaturase [61], and methylation of hepatic fatty acid desaturase 2 occurred in maternal obesity-exposed rats [62]. Additionally, the histone activities of regulators of fat metabolism and storage, adiponectin and leptin, were altered in high-fat exposed animals [63]. Second, mitochondrial ultrastructure is impaired in these models, leading to decreased mitochondrial density and membrane potential [61]. Studies from our laboratory and others have demonstrated ultrastructural defects in mitochondria of high-fat diet-exposed oocytes [64]. Epigenetic patterning and mitochondrial function are disrupted by maternal diet leading to cardiometabolic defects; the most common NCD manifestation of maternal obesity on offspring.

Maternal Energy Balance and Pregnancy

A healthy pregnancy is influenced by extrinsic factors. One well-known extrinsic factor is maternal smoking, which increases risks for low fetal birth weight, placental abnormalities, and chronic hypertension and death of the fetus [65]. As early as 1945, maternal energetics has been recognized as another extrinsic factor that contributes to the health of the pregnant woman and child [66–68]. Although the relevance of maternal underweight (defined by the Institute of Medicine (IOM) as a BMI of less than 19.9 kg/m^2) is less clear, maternal overweight (BMI: 25.0–29.9) and obesity (BMI $\geq$30.0) are consistently correlated with worsened pregnancy outcomes and increased perinatal death [69, 70]. In 2010, the World Health Organization estimated that 25 % of all reproductive-age (18–44) women globally were obese [71], and recently, the American College of Obstetricians and Gynecologists and the Centers for Disease Control estimated that 60 % of women are overweight or obese at the beginning of pregnancy [72, 73]. Furthermore, more than 70 % of women gain more than the IOM-recommended amount during pregnancy [70], putting them at risk of worsened maternal and fetal health [74]. Thus, perinatal overweight and obesity are serious global health concerns.

Pre- and perinatal obesity are associated with a number of conditions that put the health of both the mother and the fetus at risk. Maternal overweight and obesity are positively associated with risks for gestational diabetes mellitus, hyperglycemia, gestational hypertension, preeclampsia, caesarean section, spontaneous abortion, and postpartum weight retention [72, 75, 76]. During gestation, obesity-related complications include: difficultly with monitoring of fetal health, wound infection, thromboembolic disorders, and urinary tract and kidney infections [66, 72, 77]. Development of one of these conditions frequently correlates with the codevelopment of other morbidities. For example, obesity is a risk factor for maternal hypertension, and both obesity and hypertension are independent risk factors for preeclampsia and gestational diabetes [73]. Thus, pregnancies can more easily become very high risk in the presence of maternal obesity. Additionally, obese women experience complications during and after birth such as early induction of labor, preterm delivery, impaired epidural administration, postpartum hemorrhage, poor recovery, increased neonatal intensive care unit admission, and decreased likelihood of breastfeeding [72, 78, 79]. Finally, collective studies associate maternal obesity with miscarriage and deaths occurring from the perinatal to infant periods (gestational week 20 through 1 year old) [69].

Infants born to obese mothers also exhibit a variety of disorders and complications. These infants are at increased risk of growth disorders including either low or high birth weight [80]. In obese women, fetal weight is difficult to estimate with the current technologies, so intrapartum macrosomia (birthweight >4000 g) often goes undiagnosed [72]. Large-for-gestational-age infants have increased the risk of physical injuries during labor and delivery, such as shoulder dystocia [80]. In fetal and neonatal stages, macrosomia is a risk factor for hyperglycemia, hyperinsulinemia, and unbalanced electrolyte and metabolite concentrations [80]. Finally, the macrosomic neonate is at risk of obesity as a child, adolescent, and young adult, which in turn is a risk factor for cardiovascular and metabolic disorders in adulthood [77].

Congenital defects are also a concern when mothers are obese [72, 81]. In a comprehensive meta-analysis, Stothard et al. found that the infants of obese mothers were more likely than the infants of normal-weight mothers to present with a variety of structural complications including the neural tube defects such as spina bifida, cleft lip and palate, anorectal atresia, hydrocephaly, and cardiovascular, septal, or limb reduction anomalies [81]. In 2010, 0.6 % of the births resulted in mortality [82], and of those, the leading cause of death (21 %) was congenital malformations. It is important to note that these defects may go undetected in fetuses of obese mothers because of the limitations in ultrasonography [72].

Children born to mothers suffering from obesity and associated comorbidities grow into adults that face higher incidence of many chronic diseases. The best-studied conditions are those that follow logically from maternal obesity: cardiovascular and metabolic disorders [79, 83]. Briefly, maternal obesity is a very strong indicator of infant, adolescent, and adult overweight, obesity, and increased adiposity. Offspring of obese mothers have an increased likelihood of deregulated insulin resistance and impaired glucose/insulin homeostasis at all ages through young adulthood [22, 84]. They also exhibit hypertension, hyperlipidemia, and decreased concentrations of high-density lipoproteins [85]. Neonatal offspring of high BMI mothers exhibit increased intrahepatocellular lipid content, which may associate with childhood nonalcoholic fatty liver disease [86]. Furthermore, maternal-obesity-exposed children exhibit chronic inflammation, which is a risk factor associated with cardiometabolic disorders [87].

The transmission of chronic cardiometabolic disorders from mother to child could have multiple causes. First, as the presence of obesity in the offspring is itself a risk factor for cardiovascular disease, hypertension, insulin resistance, diabetes, and hyperglycemia (collectively called the "metabolic syndrome"), generational transmission may be secondary to obesity that develops in offspring as a result of maternal-induced changes in eating behaviors, fat storage, and metabolism. This idea is supported by the finding in animal models that maternal obesity alters the ratio of appetite/satiety genes expressed in the hypothalamus, which controls appetite, and changes the responses of hypothalamic neurons to satiety-stimulating hormones such as leptin [88]. This idea leads to a second possible explanation: that impaired hypothalamus development directly controls the onset of hypertension. Evidence for this comes from data showing that leptin signaling to the hypothalamus is a causal factor of obesity-related hypertension because it regulates the activity of the efferent sympathetic neural pathways [89].

Third, germ-line or tissue-specific epigenetic mechanisms have been implicated as a mechanism by which metabolic phenotypes could be transmitted from mother to offspring [90]. For example, placental tissues and cord blood collected from insulin-dependent diabetic, non-insulin dependent diabetic, and nondiabetic mothers had differential DNA methylation of growth- and glucose homeostasis-related genes [91]. Additionally, inflammatory pathway genes are differentially methylated in maternal obesity-exposed offspring. In fact, a genome-wide methylation study of placenta and cord blood collected from diabetic mothers reported that 11 % of the genes that were differentially methylated were involved in metabolic regulation [92]. Fourth, the environment in which the child is raised during early life and adolescence, such as the availability of energy dense foods, absence of safe play areas, and parental regulation of eating habits, contributes significantly to the risk of obesity and obesity-related comorbidities in adults [93]. These four explanations are not mutually exclusive. Likely the increased prevalence of the metabolic syndrome in Westernized populations is attributed to many or all of these factors in each affected individual.

Maternal Energy Balance and Cancer

In addition to the onset of metabolic syndrome, maternal gestational obesity influences the risk of several other chronic conditions in offspring, including neurodevelopmental disorders (e.g., attention deficit hyperactivity disorder and autism spectrum disorders) [94], impaired circadian rhythms of the hypothalamus and adipose tissues [95, 96], and asthma [97]. However, each of these could be the subject of an entire chapter; we focus here on the link between maternal energetics and cancer predisposition.

Many cancers are initiated by gene expression changes in cells, environmental modifications that provide fitness to cells in particular niches, or both. Additionally, most cancers are undetectable in early stages. Thus, cancer initiation and outcome are particularly difficult to predict. Despite these difficulties, some shared hallmarks of many types of cancers have been identified. For example, most cancers undergo tissue-of-origin-specific changes in gene expression, leading to subsequent alterations that make transformed cells more robust than their unaffected neighbors [98]. Given that the patterning of tissue expression patterns can be negatively influenced by maternal energetics during periconception, maternal energetic imbalance could also lead to changes, such as epigenetic and metabolic defects, that predispose tissues to becoming cancerous.

Few studies have investigated maternal obesity as a risk factor for the development of cancers in offspring despite the fact that patient's BMI influences the outcomes of many cancers, including advanced prostate adenocarcinoma. Additionally, identification of modifiable risk factors for these malignancies is critical so as to develop preventative interventions. As early development is essential for the patterning of tissue behaviors that will persist throughout the life of the offspring (e.g., inflammation and metabolism), this time window is a prime candidate. However, prospective studies of early developmental exposures that may increase risk for cancer in adulthood would be extremely costly and challenging in humans. Retrospective studies are feasible, but they suffer from the fact that recall (e.g., asking 60-year-old prostate cancer patients how much their mothers weighed during pregnancy) at the time of diagnosis is unreliable. Nonetheless, some studies have been done in human and animal models to address whether the maternal energetics is a modifiable risk factor for cancer and which time window is the appropriate target for potential interventions.

Breast Adenocarcinoma

Breast cancer incidence has long been associated with maternal exposures [99], but the data regarding maternal energetics and generationally transmitted breast cancer risk in humans are mixed. For example, a study combining two small population datasets showed that prepregnancy weight gain, but not prepregnancy BMI, was positively associated with breast cancer risk in offspring [100]. Likewise, data from a study out of the Nurses' Health Study Cohorts I and II showed no evidence of correlation between maternal prepregnancy BMI or gestational weight gain and breast cancer risk in daughters [101]. On the other hand, a different large population study from the Nurses' Health Study Cohorts [102] and a meta-analysis [103] showed positive linear associations between increased birth weight and breast cancer risk, suggesting that metabolic conditions influenced by gestational obesity may in turn promote development of breast cancer in offspring. Unfortunately, such retrospective studies in humans are limited in their ability to precisely account for macromolecule consumption rates and precise time windows of exposure; they, therefore, provide conflicting results when evaluating dietary intake and cancer risk.

In rodent models, maternal obesity has a strong, consistent, positive correlation with generational transmission of mammary tumors. Rats spontaneously develop tumors after 18 months aging and are, thus, a useful model to study breast cancer. Additionally, rats can readily develop tumors earlier in genetic models or when treated with carcinogens such as 7,12-dimethylbenz[*a*]anthracene (DMBA) [104].

In these models, the effects of maternal obesity on the incidence of induced mammary tumors in offspring have been extensively evaluated, and the models have proven useful for mechanistic elucidation [105]. In 1997, Hilakivi-Clarke et al. demonstrated that feeding rat dams a high-fat diet (consisting of 46 % calories from the ω-6 polyunsaturated fatty acid [PUFA] linoleic acid) increased circulating estradiol levels in the dam and caused a twofold increase in DMBA-induced mammary tumor incidence in offspring exposed to chow diet after weaning [106]. Additional studies showed that transmission of mammary tumor risk extended to both the daughter and granddaughter generations [107], and that perinatal diet exposure was more important to the phenotype than postnatal exposure [108]. Interestingly, the effect of maternal high-fat diet exposure may be specific to plant-derived ω-6 PUFAs; the tumorigenic effects were not observed when the mothers consumed an animal-fat-based diet or when the mothers also consumed ω-3 PUFAs [109, 110]. Thus, during the perinatal time window, patterning of adult breast tissue may be exquisitely sensitive to exposure to certain fatty acids.

Transmission of mammary tumor risk to offspring may be due to both epigenetic deregulation of gene expression and alteration in tissue metabolism. For example, in the MMTV-Wnt1-transgenic mouse mammary tumor model, maternal high-fat diet exposure increased tumor number and size and affected insulin levels (hence altered metabolism) and the expression of oxidative stress markers [111]. Changes were also observed in the expression of tumor suppressor genes and oncogenes, and studies indicate that increased insulin sensitivity coordinates with the gene expression changes of the tumor regulating genes. Another study found that mammary tumors in offspring of obese mice overexpressed DNMT1, DNMT3a, and DNMT3b, suggesting that the DNA methylation machinery was deregulated [107]. Additional studies have demonstrated that the estrogen receptor gene was hypomethylated and subsequently overexpressed in offspring of obese mice, whereas the gene was nearly completely methylated in offspring of mice exposed to a low-fat diet [104]. As estrogen receptor signaling is a critical factor regulating the development of breast tissue and adenocarcinomas in rodents and humans [112], deregulated epigenetic patterning of the estrogen receptor gene and other related genes is a likely mechanism by which maternal diet influences breast cancer development in offspring.

Prostate Hyperplasia and Prostatic Adenocarcinoma

Many studies have investigated whether body weight at the time of diagnosis correlates with prostate cancer incidence in humans. Although these collective studies loosely suggest that body weight at diagnosis has a modest positive correlation with worsened prostate cancer outcomes, they have provided few definitive conclusions [11]. Studies have also evaluated the weights of prostate cancer patients during adolescence and have achieved only mixed results [20]. In several cohorts, increased cancer incidence was observed among males with a higher young adult BMI [113–115]. Conversely, BMI from ages 5 through 30 has been inversely associated with both prostate cancer development and secretion of early cancer markers such as prostate specific antigen [116–118]. Sutcliffe and Colditz recently summarized this body of literature and noted that few studies investigated the same nutrients, and many relied on recall of diet several decades prior to diagnosis, which is inaccurate [20]. Thus, more controlled studies are required to reach solid conclusions about the effects of adolescent obesity.

It is of little surprise that even fewer epidemiologic studies have attempted to investigate links between maternal obesity, gestational health, and prostate cancer outcomes in offspring. A Finnish cohort study recently found that maternal intercristal pelvic diameter, used as a proxy for maternal pubertal growth and subsequent circulating estradiol levels, positively correlated with prostate cancer outcomes in offspring [119]. Another group found that whereas birth size indicators, maternal age, socioeconomic status, and parity were not associated with prostate cancer risk, increased length of gestation was slightly protective, suggesting that events occurring at late gestation are important regulators of prostate health [120]. Similarly, another study suggested that prenatal factors that correlate with pregnancy hormone levels (e.g., prematurity and placental weight) may be relevant to prostate cancer incidence in offspring [121]. However, neither of these studies considered maternal BMI. Longitudinal studies covering the entire life course are greatly needed to determine whether the perinatal period is a critical developmental time window for prostate health in human offspring.

A growing body of literature has investigated the effects of maternal exposures during gestation on development of prostate cancer in rodents. When rat dams were exposed to the carcinogen 2-amino-1-methyl-6-phenylimidazo[4,5-b]pyridine (PhIP) during gestation, offspring developed preneoplastic prostate lesions, even if the offspring were not exposed to PhIP throughout adult life [122]. This indicates that the perigestational time window was critical to developing prostate tissue, and that the prostate was susceptible to long-term effects of carcinogen exposure. Similarly, high-fat diet exposure from periconception through weaning has resulted in worsened prostate health outcomes. ACI/Seg rats, which spontaneously develop atypical hyperplasia lesions and carcinoma in situ, and that were exposed to high-fat diet both maternally and throughout their lives, had significantly more lesions than either nonexposed animals or animals only exposed to the high-fat diet after weaning [123]. This finding suggests that the perinatal/early postnatal period is important for long-term prostate health [124]. Furthermore, our recent studies showed that maternal high-fat diet (consisting of 59 % kcal from palmitate-based fat) given from 1 month before gestation through weaning stimulated prostate hyperplasia and nuclear atypia in offspring, as determined by three independent quantification methods. The prostates of these animals exhibited posttranslationally hyperactivated protein kinase B (Akt) signaling due to permissive deactivation of the Akt antagonist, phosphatase, and tensin homolog (Pten). We also observed concomitant elevation in apoptosis that appeared to balance the hyperproliferative response. Importantly, changes in Pten/Akt signaling are hallmarks of human and rodent prostate cancers, suggesting that this hyperplasia could predispose offspring to cancer after additional genetic or carcinogenic hits. This study was the first to switch weaned offspring to a control chow diet for the remaining life, definitively narrowing the relevant window of exposure to the perinatal and lactation periods [125]. To our knowledge, these studies are the whole of the rodent literature on maternal high-fat diet exposures and generational transmission of prostate health.

A few studies have investigated prostate health in the context of maternal protein restriction. One found that maternal protein restriction delays prostate development but led to higher incidence of dysplasia and prostatitis in aged offspring [126]. Another group found that maternal protein restriction during lactation altered the concentrations of sex hormones, which are important for normal prostate growth [127]. Thus, the perinatal time period is important for prostate health in rodents, but more studies are required to confirm these findings and elucidate the underlying mechanisms.

Other Cancers

A recent study from the Helsinki Birth Cohort found that maternal obesity positively correlates with all cancer outcomes in human offspring [128]. Additionally, Walker and colleagues demonstrated that mice born to the mothers that were exposed to a high-fat diet had significantly more reproductive and pituitary tumors than controls [129]. However, no studies have yet been done to address the mechanism by which maternal obesity contributes to other cancers in rodents or humans. At best, maternal obesity can only be linked to outcomes of other cancers through secondary mechanisms. For example, a recent review proposed that maternal obesity influences obesity and weight gain in offspring and thereby predisposes offspring to colorectal cancer (obesity of the patient at diagnosis and weight gain in adult life are risk factors for colorectal cancer) [130]. These authors also suggested a similar mechanism for hepatic carcinoma. Clearly, more work is needed.

Maternal Energetics and Oocyte Health: A Foundation for Generational Disease Transmission

The information presented above suggests that some aspect of the in utero environment is highly sensitive to maternal energetics and plays a critical role in determining offspring health, but when is the critical period in gestation? Is there a single stage in which the type and extent of damaging exposure could produce a spectrum of effects in offspring and lead to the above-discussed NCDs? What mechanisms are responsible for long-term health outcomes of offspring, and how are they impaired in an obese pregnancy?

Evidence from our group and others have demonstrated that the health of the oocyte (the unfertilized female gamete) sets the foundation for the health of the offspring and is impaired by maternal diabetes and obesity [131–133]. Here, we will (1) briefly describe the stages of early embryonic development in human and mouse, (2) provide evidence that maternal obesity endangers offspring health as early as the oocyte maturation process and persisting throughout the pregnancy to affect outcomes after birth, and (3) discuss two unifying mechanisms—epigenetics and metabolic function—that in the gamete are sensitive to alteration by maternal energetics and may program chronic diseases such as prostate cancer in offspring.

The Stages of Folliculogenesis and Early Embryonic Development in Mammals

In the human female, oogenesis, or the development of oocytes, begins with primordial germ cell (PGC) formation around the fourth week of embryogenesis [36, 134]. PGCs migrate to the genital area to form the genital ridge, and by the third month, ovaries have formed. The diploid PGCs undergo rapid proliferation, which reduces their size, and differentiate into oogonia, which then undergo meiosis until they arrest in prophase I. At this point, a primordial follicle, a layer of supportive cells from the germinal epithelium, surrounds each primary oocyte. These groups of cells remain quiescent until the female undergoes puberty. At puberty, a primordial follicle is stimulated to mature into a preovulatory follicle in a 375-day process called folliculogenesis. During folliculogenesis, the growing follicle passes through stages that include division of the follicular granulosa cells, increased follicle size, and formation of the fluid-filled antrum. Once a preovulatory follicle has developed, a surge of hormones stimulates oocyte maturation, which consists of meiotic divisions progressing the oocyte from prophase I through metaphase II. At this stage, the nuclear envelope disappears in a process termed germinal vesicle breakdown [135]. After ovulation, the mature (MII) oocyte, surrounded by a supportive layer of cumulus cells, enters the oviduct and will only finish meiosis II upon being fertilized by a sperm, which should occur in the fallopian tube [36, 134].

After fertilization, the diploid zygote divides to form an embryo composed of two blastomeres. Equal division of the blastomeres continue for 3 days after fertilization as the embryo travels down the fallopian tubes to the uterine cavity. By the time the embryo enters the uterine cavity, it has formed into a densely packed group of approximately 32 cells, called an early morula. Next, fluid accumulates between the morula cells to form a central cavity; the embryo at this stage is called a blastocyst and contains approximately 100 cells. Over the next 5 days, the blastocyst implants into the endometrial epithelium of the uterus. From this point, differentiation of the embryonic cells begins. Organogenesis progresses throughout the first trimester, and the heart is beating by 4 weeks. During the second and third trimesters, the fetus increases in size until parturition, approximately 38 weeks after fertilization [36, 134].

Many aspects of the rodent estrous cycle and gestation are similar to those of humans, but there are some distinct differences. During the rodent estrous cycle, hormone signaling stimulates a group of primordial follicles to differentiate into preovulatory follicles. The oocytes increase in size and undergo nuclear maturation through meiosis I until being arrested in metaphase II until fertilization, when multiple oocytes are asynchronously released from the ovary. The oocytes move into the oviduct where they are fertilized, and after 5 days are implanted in the uterus through peristaltic movements that evenly space blastocysts along the uterine wall. The generalized stages of rodent embryonic development, from one cell zygote through organogenesis and growth, are similar to those of human development [136]. Because of these similarities and powerful rodent-based genetic tools, rodent models are frequently used to test the effect of exposures on gestation and health outcomes in offspring.

Evidence That Aberrant Maternal Metabolism Impairs Oocyte Quality and Offspring Health

Initial studies investigating the effects of maternal metabolic exposures on oocyte quality used a murine model of drug-induced type-I diabetes, and later studies observed similar phenotypes in the hyperglycemic Akita genetic model [137]. In these models, fewer oocytes reached germinal vesicle breakdown, and embryos resulting from the remaining mature oocytes exhibited developmental delays at the two-cell embryo, morula, and blastocyst stages; all of these outcomes were improved by insulin treatment [138, 139]. Culturing of wild-type two-cell embryos in specific components present in the diabetic milieu demonstrated that hyperglycemia and hyperketonemia, conditions found in uncontrolled diabetes, caused developmental delays in embryogenesis [140]. Furthermore, diabetes-exposed preimplantation embryos, as early as the preovulatory oocyte stage, were smaller than nonexposed embryos and exhibited delayed maturation and aberrant apoptosis of the supportive granulosa cells [131, 137, 141]. Preimplantation blastocysts derived from diabetic mothers also had altered metabolic profiles, including alterations in components of the AMP-activated protein kinase (AMPK) signaling pathway, decreased available ATP, and impaired glucose uptake and utilization [141–144]. The long-term outcome for oocytes exposed to uncontrolled type-I diabetes were growth retardation and congenital defects [145]. Importantly, one-cell zygotes or blastocysts that came from diabetic mothers and were transferred into nondiabetic mothers still demonstrated growth and malformation defects, suggesting that the effect occurs acutely in the oocyte. Indeed, diabetes-exposed oocytes had more mitochondria that were misshapen and abnormally distributed than did control oocytes [64]. Oocytes from diabetic mothers also demonstrated severe spindle defects and chromosome misalignments. Mitochondrial defects and increased apoptosis were also found in the oocyte-supporting cumulus cells [146]. Finally, glucose uptake by cumulus cells for utilization in the oocyte was impaired in a hyperglycemic environment, further contributing to impaired oocyte quality [147, 148].

More recent studies have investigated oocyte quality in a high-fat diet model of murine obesity. Dams exposed to a high-fat diet starting at 4 weeks of age develop obesity, hyperglycemia, and hyperlipidemia [133, 149]. After 16 weeks on the diet, ovaries from obese dams have altered estrous cycle patterning [150], more apoptotic ovarian follicles, and fewer and smaller MII oocytes [149]. With as short as 4 weeks of exposure to a high-fat diet, one-cell zygotes from obese dams exhibit developmental delays and reduced blastocyst survival rates [132, 133]. Oocytes from obese mothers also exhibit impaired chromosome alignment and spindle structures, as well as abnormally shaped mitochondria [132]. When the mother is obese, metabolic function of oocytes is impaired [151], and uteri and blastocysts exhibit pro-inflammatory and hypermetabolic RNA signatures [152]. Late embryonic and early postnatal offspring are smaller than those from nonobese mothers, but the embryos demonstrate catch-up growth and eventually surpass control-exposed littermates by early adulthood. This effect was also observed in blastocysts cultured in saturated fatty acid and transferred to control dams for the remainder of gestation [149, 153]. Offspring of high-fat-diet-exposed mothers exhibited fetal growth retardation, structural defects in the ventricle and choroid plexus in the brain, glucose intolerance, and increased cholesterol and body fat, despite exposure to a control diet after weaning [132]. Interestingly, maternal undernutrition also significantly reduces primordial and secondary follicle numbers in offspring [154], emphasizing the importance of balanced nutrition during gestation.

The mechanistic alterations observed in murine oocytes from diabetic and obese mothers might have strong implications for health outcomes of pregnancies in obese women. Obese women who undergo in vitro fertilization (IVF) with autologous oocytes have higher failure rates than nonobese women [155]. However, if oocytes from nonobese donors are used, then IVF success does not differ between obese and lean women [156]. Also, oocytes that fail to fertilize in obese IVF patients exhibit significantly higher numbers of spindle and chromosome alignment abnormalities than do failed oocytes from lean women [157]. Taken together, these results indicate that maternal obesity in rodents and humans dramatically affects the health of the oocyte, which manifests in metabolic, gene expression, and chromosomal defects.

Perspective—Maternal Energetics Influences Epigenetics and Metabolism to Program Prostate Cancer Risk in Offspring

We will now consider whether maternal obesity is a risk factor for prostate cancer. It is important to note that, given the small number of studies addressing this question, definitive answers cannot be reached. Our data demonstrate that maternal obesity did not lead to prostate adenocarcinoma in situ in a wild-type mouse model, which is not surprising given that mice do not spontaneously develop prostate cancer [125]. Similarly, most murine models of breast cancer incorporate additional carcinogenic hits to stimulate mammary tumor growth [158]; nevertheless, maternal obesity worsened the precancer phenotype in these models. Furthermore, we found that some high-fat diet-induced gene expression changes in the prostate mirrored those observed in prostate cancer in both humans and rodents, suggesting that maternal obesity predisposes offspring to prostate cancer. These results provide a

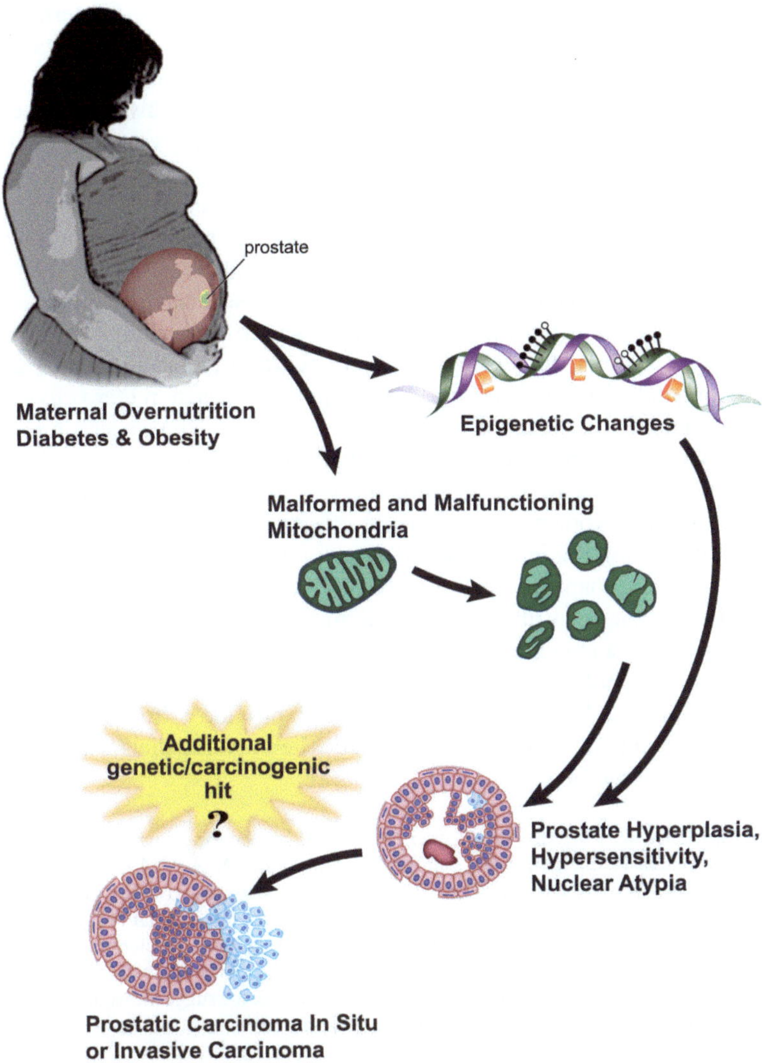

Fig. 11.4 A schematic of possible maternal influences on prostate health in offspring. In maternal obesogenic environment, offspring prostate cells could inherit malfunctioning mitochondria and/or deregulated epigenetic signatures. These alterations contribute to a hypersensitive prostatic environment, characterized by hyperplasia and nuclear atypia. With additional genetic or carcinogenic hits this predisposed phenotype may lead to prostate cancer outcomes in exposed offspring

promising foundation for future studies using multi-hit animal models and investigating longitudinal population data to determine whether maternal obesity is truly a risk factor for prostate cancer.

If maternal obesity predisposes offspring to prostate cancer, then what is the underlying mechanism? A schematic depicting two possible mechanisms is presented in Fig. 11.4. Collective data indicate that the oocyte is altered by maternal dietary exposures prior to conception exhibiting changes in gene expression patterns, chromosomal structure, and metabolism. Embryo transfer experiments suggest that epigenetic alterations and impaired mitochondrial function (directly transmitted from the mother) are heritable contributions to these outcomes. It is very well established that lipid biogenesis and metabolism are both dramatically deregulated in prostate cancer cells and tissue [159]. Additionally, all prostate cancers exhibit alterations in epigenetic patterning that are used as cancer signatures in the clinic, but the cause of these alterations is unknown [160]. We propose that the oocyte sets the foundation for gene expression and metabolic function in tissues such as the prostate. In an obese maternal environment, this foundation patterns the prostate environment in a pro-cancer manner, for example, by provoking aberrant hormone signaling, inflammation, and proliferation (Fig. 11.4). Importantly, each in utero environment is unique depending on the extent and type of exposures. Thus, specific responses will differ in individual prostate tissues, and this variability may explain the current unpredictability of prostate cancer initiation and outcomes. Additionally, postnatal diet could either worsen or improve the effects of maternal dietary exposures, potentially explaining the inconsistent data around adult BMI and prostate cancer incidence. These mechanisms are generalizable; they are altered in many offspring tissues and could thus contribute to the spectrum of other NCDs manifested in exposed offspring. Taken together, we propose that maternal obesity patterns systemic epigenetic and metabolic defects in the oocyte that present, in part, as prostate hyperplasia in offspring. Such an environment could provide a "first hit" for prostate cancer development and should be considered when making dietary recommendations for expectant mothers.

Acknowledgements K.H.M. and E.C.B. were supported by the Washington University Transdisciplinary Research in Energetics and Cancer Program Grant (NCI; U54 CA 155496). E.C.B. was funded by the Washington University Department of Pathology Training Grant (5T32 DK 7296-33) and the Washington University Institute of Clinical and Translational Sciences grant UL1 TR000448 from the National Center for Advancing Translational Sciences. The content is solely the responsibility of the authors and does not necessarily represent the official views of the National Institute of Health (NIH). We thank Dr. Deborah Frank for suggestions on the manuscript and Michaela Reid for assistance with bibliography generation.

References

1. Kleinsmith LJ, Kerrigan D, Kelly J, Hollen B. What is Cancer? National Cancer Institute. 2009. http://www.cancer.gov/cancertopics/understandingcancer/cancer/page1. Accessed 10 June 2014.
2. Howlader N, Noone A, Krapcho M, Garshell J, Neyman N, Altekruse SF, Kosary CL, Yu M, Ruhl J, Tatalovich Z, Cho H, Mariotto A, Lewis DR, Chen HS, Feuer EJ, Cronin KA. SEER Cancer Statistics Review, 1975–2011. National Cancer Institute. 2014. http://seer.cancer.gov/csr/1975_2011/. Accessed 20 April 2014.
3. Cancer Trends Progress Report—2011/2012. Update National Cancer Institute, NIH, DHHS. 2012. http://progressreport.cancer.gov.

4. Klein EA, Thompson IM Jr, Tangen CM, Crowley JJ, Lucia MS, Goodman PJ, Minasian LM, Ford LG, Parnes HL, Gaziano JM, Karp DD, Lieber MM, Walther PJ, Klotz L, Parsons JK, Chin JL, Darke AK, Lippman SM, Goodman GE, Meyskens FL Jr, Baker LH. Vitamin E and the risk of prostate cancer: the Selenium and Vitamin E Cancer Prevention Trial (SELECT). JAMA. 2011;306(14):1549–56.
5. Figueiredo JC, Grau MV, Haile RW, Sandler RS, Summers RW, Bresalier RS, Burke CA, McKeown-Eyssen GE, Baron JA. Folic acid and risk of prostate cancer: results from a randomized clinical trial. J Natl Cancer Inst. 2009;101(6):432–5.
6. Rycyna KJ, Bacich DJ, O'Keefe DS. Opposing roles of folate in prostate cancer. Urology. 2013;82(6):1197–203.
7. Chan JM, Giovannucci EL. Dairy products, calcium, and vitamin D and risk of prostate cancer. Epidemiol Rev. 2001;23(1):87–92.
8. Skinner HG, Schwartz GG. Serum calcium and incident and fatal prostate cancer in the National Health and Nutrition Examination Survey. Cancer Epidemiol Biomarkers Prev (a publication of the American Association for Cancer Research, cosponsored by the American Society of Preventive Oncology). 2008;17(9):2302–5.
9. Van Hemelrijck M, Hermans R, Michaelsson K, Melvin J, Garmo H, Hammar N, Jungner I, Walldius G, Holmberg L. Serum calcium and incident and fatal prostate cancer in the Swedish AMORIS study. Cancer Causes Control. 2012;23(8):1349–58.
10. Wynder EL. Nutrition and cancer. Fed Proc. 1976;35(6):1309–15.
11. Holmberg L. Obesity, nutrition, and prostate cancer: insights and issues. Eur Urol. 2013;63(5):821–2.
12. Discacciati A, Orsini N, Wolk A. Body mass index and incidence of localized and advanced prostate cancer—a dose-response meta-analysis of prospective studies. Ann Oncol. 2012;23(7):1665–71.
13. Hsing AW, Sakoda LC, Chua S Jr. Obesity, metabolic syndrome, and prostate cancer. Am J Clin Nutr. 2007;86(3):843–57.
14. Cao Y, Ma J. Body mass index, prostate cancer-specific mortality, and biochemical recurrence: a systematic review and meta-analysis. Cancer Prev Res. 2011;4(4):486–501.
15. Hu MB, Xu H, Bai PD, Jiang HW, Ding Q. Obesity has multifaceted impact on biochemical recurrence of prostate cancer: a dose-response meta-analysis of 36,927 patients. Med Oncol. 2014;31(2):829.
16. Allott EH, Masko EM, Freedland SJ. Obesity and prostate cancer: weighing the evidence. Eur Urol. 2013;63(5):800–9.
17. Humphrey PA. Prostate pathology. Chicago: American society for clinical pathology; 2003. pp. 1–13.
18. Abate-Shen C, Shen MM. Molecular genetics of prostate cancer. Genes Dev. 2000;14(19):2410–34.
19. Moller E, Adami HO, Mucci LA, Lundholm C, Bellocco R, Johansson JE, Gronberg H, Balter K. Lifetime body size and prostate cancer risk in a population-based case-control study in Sweden. Cancer Causes Control. 2013;24(12):2143–55.
20. Sutcliffe S, Colditz GA. Prostate cancer: is it time to expand the research focus to early-life exposures? Nature Rev Cancer 2013;13(3):208–518.
21. Barker DJ. The fetal and infant origins of adult disease. BMJ. 1990;301(6761):1111.
22. Drake AJ, Reynolds RM. Impact of maternal obesity on offspring obesity and cardiometabolic disease risk. Reproduction. 2010;140(3):387–98.
23. Ma RC, Chan JC, Tam WH, Hanson MA, Gluckman PD. Gestational diabetes, maternal obesity, and the NCD burden. Clin Obstet Gynecol. 2013;56(3):633–41.
24. Galliano D, Bellver J. Female obesity: short- and long-term consequences on the offspring. Gynecol Endocrinol. 2013;29(7):626–31.
25. Williams L, Seki Y, Vuguin PM, Charron MJ. Animal models of in utero exposure to a high fat diet: a review. Biochim Biophys Acta. 2014;1842(3):507–19.
26. Waddington CH. Epigenetics and evolution. Symp Soc Exp Biol. 1953;7:186–99.

27. Fukushige S, Horii A. DNA methylation in cancer: a gene silencing mechanism and the clinical potential of its biomarkers. Tohoku J Exp Med. 2013;229(3):173–85.
28. Baylin SB, Jones PA. A decade of exploring the cancer epigenome—biological and translational implications. Nature Rev Cancer. 2011;11(10):726–34.
29. Holliday R, Pugh JE. DNA modification mechanisms and gene activity during development. Science. 1975;187(4173):226–32.
30. Schapira F. DNA methylation and gene expression. Biomed Pharmacother. 1983;37(4):173–5.
31. Christman JK, Price P, Pedrinan L, Acs G. Correlation between hypomethylation of DNA and expression of globin genes in Friend erythroleukemia cells. Eur J Biochem. 1977;81(1):53–61.
32. Robertson KD. DNA methylation and human disease. Nature Rev Genet. 2005;6(8):597–610.
33. Waterland RA, Jirtle RL. Early nutrition, epigenetic changes at transposons and imprinted genes, and enhanced susceptibility to adult chronic diseases. Nutrition. 2004;20(1):63–8.
34. Williams K, Christensen J, Helin K. DNA methylation: TET proteins-guardians of CpG islands? EMBO Rep. 2012;13(1):28–35.
35. Liu R, Leslie KL, Martin KA. Epigenetic regulation of smooth muscle cell plasticity. Biochim Biophys Acta. 2014;1849:448–53.
36. Campbell NA, Reece JB. Biology. 6th edn. San Francisco: Benjamin Cummings; 2002. p. 87, 94–95, 125–7, 160, 172, 363–4, 984–92.
37. Bannister AJ, Kouzarides T. Regulation of chromatin by histone modifications. Cell Res. 2011;21(3):381–95.
38. Jenuwein T, Allis CD. Translating the histone code. Science. 2001;293(5532):1074–80.
39. Song N, Liu J, An S, Nishino T, Hishikawa Y, Koji T. Immunohistochemical analysis of histone H3 modifications in germ cells during mouse spermatogenesis. Acta Histochem Cytochem. 2011;44(4):183–90.
40. Kouzarides T. Chromatin modifications and their function. Cell. 2007;128(4):693–705.
41. Barski A, Cuddapah S, Cui K, Roh TY, Schones DE, Wang Z, Wei G, Chepelev I, Zhao K. High-resolution profiling of histone methylations in the human genome. Cell. 2007;129(4):823–37.
42. Strahl BD, Allis CD. The language of covalent histone modifications. Nature. 2000;403(6765):41–5.
43. He L, Hannon GJ. MicroRNAs: small RNAs with a big role in gene regulation. Nature Rev Genet. 2004;5(7):522–31.
44. Morozova N, Zinovyev A, Nonne N, Pritchard LL, Gorban AN, Harel-Bellan A. Kinetic signatures of microRNA modes of action. RNA. 2012;18(9):1635–55.
45. Lim LP, Lau NC, Garrett-Engele P, Grimson A, Schelter JM, Castle J, Bartel DP, Linsley PS, Johnson JM. Microarray analysis shows that some microRNAs downregulate large numbers of target mRNAs. Nature. 2005;433(7027):769–73.
46. Yao B, Jin P. Unlocking epigenetic codes in neurogenesis. Genes Dev. 2014;28(12):1253–71.
47. Joh RI, Palmieri CM, Hill IT, Motamedi M. Regulation of histone methylation by noncoding RNAs. Biochim Biophys Acta. 2014;1839:1385–94.
48. Fall CH. Fetal programming and the risk of noncommunicable disease. Ind J Pediatr. 2013;80(Suppl 1):13–20.
49. Edwards TM, Myers JP. Environmental exposures and gene regulation in disease etiology. Environ Health Perspect. 2007;115(9):1264–70.
50. Plasschaert RN, Bartolomei MS. Genomic imprinting in development, growth, behavior and stem cells. Development. 2014;141(9):1805–13.
51. Reynolds EH. The neurology of folic acid deficiency. Handb Clin Neurol. 2014;120:927–43.
52. Zhang S, Rattanatray L, McMillen IC, Suter CM, Morrison JL. Periconceptional nutrition and the early programming of a life of obesity or adversity. Prog Biophysics Mol Biol. 2011;106(1):307–14.

53. Damaschke NA, Yang B, Bhusari S, Svaren JP, Jarrard DF. Epigenetic susceptibility factors for prostate cancer with aging. Prostate. 2013;73(16):1721–30.
54. Bergman D, Halje M, Nordin M, Engstrom W. Insulin-like growth factor 2 in development and disease: a mini-review. Gerontology. 2013;59(3):240–49.
55. Grindler NM, Moley KH. Maternal obesity, infertility and mitochondrial dysfunction: potential mechanisms emerging from mouse model systems. Mol Hum Reprod. 2013;19(8):486–94.
56. Hartil K, Vuguin PM, Kruse M, Schmuel E, Fiallo A, Vargas C, Warner MJ, Durand JL, Jelicks LA, Charron MJ. Maternal substrate utilization programs the development of the metabolic syndrome in male mice exposed to high fat in utero. Pediatr Res. 2009;66(4):368–73.
57. Barbour LA. Changing perspectives in pre-existing diabetes and obesity in pregnancy: maternal and infant short and long-term outcomes. Curr Opin Endocrinol Diabetes Obes. 2014;21:257–63.
58. Klimek P, Leitner M, Kautzky-Willer A, Thurner S. Effect of fetal and infant malnutrition on metabolism in older age. Gerontology. 2014;60:502–7.
59. Li CC, Young PE, Maloney CA, Eaton SA, Cowley MJ, Buckland ME, Preiss T, Henstridge DC, Cooney GJ, Febbraio MA, Martin DI, Cropley JE, Suter CM. Maternal obesity and diabetes induces latent metabolic defects and widespread epigenetic changes in isogenic mice. Epigenetics. 2013;8(6):602–11.
60. Heerwagen MJ, Miller MR, Barbour LA, Friedman JE. Maternal obesity and fetal metabolic programming: a fertile epigenetic soil. Am J Physiol Regul Integr Comp Physiol. 2010;299(3):R711–22.
61. Turdi S, Ge W, Hu N, Bradley KM, Wang X, Ren J. Interaction between maternal and postnatal high fat diet leads to a greater risk of myocardial dysfunction in offspring via enhanced lipotoxicity, IRS-1 serine phosphorylation and mitochondrial defects. J Mol Cell Cardiol. 2013;55:117–29.
62. Hoile SP, Irvine NA, Kelsall CJ, Sibbons C, Feunteun A, Collister A, Torrens C, Calder PC, Hanson MA, Lillycrop KA, Burdge GC. Maternal fat intake in rats alters 20:4n-6 and 22:6n-3 status and the epigenetic regulation of *Fads2* in offspring liver. J Nutr Biochem. 2013;24(7):1213–20.
63. Masuyama H, Hiramatsu Y. Effects of a high-fat diet exposure in utero on the metabolic syndrome-like phenomenon in mouse offspring through epigenetic changes in adipocytokine gene expression. Endocrinology. 2012;153(6):2823–30.
64. Wang Q, Ratchford AM, Chi MM, Schoeller E, Frolova A, Schedl T, Moley KH. Maternal diabetes causes mitochondrial dysfunction and meiotic defects in murine oocytes. Mol Endocrinol. 2009;23(10):1603–12.
65. Phelan S. Smoking cessation in pregnancy. Obstet Gynecol Clin North Am. 2014;41(2):255–66.
66. Bhattacharya S, Campbell DM, Liston WA, Bhattacharya S. Effect of body mass index on pregnancy outcomes in nulliparous women delivering singleton babies. BMC Public Health. 2007;7:168.
67. Watkins ML, Rasmussen SA, Honein MA, Botto LD, Moore CA. Maternal obesity and risk for birth defects. Pediatrics. 2003;111(5 Pt 2):1152–58.
68. Odell LD, Mengert WF. The overweight obstetric patient. JAMA. 1945;128(2):87–90.
69. Aune D, Saugstad OD, Henriksen T, Tonstad S. Maternal body mass index and the risk of fetal death, stillbirth, and infant death: a systematic review and meta-analysis. JAMA. 2014;311(15):1536–46.
70. Rasmussen KM, Yaktine AL, editors. Weight gain during pregnancy: reexamining the guidelines. The National Academies Collection: Reports funded by National Institutes of Health. Washington, DC. 2009.
71. James WP. WHO recognition of the global obesity epidemic. Int J Obes. 2008;32(Suppl 7):120–6.
72. American College of Obstetricians and Gynecologists. ACOG Committee opinion no. 549: obesity in pregnancy. Obstet Gynecol. 2013;121(1):213–7.

73. Prevention CfDCa. Pregnancy complications. 2014. http://www.cdc.gov/reproductivehealth/maternalinfanthealth/pregcomplications.htm. Accessed 05 April 2014.
74. Thompson ML, Ananth CV, Jaddoe VW, Miller RS, Williams MA. The association of maternal adult weight trajectory with preeclampsia and gestational diabetes mellitus. Paediatr Prenat Epidemiol. 2014;28(4):287–96.
75. Elhddad AS, Fairlie F, Lashen H. Impact of gestational weight gain on fetal growth in obese normoglycemic mothers: a comparative study. Acta Obstet Gynecol Scand. 2014;98:771–7.
76. Lashen H, Fear K, Sturdee DW. Obesity is associated with increased risk of first trimester and recurrent miscarriage: matched case-control study. Hum Reprod. 2004;19(7):1644–6.
77. Sebire NJ, Jolly M, Harris JP, Wadsworth J, Joffe M, Beard RW, Regan L, Robinson S. Maternal obesity and pregnancy outcome: a study of 287,213 pregnancies in London. Int J Obes Relat Metab Disord. 2001;25(8):1175–82.
78. Li R, Jewell S, Grummer-Strawn L. Maternal obesity and breast-feeding practices. Am J Clin Nutr. 2003;77(4):931–6.
79. Tenenbaum-Gavish K, Hod M. Impact of maternal obesity on fetal health. Fetal Diagn Ther. 2013;34(1):1–7.
80. Walsh JM, McAuliffe FM. Prediction and prevention of the macrosomic fetus. Eur J Obstet Gynecol Reprod Biol. 2012;162(2):125–30.
81. Stothard KJ, Tennant PW, Bell R, Rankin J. Maternal overweight and obesity and the risk of congenital anomalies: a systematic review and meta-analysis. JAMA. 2009;301(6):636–50.
82. Matthews TJ, MacDorman MF. Infant mortality statistics from the 2010 period linked birth/infant death data set. Natl Vital Stat Rep (from the Centers for Disease Control and Prevention, National Center for Health Statistics, National Vital Statistics System). 2013;62(8):1–26.
83. Patro B, Liber A, Zalewski B, Poston L, Szajewska H, Koletzko B. Maternal and paternal body mass index and offspring obesity: a systematic review. Ann Nutr Metab. 2013;63(1–2):32–41.
84. Catalano PM, McIntyre HD, Cruickshank JK, McCance DR, Dyer AR, Metzger BE, Lowe LP, Trimble ER, Coustan DR, Hadden DR, Persson B, Hod M, Oats JJ, Group HSCR. The hyperglycemia and adverse pregnancy outcome study: associations of GDM and obesity with pregnancy outcomes. Diabetes Care. 2012;35(4):780–6.
85. Gaillard R, Steegers EA, Duijts L, Felix JF, Hofman A, Franco OH, Jaddoe VW. Childhood cardiometabolic outcomes of maternal obesity during pregnancy: the Generation R Study. Hypertension. 2014;63(4):683–91.
86. Brumbaugh DE, Tearse P, Cree-Green M, Fenton LZ, Brown M, Scherzinger A, Reynolds R, Alston M, Hoffman C, Pan Z, Friedman JE, Barbour LA. Intrahepatic fat is increased in the neonatal offspring of obese women with gestational diabetes. J Pediatr. 2013;162(5):930–6.e1.
87. Guenard F, Tchernof A, Deshaies Y, Cianflone K, Kral JG, Marceau P, Vohl MC. Methylation and expression of immune and inflammatory genes in the offspring of bariatric bypass surgery patients. J Obes. 2013;2013:492170.
88. Desai M, Gayle D, Han G, Ross MG. Programmed hyperphagia due to reduced anorexigenic mechanisms in intrauterine growth-restricted offspring. Reprod Sci. 2007;14(4):329–37.
89. Taylor PD, Samuelsson AM, Poston L. Maternal obesity and the developmental programming of hypertension: a role for leptin. Acta Physiol. 2014;210(3):508–23.
90. Bird A. Perceptions of epigenetics. Nature. 2007;447(7143):396–8.
91. Ge ZJ, Zhang CL, Schatten H, Sun QY. Maternal diabetes mellitus and the origin of non-communicable diseases in offspring: the role of epigenetics. Biol Reprod. 2014;90:139.
92. Ruchat SM, Houde AA, Voisin G, St-Pierre J, Perron P, Baillargeon JP, Gaudet D, Hivert MF, Brisson D, Bouchard L. Gestational diabetes mellitus epigenetically affects genes predominantly involved in metabolic diseases. Epigenetics. 2013;8(9):935–43.
93. Robinson S, Yardy K, Carter V. A narrative literature review of the development of obesity in infancy and childhood. J Child Health Care. 2012;16(4):339–54.

94. Sullivan EL, Nousen EK, Chamlou KA. Maternal high fat diet consumption during the perinatal period programs offspring behavior. Physiol Behav. 2014;123:236–42.
95. Breton C. The hypothalamus-adipose axis is a key target of developmental programming by maternal nutritional manipulation. J Endocrinol. 2013;216(2):R19–31.
96. Lukaszewski MA, Eberle D, Vieau D, Breton C. Nutritional manipulations in the perinatal period program adipose tissue in offspring. Am J Physiol Endocrinol Metab. 2013;305 (10):E1195–207.
97. O'Reilly JR, Reynolds RM. The risk of maternal obesity to the long-term health of the offspring. Clin Endocrinol. 2013;78(1):9–16.
98. Suburu J, Chen YQ. Lipids and prostate cancer. Prostaglandins Other Lipid Mediat. 2012;98(1–2):1–10.
99. Sanderson M, Williams MA, Malone KE, Stanford JL, Emanuel I, White E, Daling JR. Perinatal factors and risk of breast cancer. Epidemiology. 1996;7(1):34–7.
100. Sanderson M, Williams MA, Daling JR, Holt VL, Malone KE, Self SG, Moore DE. Maternal factors and breast cancer risk among young women. Paediatr Perinat Epidemiol. 1998;12(4):397–407.
101. Wilson KM, Willett WC, Michels KB. Mothers' pre-pregnancy BMI and weight gain during pregnancy and risk of breast cancer in daughters. Breast Cancer Res Treat. 2011;130(1):273–9.
102. Michels KB, Trichopoulos D, Robins JM, Rosner BA, Manson JE, Hunter DJ, Colditz GA, Hankinson SE, Speizer FE, Willett WC. Birthweight as a risk factor for breast cancer. Lancet. 1996;348(9041):1542–46.
103. Park SK, Kang D, McGlynn KA, Garcia-Closas M, Kim Y, Yoo KY, Brinton LA. Intrauterine environments and breast cancer risk: meta-analysis and systematic review. Breast Cancer Res. 2008;10(1):R8.
104. Yenbutr P, Hilakivi-Clarke L, Passaniti A. Hypomethylation of an exon I estrogen receptor CpG island in spontaneous and carcinogen-induced mammary tumorigenesis in the rat. Mech Ageing Dev. 1998;106(1–2):93–102.
105. La Merrill M, Harper R, Birnbaum LS, Cardiff RD, Threadgill DW. Maternal dioxin exposure combined with a diet high in fat increases mammary cancer incidence in mice. Environ Health Perspect. 2010;118(5):596–601.
106. Hilakivi-Clarke L, Clarke R, Onojafe I, Raygada M, Cho E, Lippman M. A maternal diet high in n−6 polyunsaturated fats alters mammary gland development, puberty onset, and breast cancer risk among female rat offspring. Proc Natl Acad Sci U S A. 1997;94(17):9372–7.
107. de Assis S, Warri A, Cruz MI, Laja O, Tian Y, Zhang B, Wang Y, Huang TH, Hilakivi-Clarke L. High-fat or ethinyl-oestradiol intake during pregnancy increases mammary cancer risk in several generations of offspring. Nat Commun. 2012;3:1053.
108. Lo CY, Hsieh PH, Chen HF, Su HM. A maternal high-fat diet during pregnancy in rats results in a greater risk of carcinogen-induced mammary tumors in the female offspring than exposure to a high-fat diet in postnatal life. Int J Cancer. 2009;125(4):767–73.
109. Su HM, Hsieh PH, Chen HF. A maternal high n-6 fat diet with fish oil supplementation during pregnancy and lactation in rats decreases breast cancer risk in the female offspring. J Nutr Biochem. 2010;21(11):1033–7.
110. de Oliveira Andrade F, Fontelles CC, Rosim MP, de Oliveira TF, de Melo Loureiro AP, Mancini-Filho J, Rogero MM, Moreno FS, de Assis S, Barbisan LF, Hilakivi-Clarke L, Ong TP. Exposure to lard-based high-fat diet during fetal and lactation periods modifies breast cancer susceptibility in adulthood in rats. J Nutr Biochem 2014;25(6):613–22.
111. Montales MT, Melnyk SB, Simmen FA, Simmen RC. Maternal metabolic perturbations elicited by high-fat diet promote Wnt-1-induced mammary tumor risk in adult female offspring via long-term effects on mammary and systemic phenotypes. Carcinogenesis. 2014;35:2012–12.
112. Kerdivel G, Flouriot G, Pakdel F. Modulation of estrogen receptor alpha activity and expression during breast cancer progression. Vitam Horm. 2013;93:135–60.

113. Burton A, Martin R, Galobardes B, Davey Smith G, Jeffreys M. Young adulthood body mass index and risk of cancer in later adulthood: historical cohort study. Cancer Causes Control. 2010;21(12):2069–77.
114. Dal Maso L, Zucchetto A, La Vecchia C, Montella M, Conti E, Canzonieri V, Talamini R, Tavani A, Negri E, Garbeglio A, Franceschi S. Prostate cancer and body size at different ages: an Italian multicentre case-control study. Br J Cancer. 2004;90(11):2176–80.
115. Schuurman AG, Goldbohm RA, Dorant E, van den Brandt PA. Anthropometry in relation to prostate cancer risk in the Netherlands cohort study. Am J Epidemiol. 2000;151(6):541–9.
116. Giovannucci E, Rimm EB, Stampfer MJ, Colditz GA, Willett WC. Height, body weight, and risk of prostate cancer. Cancer Epidemiol Biomarkers Prev (a publication of the American Association for Cancer Research, cosponsored by the American Society of Preventive Oncology). 1997;6(8):557–63.
117. Robinson WR, Stevens J, Gammon MD, John EM. Obesity before age 30 years and risk of advanced prostate cancer. Am J Epidemiol. 2005;161(12):1107–14.
118. Discacciati A, Orsini N, Andersson SO, Andren O, Johansson JE, Wolk A. Body mass index in early and middle-late adulthood and risk of localised, advanced and fatal prostate cancer: a population-based prospective study. Br J Cancer. 2011;105(7):1061–68.
119. Barker DJ, Osmond C, Thornburg KL, Kajantie E, Eriksson JG. A possible link between the pubertal growth of girls and prostate cancer in their sons. Am J Hum Biol. 2012;24(4):406–10.
120. Ekstrom S, Magnusson J, Kull I, Lind T, Almqvist C, Melen E, Bergstrom A. Maternal BMI in early pregnancy and offspring asthma, rhinitis and eczema up to 16 years of age. Clin Exp Allergy. 2014;45:283–91.
121. Ekbom A, Hsieh CC, Lipworth L, Wolk A, Ponten J, Adami HO, Trichopoulos D. Perinatal characteristics in relation to incidence of and mortality from prostate cancer. BMJ. 1996;13(7053):337–41.
122. Ikeda Y, Cho YM, Takahashi S, Tang M, Asamoto M, Ogawa K, Shirai T. Equivocal impact of transplacental and lactational exposure to a food-derived carcinogen, 2-amino-1-methyl-6-phenylimidazo[4,5-b]pyridine, on prostate and colon lesion development in F344 rats. Cancer Lett. 2005;224(1):23–30.
123. Kondo Y, Homma Y, Aso Y, Kakizoe T. Promotional effect of two-generation exposure to a high-fat diet on prostate carcinogenesis in ACI/Seg rats. Cancer Res. 1994;54(23):6129–32.
124. Tobisu K, Kakizoe T, Tanaka Y, Takai K, Shirai T. Influence of dietary fat and total calorie intake on the incidence of spontaneous prostatic cancer. Proceeding of the First Workshop on Prostatic Cancer; 1990; Shalom Co., Tokyo.
125. Benesh EC, Humphrey PA, Wang Q, Moley KH. Maternal high-fat diet induces hyperproliferation and alters Pten/Akt signaling in prostates of offspring. Sci Rep. 2013;3:3466.
126. Rinaldi JC, Justulin LA Jr, Lacorte LM, Sarobo C, Boer PA, Scarano WR, Felisbino SL. Implications of intrauterine protein malnutrition on prostate growth, maturation and aging. Life Sci. 2013;92(13):763–74.
127. Ramos Cda F, Babinski MA, Costa WS, Sampaio FJ. The prostate of weaned pups is altered by maternal malnutrition during lactation in rats. Asian J Androl. 2010;12(2):180–5.
128. Eriksson JG, Sandboge S, Salonen MK, Kajantie E, Osmond C. Long-term consequences of maternal overweight in pregnancy on offspring later health: findings from the Helsinki Birth Cohort Study. Ann Med. 2014;46:434–8.
129. Walker BE. Tumors in female offspring of control and diethylstilbestrol-exposed mice fed high-fat diets. J Natl Cancer Inst. 1990;82(1):50–4.
130. Simmen FA, Simmen RC. The maternal womb: a novel target for cancer prevention in the era of the obesity pandemic? Eur J Cancer Prev (the official journal of the European Cancer Prevention Organisation). 2011;20(6):539–48.
131. Colton SA, Pieper GM, Downs SM. Altered meiotic regulation in oocytes from diabetic mice. Biol Reprod. 2002;67(1):220–31.

132. Luzzo KM, Wang Q, Purcell SH, Chi M, Jimenez PT, Grindler N, Schedl T, Moley KH. High fat diet induced developmental defects in the mouse: oocyte meiotic aneuploidy and fetal growth retardation/brain defects. PloS One. 2012;7(11):e49217.
133. Minge CE, Bennett BD, Norman RJ, Robker RL. Peroxisome proliferator-activated receptor-gamma agonist rosiglitazone reverses the adverse effects of diet-induced obesity on oocyte quality. Endocrinology. 2008;149(5):2646–56.
134. Cunningham FG, Williams JW. Williams obstetrics. 21st edn. New York: McGraw-Hill; 2001. p. 86–89, 130–134.
135. Racki WJ, Richter JD. CPEB controls oocyte growth and follicle development in the mouse. Development. 2006;133(22):4527–37.
136. Behringer R. Manipulating the mouse embryo: a laboratory manual. 4th edn. New York: Cold Spring Harbor Laboratory Press, Cold Spring Harbor; 2014.
137. Chang AS, Dale AN, Moley KH. Maternal diabetes adversely affects preovulatory oocyte maturation, development, and granulosa cell apoptosis. Endocrinology. 2005;146(5):2445–53.
138. Diamond MP, Moley KH, Pellicer A, Vaughn WK, DeCherney AH. Effects of streptozotocin- and alloxan-induced diabetes mellitus on mouse follicular and early embryo development. J Reprod Fertil. 1989;86(1):1–10.
139. Moley KH, Vaughn WK, DeCherney AH, Diamond MP. Effect of diabetes mellitus on mouse pre-implantation embryo development. J Reprod Fertil. 1991;93(2):325–32.
140. Moley KH, Vaughn WK, Diamond MP. Manifestations of diabetes mellitus on mouse pre-implantation development: effect of elevated concentration of metabolic intermediates. Hum Reprod. 1994;9(1):113–21.
141. Ratchford AM, Chang AS, Chi MM, Sheridan R, Moley KH. Maternal diabetes adversely affects AMP-activated protein kinase activity and cellular metabolism in murine oocytes. Am J Physiol Endocrinol Metab. 2007;293(5):E1198–206.
142. Chi MM, Hoehn A, Moley KH. Metabolic changes in the glucose-induced apoptotic blastocyst suggest alterations in mitochondrial physiology. Am J Physiol Endocrinol Metab. 2002;283(2):E226–32.
143. Moley KH, Chi MM, Knudson CM, Korsmeyer SJ, Mueckler MM. Hyperglycemia induces apoptosis in pre-implantation embryos through cell death effector pathways. Nature Med. 1998;4(12):1421–4.
144. Moley KH, Chi MM, Mueckler MM. Maternal hyperglycemia alters glucose transport and utilization in mouse preimplantation embryos. Am J Physiol. 1998;275(1 Pt 1):E38–47.
145. Wyman A, Pinto AB, Sheridan R, Moley KH. One-cell zygote transfer from diabetic to non-diabetic mouse results in congenital malformations and growth retardation in offspring. Endocrinology. 2008;149(2):466–9.
146. Wang Q, Frolova AI, Purcell S, Adastra K, Schoeller E, Chi MM, Schedl T, Moley KH. Mitochondrial dysfunction and apoptosis in cumulus cells of type I diabetic mice. PloS One. 2010;5(12):e15901.
147. Wang Q, Chi MM, Moley KH. Live imaging reveals the link between decreased glucose uptake in ovarian cumulus cells and impaired oocyte quality in female diabetic mice. Endocrinology. 2012;153(4):1984–9.
148. Wang Q, Chi MM, Schedl T, Moley KH. An intercellular pathway for glucose transport into mouse oocytes. Am J Physiol Endocrinol Metab. 2012;302(12):E1511–8.
149. Jungheim ES, Schoeller EL, Marquard KL, Louden ED, Schaffer JE, Moley KH. Diet-induced obesity model: abnormal oocytes and persistent growth abnormalities in the offspring. Endocrinology. 2010;151(8):4039–46.
150. Bermejo-Alvarez P, Rosenfeld CS, Roberts RM. Effect of maternal obesity on estrous cyclicity, embryo development and blastocyst gene expression in a mouse model. Hum Reprod. 2012;27(12):3513–22.
151. Igosheva N, Abramov AY, Poston L, Eckert JJ, Fleming TP, Duchen MR, McConnell J. Maternal diet-induced obesity alters mitochondrial activity and redox status in mouse oocytes and zygotes. PloS One. 2012;5(4):e10074.

152. Shankar K, Zhong Y, Kang P, Lau F, Blackburn ML, Chen JR, Borengasser SJ, Ronis MJ, Badger TM. Maternal obesity promotes a proinflammatory signature in rat uterus and blastocyst. Endocrinology. 2011;152(11):4158–70.
153. Jungheim ES, Louden ED, Chi MM, Frolova AI, Riley JK, Moley KH. Preimplantation exposure of mouse embryos to palmitic acid results in fetal growth restriction followed by catch-up growth in the offspring. Biol Reprod. 2011;85(4):678–83.
154. Bernal AB, Vickers MH, Hampton MB, Poynton RA, Sloboda DM. Maternal undernutrition significantly impacts ovarian follicle number and increases ovarian oxidative stress in adult rat offspring. PloS One. 2010;5(12):e15558.
155. Luke B, Brown MB, Missmer SA, Bukulmez O, Leach R, Stern JE, Society for Assisted Reproductive Technology writing group. The effect of increasing obesity on the response to and outcome of assisted reproductive technology: a national study. Fertil Steril. 2011;96(4):820–85.
156. Jungheim ES, Schon SB, Schulte MB, DeUgarte DA, Fowler SA, Tuuli MG. IVF outcomes in obese donor oocyte recipients: a systematic review and meta-analysis. Hum Reprod. 2013;28(10):2720–7.
157. Machtinger R, Combelles CM, Missmer SA, Correia KF, Fox JH, Racowsky C. The association between severe obesity and characteristics of failed fertilized oocytes. Hum Reprod. 2012;27(11):3198–207.
158. Hilakivi-Clarke L, Cho E, deAssis S, Olivo S, Ealley E, Bouker KB, Welch JN, Khan G, Clarke R, Cabanes A. Maternal and prepubertal diet, mammary development and breast cancer risk. J Nutr. 2001;131(1):154S–7S.
159. Zadra G, Photopoulos C, Loda M. The fat side of prostate cancer. Biochim Biophys Acta. 2013;1831(10):1518–32.
160. Chiam K, Ricciardelli C, Bianco-Miotto T. Epigenetic biomarkers in prostate cancer: current and future uses. Cancer Lett. 2014;342(2):248–56.

Chapter 12
Mouse Models to Study the Effect of Natural Products on Obesity-Associated NAFLD/NASH

Dipali Sharma and Neeraj K. Saxena

Abstract Obesity and metabolic syndrome have escalated to the level of an epidemic in the last three decades. Concomitant with obesity and metabolic syndrome epidemic, there is a rapid increase in the prevalence of nonalcoholic fatty liver disease (NAFLD), the hepatic manifestation of metabolic syndrome, in the world especially in developed and developing countries. NAFLD is a benign condition characterized by excessive fat deposition (steatosis) in the form of triglycerides (TG) (histologically >5 % of hepatocytes with fat). NAFLD can stay overlooked for a long duration (multiple years), but a subset may progress to variable degree of hepatic necro-inflammation (lobular inflammation, hepatocyte injury, and ballooning), a condition defined as nonalcoholic steatohepatitis (NASH) which may further develop into fibrosis, nonreversible cirrhosis and hepatocellular carcinoma (HCC) to end-stage liver failure. Modern lifestyle and high-fat diet (HFD) are considered to be the major factors for the development of NAFLD. Availability of suitable in vivo mouse models to study diet-obesity-induced NAFLD and NASH are limited. A number of mouse models have been used to get important insights into the pathology of NAFLD/NASH. These mouse models mimic and reflect hepatic-histopathology and pathophysiology of NAFLD/NASH, but no single model represent the full range of human disease. Mouse models of NAFLD/NASH can be grouped into two categories, genetic and dietary. The purpose of this review is to elucidate vari-

N. K. Saxena (✉)
Department of Medicine, University of Maryland School of Medicine, UMMC N3W62, 655 W. Baltimore Street, Baltimore, MD 21201, USA
e-mail: nsaxena@medicine.umaryland.edu

D. Sharma
Department of Oncology, The Sidney Kimmel Comprehensive Cancer Center, Johns Hopkins University School of Medicine, Baltimore, MD 21231, USA
e-mail: dsharma7@jhmi.edu

N. A. Berger (ed.), *Murine Models, Energy Balance, and Cancer*,
Energy Balance and Cancer, DOI 10.1007/978-3-319-16733-6_12

ous mouse models of NAFLD/NASH with potential therapeutic interventions using natural products.

Keywords Nonalcoholic fatty liver disease (NAFLD) · Nonalcoholic steatohepatitis · Mouse models · Natural compounds

According to "World Disasters Report" by "International Federation of Red Cross and Red Crescent Societies," published in 2011, approximately 15 % of the world population is hungry and malnourished. While approximately 1 billion people go to bed hungry every night, paradoxically ~ 1.5 billion people are overweight, around the world. A recent study shows that the prevalence of obesity has risen to the level of an epidemic in developed as well as developing countries [1–3]. As per World Health Organization (WHO), the number of overweight and obese people around the globe has tripled in the last three decades and the rate is alarming for childhood obesity [4]. In Europe, one-fifth of school-aged children are obese or overweight [5]. Within the last 30 years, there is a profound increase in the prevalence of obesity among populations in developed countries [6, 7] and developing countries like India and China [8]. Importantly, the highest proportion of overweight and obese populations reside in the USA. According to a recent report from Centers for Disease Control and Prevention, more than one-third of adults and 17 % of youth in the USA are obese [9].

Obesity is a risk factor for various diseases in adults [10–13] and children [14–16]. Obesity can cause a number of medical disorders ranging from metabolic syndrome to life-threatening diseases such as cancer. Obesity has been associated with approximately 90,000 cancer-related deaths per year in the USA [10] and is an independent risk factor for the cancer of breast, endometrium, colon, renal cell, esophagus, pancreas, and liver [11, 17–22]. Importantly, relative risk for pancreatic and liver cancer is the highest among obese men as compared to other cancers [10, 17, 23, 24]. The likelihood of liver complications increases in obese adults with a history of childhood obesity due to early accumulation of fat in the liver along with impaired glucose tolerance [25–27]. Epidemiological studies suggest a carcinogenic role for NAFLD and obesity in HCC [10, 17].

Metabolic syndrome is a disorder of energy intake and expenditure with constellation of abnormalities in high blood pressure, triglycerides, and blood sugar leading to chronic heart and liver diseases including atherosclerotic cardiovascular disease, stroke, and type 2 diabetes [28]. Metabolic syndrome confers approximately four to fivefold and twofold increase in type 2 diabetes, and cardiovascular disease, respectively [29]. Concomitant with the obesity and metabolic syndrome epidemic, there is a rapid increase in nonalcoholic fatty liver disease (NAFLD), the hepatic manifestation of metabolic syndrome [30]. Generally, NAFLD is defined by the presence of any three of the following (listed in Table 12.1): (a) abdominal obesity with increased visceral fat (waist circumference > 102 cm in men, > 88 cm in women), (b) low levels of HDL-C (high-density lipoprotein cholesterol) (< 40 mg/dl in men, < 50 mg/dl in women), (c) elevated levels of triglycerides (> 150 mg/

Table 12.1 Defining factors for metabolic syndrome leading to NAFLD

Diagnostic risk factors	Conditions defining metabolic syndrome	Demarcating levels
Abdominal obesity	Increased waist circumference	Men > 102 cm Women > 88 cm
Atherogenic Dyslipidimia	Elevated Triglycerides OR on drugs to reduce elevated levels of triglycerides	> 150 mg/dl
	Decreased HDL (high-density lipoprotein cholesterol) OR on drugs to elevate decreased levels of HDL	Men >40 mg/dl Women > 50 mg/dl
Hypertension	Raised blood pressure	
	Systolic	> 130 mm Hg
	Diastolic	> 85 mm Hg
	OR on drugs to reduce elevated blood pressure	
Impaired fasting Glucose levels	Fasting glucose OR on antidiabetic drugs	100–125 mg/dl

dl), (d) impaired fasting glucose levels (100–125 mg/dl), (e) high blood pressure (systolic blood pressure > 130 mm Hg or diastolic blood pressure > 85 mm Hg), (f) drug treatment for elevated triglycerides, (g) drug treatment for high blood pressure, (h) drug treatment for type 2 diabetes, (i) drug treatment for increasing HDL (Table 12.1) [31, 32].

The global prevalence of NAFLD ranges from 2.8 to 46 % [33] and in the USA it concurs with an increasing trend in obesity, metabolic syndrome, and type 2 diabetes [34]. According to the National Health and Nutrition Examination Surveys (NHANES), the prevalence of NAFLD in the USA has increased from 5.5 to 11 % from 1988 to 2008 [35]. Data from the NHANES III indicate that 28 % of Americans have elevated levels of serum ALT (alanine aminotransferase) as a consequence of hepatic steatosis [36] and about one fifth of these patients progress to steatohepatitis (NASH), cirrhosis, and ultimately to HCC [37, 38]. Recent studies provide evidence for association between NAFLD and atherosclerosis [39–41], insulin resistance, and type 2 diabetes [42]. NAFLD was first introduced and recognized as a chronic disease in the year 1980 and can be defined as deposition of fat in the liver cells of patients with minimal or no alcohol intake [43]. NAFLD is a benign condition characterized by excessive fat deposition (steatosis) in the form of triglycerides (histologically > 5 % of hepatocytes with fat). Over a period of time, a subset of hepatic steatotic patients develop variable degree of hepatic necro-inflammation (lobular inflammation, hepatocyte injury, and ballooning) and this condition is defined as nonalcoholic steatohepatitis (NASH), a more severe form of NAFLD [43, 44]. NAFLD-patients with simple fatty liver (hepatic steatosis) have a good long-term prognosis as compared to patients with NASH having a threefold increase in liver-related mortality [45, 46]. NAFLD can go unnoticed for a long duration (years), but may progress to NASH, fibrosis and nonreversible cirrhosis, and hepatocellular carcinoma (HCC) to end-stage liver failure [38, 43, 47–50]. NAFLD is the most common liver disease with an estimated prevalence from 20 to 30 % of the general population in Western countries [51]. Approximately 10–25 % of NAFLD patients develop NASH [52, 53] and approximately 7 % of patients diagnosed with cirrhosis due to NASH develop HCC [54]. Importantly, recent studies show that patients with cirrhosis preceded by NASH are at greater risk of developing HCC than those who develop cirrhosis from hepatitis C virus [55].

Since NAFLD is closely related with obesity, metabolic syndrome, diabetes, and hypertriglyceridemia, its pathogenesis is still unclear and not fully explained. Fat buildup in the liver is one of the main factors responsible for metabolic imbalances that are related to central obesity and insulin resistance [56]. According to some reports the "two hit" hypothesis is a strong contender [11, 52, 57]; insulin resistance, the "first" hit results from accumulation of free fatty acids (FFAs) that are esterified and stored as hepatocyte triglycerides [57]. The "second" hit involves abnormal oxidation of these fats resulting in free radical production with membrane lipid injury and toxic lipid peroxidation product formation as well as release of pro-inflammatory cytokines. Due to opaque pathogenesis of NAFLD, the disease lacks an established treatment except to make modifications in life style and diet [58]. Since the theories on the pathogenesis of NAFLD/NASH are not clear, development of preventive and therapeutic modalities are stalled [59–62]. NAFLD/NASH being chronic disease, has very slow progression in humans, therefore it becomes very important and essential to have animal model(s) to elucidate the pathogenesis and therapeutic potentials of various agents [63–67]. One of the important requirements for these models is to reflect and portray correct pathophysiology and histopathology of NAFLD/NASH. In this chapter, we first review current mouse models of NAFLD/NASH followed by concise review of various natural plant products used as therapeutic agents in these mouse models.

Mouse Models of NAFLD/NASH

Since the review is focused on hepatocarcinogenesis with NAFLD/NASH as causal factor, the mouse models discussed in this review are pertinent to obesity, metabolic syndrome, insulin resistance, diabetes, hypertrigleridemia, and dysregulated level of adipocytokines. Ideally, mouse models of NAFLD/NASH should imitate human NAFLD/NASH at pathophysiological and histopathological level and show a sequential progression of liver disease starting with hepatic steatosis followed by intralobular inflammation necroinflammation, hepatocytic ballooning finally progressing to fibrosis, cirrhosis, and HCC [68]. Till date, a number of mouse models have been reported, however, none of the single model represents full spectrum of human-liver pathophysiology and histopathology of NAFLD/NASH [58]. All the mouse models differ in the presentation of NAFLD/NASH for various parameters of the disease. For this review, different mouse models of NAFLD/NASH have been classified into two major categories. The first category includes genetically modified mice (transgenic or knockout) and the second one includes the mouse

Table 12.2 Mouse models for NAFLD/NASH with varying degrees of parameters

Category/Model	Obesity	Insulin resistance	NAFLD	NASH	Fibrosis	HCC
Genetically modified mice						
ob/ob mice	*Present*	Present	Present	Absent	*Absent*	*Absent*
db/db mice	*Present*	Present	Present	Absent	*Absent*	*Absent*
KK-A y mice	*Present*	Present	Present	Absent	*Absent*	*Absent*
MC4R $^{-/-}$	*Present*	Present	Present	Present	*Present*	*Present*
PTEN $^{-/-}$ *mice*	*Absent*	Absent	Present	Present	*Present*	*Present*
MAT1A $^{-/-}$ *mice*	*Absent*	Absent	Present	Present	*Present*	*Present*
AOX $^{-/-}$ *mice*	*Absent*	Absent	Present	Present	*Absent*	*Present*
PPAR$^{-\alpha-/-}$ (under starvation)	Absent	Absent	Present on starvation	Absent	*Absent*	*Absent*
SREBP-1c transgenic mice	Absent	Present	Present	Present	*Present*	*Absent*
Nutritionally manipulated						
HFD	*Present*	Present	Present	Present	*Present*	*Present*
HFD with fructose	*Present*	Present	Present	Present	*Present*	*Absent*
Atherogenic Diet	*Absent*	Liver specific Present	Present	Present	*Present*	*Absent*
Methionine-/Choline-Deficient (MCD) Diet	*Absent*	Liver specific Present	Present	Severely Present	*Present*	*Present*

HFD high-fat-diet, *NAFLD* nonalcoholic fatty liver disease, *NASH* nonalcoholic steatohepatitis, *HCC* hepatocellular carcinoma

models which acquire NAFLD/NASH after nutritional (dietary) manipulations. These two major categories are further subdivided based on the presence or absence of obese phenotype. Important features of the models discussed in this review are listed in Table 12.2. These different models differ in the disease presentation, as stated above, the pathological and pathophysiological characteristics of NAFLD/NASH appear with varying degrees of severity (absent to mild to severe) and course of time (initial or later stage).

Genetically Modified Mouse Models of NAFLD/NASH with Obesity

ob/ob **Mice (Leptin Deficient Mice)** *ob/ob* mice also known as leptin-deficient mice harbor a mutation in the leptin gene. Leptin, a 16 k Da protein hormone produced by white adipose tissue, regulates the feeding and energy expenditure via hypothalamus [58, 69–72]. The spontaneous mutation in leptin gene of *ob/ob mice* makes these mice leptin deficient rendering them hyperphagic, inactive, and severely obese with hyperlipidemia, hyperglycemia, and fatty liver (NAFLD) [71]. Importantly *ob/ob* mice develop fatty liver with severe hepatic steatosis spontaneously [73], but the progression from steatosis to steatohepatitis requires "second hit." The steatotic livers of these mice are vulnerable to progression toward severe steatohepatitis and acute mortality after secondary injury from Lipopolysaccharide (LPS), ischemia-reperfusion, methionine-/choline-deficient diet, or ethanol feeding [74–77]. Mechanism for the development of hepatic steatosis is not very clear; however, it is reported that *ob/ob* mice have increased lipogenesis with elevated concentration of FFAs and pro-inflammatory TNF-α in adipose tissue. The elevated level of serum FFAs flow to liver thereby making it steatotic [71]. Quite intriguingly *ob/ob* mice are immune to liver fibrosis thereby making leptin as an essential profibrogenic adipocytokine required for the development of liver fibrosis [78–82]. Despite being severely obese, *ob/ob* mouse model does not fully represent human NAFLD/NASH due to the absence of leptin.

***db/db* Mice (Leptin-Resistant Mice)** A naturally occurring congenic mutation in the leptin receptor (*Ob-Rb)* gene results in nonfunctional leptin signaling [83] making a leptin-resistant mouse model. Phenotypically *db/db* mice are similar to *ob/ob* mice except they have normal to elevated levels of leptin due to deficiency in leptin receptor [84]. *db/db* mice are leptin resistant, obese, insulin resistant, and diabetic with hepatic steatosis. Hepatic steatotic *db/db* mice need a second-hit from MCD diet to develop steatohepatitis which progresses to severe fibrosis [85, 86]. *ob/ob* and *db/db* mice are ideal representation of human metabolic syndrome, but they lack spontaneous development of steatohepatitis and fibrosis.

KK-Ay Mice A genetic model with higher appetite, KK-Ay mice have a heterozygous mutation of the agouti gene (KK-Ay/a) resulting in loss of melanocortin. These mice have obese phenotype due to highly increased appetite as they have compromised hypothalamic control on their appetite [87]. KK-Ay are also called lethal yellow mice. KK-Ay mice are obese with most of the characteristic features of metabolic syndrome like hyperglycemia, hyperinsulinemia, hypertriglyceridemia, hypercholesterolemia, insulin resistance, and most importantly altered levels of adipocytokines. These mice are susceptible to MCD diet-induced NASH, develop hepatic steatosis which progresses to steatohepatitis because of the antagonism between melanocortin receptor 4 (MC4R) and the ectopic expression of the agouti protein and hypoadiponectinemia [88].

MC4R Null Mice MC4R maintains energy balance by regulating food consumption as well as fat metabolism [89]. High-frequency MC4R mutations have been reported to be one of the common causes of morbid obesity in humans [90]. Targeted disruption of MC4R results in obese-mice with hyperphagia, hyperglycemic, and insulin resistance [91]. Upon high-fat-diet (HFD) administration, the MC4R-deficient mice develop hepatic steatosis, steatohepatitis, liver fibrosis, and even hepatocellular carcinoma within 1 year of HFD feeding [92–94].

Genetically Modified Mouse models of NAFLD/NASH Without Obesity

PTEN$^{-/-}$ Mice Phosphatase and tensin homolog (PTEN) is a multifunctional phosphatase whose substrate is phosphatidylinositol-3,4,5- triphosphate (PIP3) [95, 96]. PTEN acts as a tumor suppressor gene and downregulates phosphatidyl inositol 3-kinase (PI3K) and serine-threonine protein kinase B (PKB or Akt) thereby regulating cell apoptosis, proliferation, and tumor formation [97]. Liver-specific PTEN knockout mice show spontaneous and extensive steatohepatitis with histopathology quite comparable to human NASH [98]. PTEN-null mice show hyperlipidemia with increased concentrations of triglycerides and cholesterol with micro- as well as macrovesicular steaotosis within 10 weeks of age. Hepatic steatosis in PTEN-null mice progresses to steatohepatitis by the age of 40 weeks exhibiting macrovesicular steatosis with hepatocyte ballooning, lobular inflammatory cell infiltration, mallory bodies, and sinusoidal fibrosis [99, 100]. By 44 weeks, livers of 47% of PTEN-knockout mice show hepatocellular adenomas and by 80 weeks of age all the livers become positive for adenomas with 66% having HCC [99]. Evidently, PTEN is important for the prevention of adipogenic and tumorigenic transformation, and PTEN-knockout mice are a novel model for NASH and especially for HCC. This model does not exhibit high levels of obesity which is one of the important factors of human NASH.

MAT1A$^{-/-}$ Mice Methionine adenosyltransferase-1A (MAT1A) is a liver-specific rate-limiting enzyme responsible for the synthesis of S-adenosylmethionine (SAM) in adult liver. SAM is a metabolite which affects a large spectrum of biological processes, including gene expression, proliferation, differentiation, and apoptosis [101]. MTA1A-null mice have decreased levels of hepatic SAM which renders them susceptible to spontaneous development of hepatic steatosis (by the age of 8 weeks) with increased rate of hepatocyte proliferation followed by steatohepatitis and hepatocellular carcinoma [102, 103].

AOX$^{-/-}$ Mice Acyl-coenzyme A oxidase (AOX) is the rate-limiting enzyme that catalyzes the reaction of peroxisomal fatty acid β-oxidation of very long chain fatty acids (VLCFA). Deficiency of AOX in *AOX* null mice results in defective peroxisomal β-oxidation of VLCFA, thereby accumulation of VLCFA in the liver. *AOX*$^{-/-}$ mice begin to exhibit severe fatty liver progression with microvesicular steatosis of hepatocytes in zones 2 and 3 of liver lobules at 7 days of age [104]. This early stage hepatic steatosis spontaneously results in steatohepatitis (1 month of age) with hepatic peroxisomal proliferation (4–5 months of age), increased peroxisomal fatty acid oxidation (6–7 months of age) and HCC (15 months of age) [59]. Absence of AOX results in a mouse model with hepatic inflammation, hepatocyte proliferation, and hepatocarcinogenesis [105] without fibrosis.

PPAR-α$^{-/-}$ Mice with Conditional Feeding Peroxisome proliferator-activated receptor-α (PPAR-α) is a member of steroid/nuclear receptor superfamily, which acts as a key regulator of genes involved in peroxisomal, mitochondrial, and microsomal fatty acid oxidation systems in the liver. Importantly, levels of PPAR-α decrease significantly in HFD-fed normal mice [106]. Fatty acid oxidation is inhibited in homozygous knocked out mice for PPAR-α, when subjected to fasting thereby developing severe hepatic steatosis [107].

SREBP-1c Transgenic Mice Sterol regulatory element-binding proteins (SREBPs) are a family of membrane-bound transcription factors that principally regulate lipid synthesis and are involved in adipocyte differentiation [108, 109]. Transgenic mice for this gene overexpress SREBP-1c in the adipose tissue creating a model of congenital lipodystrophy under the control of aP2 promoter resulting in severe insulin resistance, hyperglycemia with hepatic steatosis and steatohepatitis [110]. Importantly, these transgenic mice do not develop obesity due to decreased fat tissue, however a marked fatty liver with hepatic steatosis can be observed with 1–2 weeks of age due to increased lipid accumulation in the liver. Although these mice are not obese they exhibit insulin resistance, hypertriglyceridemia, elevated levels of transaminases, cholesterol, and inflammation with concomitant lower levels of adipocytokines, leptin, and adiponectin, by the age of 20 weeks [111]. At 30 weeks of age, overexpression of *SREBP-1c* in adipose tissue of these mice results in intralobular inflammation with hepatocyte ballooning and mallory hyaline bodies leading to pericellular fibrosis as encountered in steatohepatitis [112].

Nutritionally Manipulated Mouse Models of NAFLD/NASH with Obesity

HFD-induced Mouse Models of NAFLD/NASH HFD is often used to induce hepatic steatosis, steatohepatitis, fibrosis, and hepatocellular carcinoma in experimental animal models including mice. HFD significantly increases triglycerides as well as FFAs, resulting in obesity, hyperglycemia, and NAFLD/NASH [113]. The outcome of HFD-induced NAFLD/NASH (extent of hepatic steatosis, steatohepatitis, and fibrosis) varies with the composition of dietary fat, total fat content, type of fat in the diet, treatment duration as well as the strain of mice. Study conducted on two different strains of mice show that BALB/c male mice accumulates more lipid in the liver as compared to C57BL/6J mice when fed same HFD [114]. Gradual body fat accumulation with elevated leptin levels and 15 % increased body weight is observed in 4-week-old C57BL6/J male mice fed HFD containing 45 % of fat [115]. Chronic HFD feeding of male C57BL6/J mice exhibit impaired glucose tolerance with mild steatohepatitis [116]. Although, C57BL6 and 129S6/SvEvTac strains of mice are resistant to high-fat-induced obesity, expression of hepatic lipid synthesis and the gene profile of *SREBP1c*, *SREBP-2*, and Stearoyl-coA desaturase 1 increases by HFD [117, 118]. Intragastric forced overfeeding of a liquid HFD is a difficult method, requires equipment and expertise, but results in NASH with mild fibrosis, similar to humans [119]. Importantly, in a recent study on C57BL/6J mice, it was reported that modulation of diet from high fat to low fat can insinuate the prevention of the progression of NASH to HCC [120]. In this study, it was reported that changing diet from high fat to low fat, progression of liver pathologies toward HCC not only stops but gets reversed toward a condition without NASH and HCC [120]. Recently, in a combination study, HFD administration for 12 weeks after gold thioglucose (GTG) (i.p.) resulted in obesity with increased abdominal adiposity, glucose intolerance, insulin resistance, steatohepatitis, and fibrosis [121]. These models of HFD are useful as they recapitulate various parameters similar to hepatic steatosis, steatohepatitis (mild), fibrosis, and in some cases progression toward HCC [120].

HFD with Fructose-induced Mouse Models of NAFLD/NASH Fructose adds insult to the injury by inducing inflammation and provides a "second hit" in the progression of NAFLD to NASH [122]. Metabolites from fructose, act as toxins and cause inflammation in the pathogenesis of NASH [122]. A "western life style" in C57BL/6 exhibit NAFLD/NASH with severe steatohepatitis by combining trans fats and saturated fatty acids (like fast food) with high fructose corn syrup (like soft drinks high in sugar) and inactivity (sedentary life style) [123]. By the end of 16 weeks, mice become obese with elevated liver enzymes, proinflammtory and profibrogenic cytokines, and develop severe hepatic steatosis with necroinflammation [123]. Addition of fructose increases calories from sugar and along with HFD changes the dietary pattern toward a sugar-rich diet with excess caloric intake with an increasing risk of diabetes and the development of NAFLD in humans [124]. Addition of fructose to HFD induced severe NAFLD and has been reported to increase gut microflora, which elevates the level of endotoxin in portal blood, causing activation of Kupffer cells in the liver through an increase in proinflammatory cytokines and oxidative stress [125–130]. Addition of fructose (55 % in drinking water) to a diet high in saturated fat administered to C57BL/6 mice for 16 weeks, increases obesity and hepatic oxidative stress with significant fibrosis [131] without any carcinoma.

Nutritionally Manipulated Mouse Models of NAFLD/NASH Without Obesity

Atherogenic Diet Containing Cholesterol and Cholate Several studies suggest that dysregulated cholesterol metabolism is associated with NAFLD/NASH [132]. Cholesterol level increases in the liver due to increased uptake and decreased removal in the converted form of bile acids [133]. Excessive cholesterol deposition in liver induces lipotoxicity leading to ER stress, activation of unfolded protein response, and apoptosis of hepatocyte [134, 135]. There is a progressive increase in the cholesterol content in the livers of NAFLD and NASH patients [132]. Atherogenic diet (containing 1.25 % of cholesterol with 0.5 % cholate) administration to C57BL6/J mice, for 24 weeks, results in hepatic steatosis followed by severe steatohepatitis with cellular ballooning and fibrosis via cholesterol-induced oxidative stress and hepatic insulin resistance [136]. Importantly, aetherogenic diet supplementation with HFD containing 60 % of fat in the form of cocoa butter aggravates oxidative stress with an accelerated steatohepatitis within 12 weeks, with 9 % loss in body weight [136].

Methionine-/Choline-Deficient (MCD) Diet Methionine and choline are essential elements for hepatic β-oxidation and VLDL synthesis [58]. MCD diet containing high sucrose (40 %), fat (10 %) without methionine and choline induces severe liver injury via impaired hepatic VLDL secretion [137], oxidative stress [138, 139], and dysregulated cytokines and adipocytokines [140]. Mice fed MCD diet develop simple steatosis followed by severe steatohepatitis, inflammation, and hepatic fibrosis [141]. MCD diet model is one of the best established models to study inflammation and fibrosis in NAFLD, however, this model does not exhibit increased FA, obesity, or IR, which are recognized features of human NASH.

Although various mouse models have been developed to study NAFLD/NASH, not a single model displays the full spectrum of the human disease rather individual models can imitate few parameters of human NAFLD/NASH. One of the major reasons for these lacunae is the vastness of this disease which ranges from obese conditions with benign simple steatosis in the liver to hepatic steatosis with inflammation (steatohepatitis or NASH) to more severe liver fibrosis to irreversible cirrhosis to hepatocellular carcinoma. The present review will help investigators to choose an appropriate mouse model with one or few parameters of NAFLD/NASH based on their specific needs and hypothesis.

Bioactive Compounds as Therapeutic Options for NAFLD/NASH

NAFLD is thought to afflict more than 70 million Americans [142] with no effective treatment, hence the development of effective therapy is a pressing concern. Current safe and effective treatment for NAFLD largely relies on making changes in diet (low caloric) along with lifestyle to reduce body weight with exercise [143]. Few drugs targeting metabolic syndrome such as insulin-sensitizers (metformin), thiazolidinediones, statins, antioxidants, and Omega-3 PUFAs have been examined for NAFLD/NASH [143, 144] with mixed results. Metformin improves liver histology and liver function in leptin-deficient *ob/ob* mice [145]. Beneficial transient effects of metformin on liver chemistry were discovered in a small open-label clinical trial [120]. Positive results have been reported from the PIVENS trial while using a combination of TZDs with vitamin E to treat NAFLD [146]. To date, agents used to treat NASH focus primarily on improving insulin sensitization, and although this goal may be achievable, it is at the risk of potential liver toxicity with safety concern from long-term use of glitazone drugs [147, 148]. Since there are no effective and established therapies for NAFLD, the development of effective therapeutic options is a pressing concern. Globally, medicinal plants have been used to treat various liver complications [149]. A number of dietary natural compounds isolated from plants have been examined for their therapeutic and preventive potential against various health issues. Owing to their chemopreventive and therapeutic potential in various diseases, there is an increased interest in searching for bioactive compounds from plants that can both prevent and effectively treat NAFLD/NASH.

In this section, we will review various plant-based natural bioactive compounds used against NAFLD/NASH in mice. Since they have low to no side effects, a variety of bioactive compounds derived from various medicinal plants have been used to deliver therapeutic and preventive benefits against NAFLD/NASH in mice [65]. Silymarin derived from *Silybum marianum*; also known as "milk thistle" [150–156], curcumin from *Curcuma longa*; also known as "turmeric," Garlic essential oil, [157], have been used in the treatment of NAFLD/NASH using various mouse models. Some combination therapies have also been reported to have synergistic effect as therapeutic agent against NAFLD/NASH [158].

Silibinin Silibinin is the main polyphenolic component of milk thistle (*Silybum marianum*) seeds, used against obese-diabetic mice [152] as well as against various liver ailments, where it acts as hepatoprotectant [150, 154, 156, 159, 160]. Clinical findings on surrogate markers in NASH patients show efficacy of silibinin on insulin resistance and liver injury [151]. Silibinin along with phosphatidylcholine and co-formulated with vitamin E shows improvement in hepatic-histology in a multicenter clinical trial enrolling NAFLD patients [153]. A recent study on MCD diet-induced steatohepatitis in *db/db* mice reveal that silibinin modulates hepatic lipid homeostasis (reduced expression and activity of stearoyl-CoA desaturase-1) with significant reduction in oxidative stress (decreased reactive oxygen species, TBARS) by inhibiting NFκB activation [155]. Eight weeks old *db/db* mice administered with MCD diet for 4 weeks and treated with silibinin (20 mg/kg daily i.p.) revealed a significant decrease in liver enzymes with improved hepatic steatosis and steatohepatitis [155].

Fucoxanthin A carotenoid pigment from brown algae, (*Undaria pinnatifida*), an edible sea weed, is reported to have antiobesity and hypoglycemic effects in HFD-induced model of obesity in C57BL/6 mice. Five weeks of Fucoxanthin-administration to HFD-fed mice (10 weeks), reduces body fat accumulation and modulates blood glucose and insulin levels [161]. Similar dose-dependent effects of fucoxanthin are reported to modulate the expression of lipogenic and fatty acid beta-oxidation enzymes with decreased levels of leptin and increased levels of adiponectin, in an HFD-induced mice model of obesity [162]. Fucoxanthin beneficially modulates plasma as well as hepatic lipid metabolism by decreasing hepatic lipogenesis and increasing β-oxidation along with lowering blood glucose levels in HFD-fed C57BL/6 mice [163]. Fucoxanthin is reported to prevent development of obesity, diabetes in hyperglycemic and obese KK-A(y) mice [164]. Fucoxanthin decreases the expression level of Stearoyl-coenzyme A desaturase-1 SCD1 and modulates fatty acid composition of the liver via regulating leptin signaling in hyperleptinemic KK-A(y) mice, but fails to affect leptin-deficient *ob/ob* mice [165].

Lycopene Member of carotenoid family, lycopene, mainly derived from tomato, watermelon, papaya, orange grape fruit, has been shown to have antioxidant properties. Lycopene treatment improves hepatic steatosis in HFD-fed C57BL/6J mice via normalizing the downregulated miRNA-21. Lycopene upregulates HFD-induced decrease in fatty acid-binding protein 7 (FABP7) via miRNA-21 and resolves hepatic steatosis [166].

Nobiletin Daily dose of nobiletin, one of the polyphenolic compounds present in the peels of citrus fruit *Citrus depressa,* improves insulin resistance, blood glucose level, homeostatic model assessment (HOMA) score and plasma adiponectin levels in obese hyperglycemic *ob/ob* mice [167]. Oral gavage of nobiletin for 5 weeks in HFD diet administered C57BL/6J mice reduces weight of white adipose tissue, plasma triglycerides, and improved hyperglycemia [168].

Baicalein Baicalein is another polyphenolic compound derived from the roots of *Scutellaria baicalensis Georgi* and has protective potential against metabolic syndrome [169]. Baicalein has been shown to ameliorate metabolic syndrome in HFD-fed C57BL/6J mice, by reducing obesity, dyslipidemia, hepatic steatosis, diabetes, and insulin resistance via inhibiting SREBP-1c and activating AMPK [169].

Quercetin A flavonol compound derived from broccoli, leafy green vegetables including onion, has been reported to have therapeutic and preventive effects against hepatic manifestation of metabolic syndrome; NAFLD. Various mouse models have been reported to show efficacy of quercetin against NAFLD. Quercetin prevents HFD-induced obesity in C57BL/6 mice by modulating genes involved in lipid accumulation [170], increasing hepatic lipid omega-oxidation with decreased circulating lipid levels [171]. In MCD diet-fed C57BL/6 mice, daily oral administration of quercetin prevents progression of steatohepatitis (NASH) by attenuating levels of proinflammtory cytokines, profibrogenic genes [172], and reducing lipoperoxidation and DNA damage in the liver [173]. In leptin resistant *db/db* mice, quercetin improves blood glucose and levels of trigylcerides and cholesterol [174].

Epigallocatechin-3-gallate (EGCG) Catechins and theaflavins are the major polyphenolic compounds mainly derived from the leaves of green tea and black tea (*Camellia sinensis*). Epigallocatechin-3-gallate (EGCG) is the most abundant (~85%) polyphenolic compound in green tea. EGCG administration, in dose-dependent manner, decreases oxidative stress and hepatotoxicity in CCL_4-treated Institute of Cancer Research (ICR) mice (http://www.criver.com/products-services/basic-research/find-a-model/cd-1-mouse) [175]. Green tea extract has also been reported to be hepatoprotective in *ob/ob* mice and ameliorates steatohepatitis [176]. Administration of EGCG to HFD-fed C57BL/6J, prevents the development of obesity, metabolic syndrome including hepatic steatosis with decreased liver weight, levels of TG, and improved liver enzymes, [177–180]. Theaflavins derived from black tea have properties similar to EGCG derived from green tea. In MCD-fed C57BL/6 mice, a mixture of theaflavins extracted and purified from black tea prevent progression of liver injury from hepatic steatosis to steatohepatitis by acting as antioxidant, anti-inflammatory, and antiapoptotic agent [181].

Resveratrol Resveratrol is one of the strongest antioxidant and anti-inflammatory agents derived from the skins of red grapes. Resveratrol has been tested in a variety of NAFLD/NASH mouse models for its therapeutic and preventive potential against progression of hepatic steatosis by reducing levels of triglycerides, cholesterol, inflammation, and insulin resistance. In HFD-fed C57BL/6 mice, administration of resveratrol along with HFD diet, reverses the deleterious effects of obesity and reduces inflammation [182]. Importantly, low dose of resveratrol is found to be more effective in ameliorating visceral adiposity and hepatic steatosis via decreasing total triglycerides and cholesterol [183]. Resveratrol has been reported to have protective effect on atherogenic diet-induced NASH in C57BL/6 mice via favorable alteration of genes involved in fat metabolism [184]. In C57BL/6 mice fed a HFD, therapeutic effect of resveratrol has been reported to improve insulin resistance only in insulin-resistant tissue such as liver and white adipose tissue without affecting skeletal muscle, indicating tissue-specific and metabolism-dependent effects of resveratrol [185]. In insulin-resistant KKA^y mice, resveratrol improves insulin sensitivity via upregulation of silent information regulator 1 (Sirt 1)-AMPK in the liver [186].

Curcumin Curcumin, a polyphenolic compound, is a yellow-colored lyophobic pigment derived mainly from the rhizomes of *Curcuma longa* commonly known as turmeric. Turmeric is extensively used as a medicinal plant in traditional medicine for the treatment of various diseases such as inflammation of joints, skin disease, infections, ulcers, jaundice, and other liver disorders [187]. The protective and therapeutic potential of curcumin against NAFLD/NASH has been reported in several mouse models [188]. Curcumin reduces body weight, serum triglycerides, improves hepatic steatosis via significant decrease in inflammation and improved mitochondrial dysfunction in HFD-fed C57BL/6 mice [189]. In a similar model of HFD-fed C57BL/6 mice along with leptin deficient *ob/ob* mice, curcumin is reported to treat diabetes by improving insulin resistance, ameliorating inflammation in white adipose tissue, and increasing adiponectin levels [190].

Garlic A recent study reports the use of garlic essential oil (GEO) and major bioactive component diallyl disulfide (DADS) against obesity-driven NAFLD in HFD-fed C57BL/6J mice [191]. GEO and DADS act as antiobesity and anti-hyperlipidemic agents in a dose-dependent manner and reduce HFD-induced lipid accumulation, inflammation, and oxidative damage. Garlic essential oil as well as the active ingredient (a) reduce HFD-induced increased-body-weight, adipose tissue weight, levels of proinflammatory cytokines in liver, (b) down-regulate gene expression for sterol regulatory element binding protein-1c (SREBP-1c), acetyl-CoA carboxylase, fatty acid synthase, and 3-hydroxy-3-methylglutaryl-coenzyme A reductase and stimulate (PPAR-α) and carnitine palmitoyltransferase-1 with improved liver enzymes [191].

Combination of Caffeine, Arginine, Soy Isoflavones, and L-Carnitine In obese, KK mice on low-fat diet, combination of Caffeine, Arginine, Soy Isoflavones, and L-Carnitine significantly inhibits hepatic-lipogenesis with an overall decrease in body weight, adipose tissue weight, and triglyceride levels in the plasma and liver [158].

Conclusions

NAFLD and NASH are the hepatic manifestations of metabolic syndrome, which encompass a wide spectrum of metabolic dysfunctions. The incidence of NAFLD/NASH has been increasing worldwide, with concomitant increased prevalence of obesity, type 2 diabetes with elevated levels of triglycerides and cholesterol. The pathogenesis of NAFLD/NASH is still not very clear and is intricately entangled with metabolic manifestations of other organs. Mouse models are very important tools to understand and elucidate the pathology, pathogenesis, and treatment approaches for NAFLD/NASH. Although, no single mouse model provides full spectrum of the disease, but cumulatively these models have provided important information about various factors (obesity, metabolic syndrome, type 2 diabetes, insulin resistance, hepatotoxicity due to fat accumulation, ER stress, hepatic inflammation, and Hepato Cellular Cancer), signaling nodes responsible for the disease. More importantly these mouse models have provided surrogate markers for NAFLD/NASH and probable therapeutic and/or preventive treatment regimens. Dietary natural compounds derived from plants have provided, "to some extent," variety of therapeutic and preventive treatment opportunities for NAFLD/NASH. Proper validation with respect to dosage, combinations with present medications and contraindications of these molecules/compounds is needed for the treatment and/or prevention of NAFLD/NASH.

References

1. Ng M, Fleming T, Robinson M, Thomson B, Graetz N, Margono C, et al. Global, regional, and national prevalence of overweight and obesity in children and adults during 1980–2013: a systematic analysis for the Global Burden of Disease Study 2013. Lancet. 2014;384(9945):766–81.
2. Kassebaum NJ, Bertozzi-Villa A, Coggeshall MS, Shackelford KA, Steiner C, Heuton KR, et al. Global, regional, and national levels and causes of maternal mortality during 1990–2013: a systematic analysis for the Global Burden of Disease Study 2013. Lancet. 2014;384(9947):980–1004.
3. Wang H, Liddell CA, Coates MM, Mooney MD, Levitz CE, Schumacher AE, et al. Global, regional, and national levels of neonatal, infant, and under-5 mortality during 1990–2013: a systematic analysis for the Global Burden of Disease Study 2013. Lancet. 2014;384(9947):957–79.
4. Rocchini AP. Childhood obesity and a diabetes epidemic. N Engl J Med. 2002;346(11):854–5.
5. Brug J, van Stralen MM, Te Velde SJ, Chinapaw MJ, De Bourdeaudhuij I, Lien N, et al. Differences in weight status and energy-balance related behaviors among schoolchildren across Europe: the ENERGY-project. PLoS One. 2012;7(4):e34742. PMCID: 3338827.
6. Flegal KM, Carroll MD, Ogden CL, Johnson CL. Prevalence and trends in obesity among US adults, 1999–2000. JAMA. 2002;288(14):1723–7.
7. Hedley AA, Ogden CL, Johnson CL, Carroll MD, Curtin LR, Flegal KM. Prevalence of overweight and obesity among US children, adolescents, and adults, 1999–2002. JAMA. 2004;291(23):2847–50.

8. Adams JP, Murphy PG. Obesity in anaesthesia and intensive care. Br J Anaesth. 2000;85(1):91–108.
9. Ogden CL, Carroll MD, Kit BK, Flegal KM. Prevalence of childhood and adult obesity in the United States, 2011–2012. JAMA. 2014;311(8):806–14.
10. Calle EE, Rodriguez C, Walker-Thurmond K, Thun MJ. Overweight, obesity, and mortality from cancer in a prospectively studied cohort of U.S. adults. N Engl J Med. 2003;348(17):1625–38.
11. Calle EE, Kaaks R. Overweight, obesity and cancer: epidemiological evidence and proposed mechanisms. Nat Rev Cancer. 2004;4(8):579–91.
12. Saunders D, Seidel D, Allison M, Lyratzopoulos G. Systematic review: the association between obesity and hepatocellular carcinoma—epidemiological evidence. Aliment Pharmacol Ther. 2010;31(10):1051–63.
13. Schlesinger S, Aleksandrova K, Pischon T, Fedirko V, Jenab M, Trepo E, et al. Abdominal obesity, weight gain during adulthood and risk of liver and biliary tract cancer in a European cohort. Int J Cancer. 2013;132(3):645–57.
14. Park MH, Falconer C, Viner RM, Kinra S. The impact of childhood obesity on morbidity and mortality in adulthood: a systematic review. Obes Rev. 2012;13(11):985–1000.
15. Reilly JJ, Kelly J. Long-term impact of overweight and obesity in childhood and adolescence on morbidity and premature mortality in adulthood: systematic review. Int J Obes (Lond). 2011;35(7):891–8.
16. Baker JL, Olsen LW, Sorensen TI. Childhood body-mass index and the risk of coronary heart disease in adulthood. N Engl J Med. 2007;357(23):2329–37. PMCID: 3062903.
17. Nair S, Mason A, Eason J, Loss G, Perrillo RP. Is obesity an independent risk factor for hepatocellular carcinoma in cirrhosis? Hepatology. 2002;36(1):150–5.
18. Chow WH, Gridley G, Fraumeni JF, Jr., Jarvholm B. Obesity, hypertension, and the risk of kidney cancer in men. N Engl J Med. 2000;343(18):1305–11.
19. Richardson LC, Thomas C, Bowman BA. Obesity and endometrial cancer: challenges for public health action. Womens Health (Lond Engl). 2009;5(6):595–7.
20. Schapira DV, Kumar NB, Lyman GH, Cox CE. Abdominal obesity and breast cancer risk. Ann Intern Med. 1990;112(3):182–6.
21. Sinicrope FA, Dannenberg AJ. Obesity and breast cancer prognosis: weight of the evidence. J Clin Oncol. 2011;29(1):4–7.
22. Sinicrope FA, Foster NR, Sargent DJ, O'Connell MJ, Rankin C. Obesity is an independent prognostic variable in colon cancer survivors. Clin Cancer Res. 2010;16(6):1884–93. PMCID: 2948494.
23. Haslam DW, James WP. Obesity. Lancet. 2005;366(9492):1197–209.
24. Regimbeau JM, Colombat M, Mognol P, Durand F, Abdalla E, Degott C, et al. Obesity and diabetes as a risk factor for hepatocellular carcinoma. Liver Transpl. 2004;10(2 Suppl 1):S69–73.
25. Cheung CR, Kelly DA. Non-alcoholic fatty liver disease in children. BMJ. 2011;343:d4460.
26. Mencin AA, Lavine JE. Nonalcoholic fatty liver disease in children. Curr Opin Clin Nutr Metab Care. 2011;14(2):151–7.
27. Sinha R, Fisch G, Teague B, Tamborlane WV, Banyas B, Allen K, et al. Prevalence of impaired glucose tolerance among children and adolescents with marked obesity. N Engl J Med. 2002;346(11):802–10.
28. Kaur J. A comprehensive review on metabolic syndrome. Cardiol Res Pract. 2014;2014:943162. PMCID: 3966331.
29. Alberti KG, Zimmet P, Shaw J. The metabolic syndrome–a new worldwide definition. Lancet. 2005;366(9491):1059–62.
30. Boppidi H, Daram SR. Nonalcoholic fatty liver disease: hepatic manifestation of obesity and the metabolic syndrome. Postgrad Med. 2008;120(2):E01–7.
31. Alberti KG, Eckel RH, Grundy SM, Zimmet PZ, Cleeman JI, Donato KA, et al. Harmonizing the metabolic syndrome: a joint interim statement of the International Diabetes Federation Task Force on Epidemiology and Prevention; National Heart, Lung, and Blood Institute;

American Heart Association; World Heart Federation; International Atherosclerosis Society; and International Association for the Study of Obesity. Circulation. 2009;120(16):1640–5.
32. Grundy SM, Cleeman JI, Daniels SR, Donato KA, Eckel RH, Franklin BA, et al. Diagnosis and management of the metabolic syndrome: an American Heart Association/National Heart, Lung, and Blood Institute Scientific Statement. Circulation. 2005;112(17):2735–52.
33. Lazo M, Hernaez R, Eberhardt MS, Bonekamp S, Kamel I, Guallar E, et al. Prevalence of nonalcoholic fatty liver disease in the United States: the Third National Health and Nutrition Examination Survey, 1988–1994. Am J Epidemiol. 2013;178(1):38–45. PMCID: 3698993.
34. Lazo M, Clark JM. The epidemiology of nonalcoholic fatty liver disease: a global perspective. Semin Liver Dis. 2008;28(4):339–50.
35. Younossi ZM, Stepanova M, Afendy M, Fang Y, Younossi Y, Mir H, et al. Changes in the prevalence of the most common causes of chronic liver diseases in the United States from 1988 to 2008. Clin Gastroenterol Hepatol. 2011;9(6):524–30. e1; quiz e60.
36. Ruhl CE, Everhart JE. Epidemiology of nonalcoholic fatty liver. Clin Liver Dis. 2004;8(3):501–19, vii.
37. Serfaty L, Lemoine M. Definition and natural history of metabolic steatosis: clinical aspects of NAFLD, NASH and cirrhosis. Diabetes Metab. 2008;34(6 Pt 2):634–7.
38. Starley BQ, Calcagno CJ, Harrison SA. Nonalcoholic fatty liver disease and hepatocellular carcinoma: a weighty connection. Hepatology. 2010;51(5):1820–32.
39. Bellentani S, Bedogni G, Tiribelli C. Liver and heart: a new link? J Hepatol. 2008;49(2):300–2.
40. Yu JH, Lee KS, Lee SY, Hong AR, Park YS. The association of cardiovascular risk factors with nonalcoholic fatty liver disease in health checkup examinees. J Prev Med Public Health. 2008;41(6):407–12.
41. Fracanzani AL, Burdick L, Raselli S, Pedotti P, Grigore L, Santorelli G, et al. Carotid artery intima-media thickness in nonalcoholic fatty liver disease. Am J Med. 2008;121(1):72–8.
42. Marchesini G, Brizi M, Bianchi G, Tomassetti S, Bugianesi E, Lenzi M, et al. Nonalcoholic fatty liver disease: a feature of the metabolic syndrome. Diabetes. 2001;50(8):1844–50.
43. Caldwell S, Argo C. The natural history of non-alcoholic fatty liver disease. Dig Dis. 2010;28(1):162–8.
44. Brunt EM, Tiniakos DG. Histopathology of nonalcoholic fatty liver disease. World J Gastroenterol. 2010;16(42):5286–96. PMCID: 2980677.
45. Soderberg C, Stal P, Askling J, Glaumann H, Lindberg G, Marmur J, et al. Decreased survival of subjects with elevated liver function tests during a 28-year follow-up. Hepatology. 2010;51(2):595–602.
46. Rafiq N, Bai C, Fang Y, Srishord M, McCullough A, Gramlich T, et al. Long-term follow-up of patients with nonalcoholic fatty liver. Clin Gastroenterol Hepatol. 2009;7(2):234–8.
47. Takamatsu S, Noguchi N, Kudoh A, Nakamura N, Kawamura T, Teramoto K, et al. Influence of risk factors for metabolic syndrome and non-alcoholic fatty liver disease on the progression and prognosis of hepatocellular carcinoma. Hepatogastroenterology. 2008;55(82–83):609–14.
48. Petta S, Craxi A. Hepatocellular carcinoma and non-alcoholic fatty liver disease: from a clinical to a molecular association. Curr Pharm Des. 2010;16(6):741–52.
49. Baffy G, Brunt EM, Caldwell SH. Hepatocellular carcinoma in non-alcoholic fatty liver disease: an emerging menace. J Hepatol. 2012;56(6):1384–91.
50. Tiniakos DG, Vos MB, Brunt EM. Nonalcoholic fatty liver disease: pathology and pathogenesis. Annu Rev Pathol. 2010;5:145–71.
51. Bedogni G, Miglioli L, Masutti F, Tiribelli C, Marchesini G, Bellentani S. Prevalence of and risk factors for nonalcoholic fatty liver disease: the Dionysos nutrition and liver study. Hepatology. 2005;42(1):44–52.
52. Day CP. Natural history of NAFLD: remarkably benign in the absence of cirrhosis. Gastroenterology. 2005;129(1):375–8.
53. Harrison SA, Torgerson S, Hayashi PH. The natural history of nonalcoholic fatty liver disease: a clinical histopathological study. Am J Gastroenterol. 2003;98(9):2042–7.

54. Sanyal AJ, Banas C, Sargeant C, Luketic VA, Sterling RK, Stravitz RT, et al. Similarities and differences in outcomes of cirrhosis due to nonalcoholic steatohepatitis and hepatitis C. Hepatology. 2006;43(4):682–9.
55. Ascha MS, Hanouneh IA, Lopez R, Tamimi TA, Feldstein AF, Zein NN. The incidence and risk factors of hepatocellular carcinoma in patients with nonalcoholic steatohepatitis. Hepatology. 2010;51(6):1972–8.
56. Takahashi Y, Fukusato T. Pediatric nonalcoholic fatty liver disease: overview with emphasis on histology. World J Gastroenterol. 2010;16(42):5280–5. PMCID: 2980676.
57. Day CP, James OF. Steatohepatitis: a tale of two "hits"? Gastroenterology. 1998;114(4):842–5.
58. Anstee QM, Goldin RD. Mouse models in non-alcoholic fatty liver disease and steatohepatitis research. Int J Exp Pathol. 2006;87(1):1–16. PMCID: 2517349.
59. London RM, George J. Pathogenesis of NASH: animal models. Clin Liver Dis. 2007;11(1):55–74, viii.
60. Wanless IR, Shiota K. The pathogenesis of nonalcoholic steatohepatitis and other fatty liver diseases: a four-step model including the role of lipid release and hepatic venular obstruction in the progression to cirrhosis. Semin Liver Dis. 2004;24(1):99–106.
61. Farrell GC, Larter CZ. Nonalcoholic fatty liver disease: from steatosis to cirrhosis. Hepatology. 2006;43(2 Suppl 1):S99–S112.
62. Cohen JC, Horton JD, Hobbs HH. Human fatty liver disease: old questions and new insights. Science. 2011;332(6037):1519–23. PMCID: 3229276.
63. Fan JG, Xu ZJ, Wang GL. Effect of lactulose on establishment of a rat non-alcoholic steatohepatitis model. World J Gastroenterol. 2005;11(32):5053–6.
64. Larter CZ. Not all models of fatty liver are created equal: understanding mechanisms of steatosis development is important. J Gastroenterol Hepatol. 2007;22(9):1353–4.
65. Koteish A, Diehl AM. Animal models of steatosis. Semin Liver Dis. 2001;21(1):89–104.
66. Imajo K, Yoneda M, Kessoku T, Ogawa Y, Maeda S, Sumida Y, et al. Rodent models of nonalcoholic fatty liver disease/nonalcoholic steatohepatitis. Int J Mol Sci. 2013;14(11):21833–57. PMCID: 3856038.
67. Takahashi Y, Soejima Y, Fukusato T. Animal models of nonalcoholic fatty liver disease/nonalcoholic steatohepatitis. World J Gastroenterol. 2012;18(19):2300–8. PMCID: 3353364.
68. Kleiner DE, Brunt EM, Van Natta M, Behling C, Contos MJ, Cummings OW, et al. Design and validation of a histological scoring system for nonalcoholic fatty liver disease. Hepatology. 2005;41(6):1313–21.
69. Campfield LA, Smith FJ, Burn P. The OB protein (leptin) pathway—a link between adipose tissue mass and central neural networks. Horm Metab Res. 1996;28(12):619–32.
70. Zhang Y, Proenca R, Maffei M, Barone M, Leopold L, Friedman JM. Positional cloning of the mouse obese gene and its human homologue. Nature. 1994;372(6505):425–32.
71. Pelleymounter MA, Cullen MJ, Baker MB, Hecht R, Winters D, Boone T, et al. Effects of the obese gene product on body weight regulation in ob/ob mice. Science. 1995;269(5223):540–3.
72. Halaas JL, Gajiwala KS, Maffei M, Cohen SL, Chait BT, Rabinowitz D, et al. Weight-reducing effects of the plasma protein encoded by the obese gene. Science. 1995;269(5223):543–6.
73. Diehl AM. Lessons from animal models of NASH. Hepatol Res. 2005;33(2):138–44.
74. Brix AE, Elgavish A, Nagy TR, Gower BA, Rhead WJ, Wood PA. Evaluation of liver fatty acid oxidation in the leptin-deficient obese mouse. Mol Genet Metab. 2002;75(3):219–26.
75. Yang SQ, Lin HZ, Lane MD, Clemens M, Diehl AM. Obesity increases sensitivity to endotoxin liver injury: implications for the pathogenesis of steatohepatitis. Proc Natl Acad Sci U S A. 1997;94(6):2557–62. PMCID: 20127.
76. Chavin KD, Yang S, Lin HZ, Chatham J, Chacko VP, Hoek JB, et al. Obesity induces expression of uncoupling protein-2 in hepatocytes and promotes liver ATP depletion. J Biol Chem. 1999;274(9):5692–700.
77. Faggioni R, Fantuzzi G, Gabay C, Moser A, Dinarello CA, Feingold KR, et al. Leptin deficiency enhances sensitivity to endotoxin-induced lethality. Am J Physiol. 1999;276(1 Pt 2):R136–42.

78. Saxena NK, Ikeda K, Rockey DC, Friedman SL, Anania FA. Leptin in hepatic fibrosis: evidence for increased collagen production in stellate cells and lean littermates of ob/ob mice. Hepatology. 2002;35(4):762–71. PMCID: 2935193.
79. Saxena NK, Saliba G, Floyd JJ, Anania FA. Leptin induces increased alpha2(I) collagen gene expression in cultured rat hepatic stellate cells. J Cell Biochem. 2003;89(2):311–20. PMCID: 2925439.
80. Saxena NK, Titus MA, Ding X, Floyd J, Srinivasan S, Sitaraman SV, et al. Leptin as a novel profibrogenic cytokine in hepatic stellate cells: mitogenesis and inhibition of apoptosis mediated by extracellular regulated kinase (Erk) and Akt phosphorylation. Faseb J. 2004;18(13):1612–4. PMCID: 2924993.
81. Leclercq IA, Farrell GC, Schriemer R, Robertson GR. Leptin is essential for the hepatic fibrogenic response to chronic liver injury. J Hepatol. 2002;37(2):206–13.
82. Ikejima K, Honda H, Yoshikawa M, Hirose M, Kitamura T, Takei Y, et al. Leptin augments inflammatory and profibrogenic responses in the murine liver induced by hepatotoxic chemicals. Hepatology. 2001;34(2):288–97.
83. Chen H, Charlat O, Tartaglia LA, Woolf EA, Weng X, Ellis SJ, et al. Evidence that the diabetes gene encodes the leptin receptor: identification of a mutation in the leptin receptor gene in db/db mice. Cell. 1996;84(3):491–5.
84. Tartaglia LA, Dembski M, Weng X, Deng N, Culpepper J, Devos R, et al. Identification and expression cloning of a leptin receptor, OB-R. Cell. 1995;83(7):1263–71.
85. Wortham M, He L, Gyamfi M, Copple BL, Wan YJ. The transition from fatty liver to NASH associates with SAMe depletion in db/db mice fed a methionine choline-deficient diet. Dig Dis Sci. 2008;53(10):2761–74. PMCID: 3991247.
86. Sahai A, Malladi P, Pan X, Paul R, Melin-Aldana H, Green RM, et al. Obese and diabetic db/db mice develop marked liver fibrosis in a model of nonalcoholic steatohepatitis: role of short-form leptin receptors and osteopontin. Am J Physiol Gastrointest Liver Physiol. 2004;287(5):G1035–43.
87. Schattenberg JM, Galle PR. Animal models of non-alcoholic steatohepatitis: of mice and man. Dig Dis. 2010;28(1):247–54.
88. Okumura K, Ikejima K, Kon K, Abe W, Yamashina S, Enomoto N, et al. Exacerbation of dietary steatohepatitis and fibrosis in obese, diabetic KK-A(y) mice. Hepatol Res. 2006;36(3):217–28.
89. Balthasar N, Dalgaard LT, Lee CE, Yu J, Funahashi H, Williams T, et al. Divergence of melanocortin pathways in the control of food intake and energy expenditure. Cell. 2005;123(3):493–505.
90. Vaisse C, Clement K, Durand E, Hercberg S, Guy-Grand B, Froguel P. Melanocortin-4 receptor mutations are a frequent and heterogeneous cause of morbid obesity. J Clin Invest. 2000;106(2):253–62. PMCID: 314306.
91. Huszar D, Lynch CA, Fairchild-Huntress V, Dunmore JH, Fang Q, Berkemeier LR, et al. Targeted disruption of the melanocortin-4 receptor results in obesity in mice. Cell. 1997;88(1):131–41.
92. Itoh M, Suganami T, Nakagawa N, Tanaka M, Yamamoto Y, Kamei Y, et al. Melanocortin 4 receptor-deficient mice as a novel mouse model of nonalcoholic steatohepatitis. Am J Pathol. 2011;179(5):2454–63. PMCID: 3204024.
93. Albarado DC, McClaine J, Stephens JM, Mynatt RL, Ye J, Bannon AW, et al. Impaired coordination of nutrient intake and substrate oxidation in melanocortin-4 receptor knockout mice. Endocrinology. 2004;145(1):243–52.
94. Sutton GM, Trevaskis JL, Hulver MW, McMillan RP, Markward NJ, Babin MJ, et al. Diet-genotype interactions in the development of the obese, insulin-resistant phenotype of C57BL/6J mice lacking melanocortin-3 or -4 receptors. Endocrinology. 2006;147(5):2183–96. PMCID: 2735103.
95. Li J, Yen C, Liaw D, Podsypanina K, Bose S, Wang SI, et al. PTEN, a putative protein tyrosine phosphatase gene mutated in human brain, breast, and prostate cancer. Science. 1997;275(5308):1943–7.

96. Maehama T, Dixon JE. The tumor suppressor, PTEN/MMAC1, dephosphorylates the lipid second messenger, phosphatidylinositol 3,4,5-trisphosphate. J Biol Chem. 1998;273(22):13375–8.
97. Stiles B, Wang Y, Stahl A, Bassilian S, Lee WP, Kim YJ, et al. Liver-specific deletion of negative regulator Pten results in fatty liver and insulin hypersensitivity [corrected]. Proc Natl Acad Sci U S A. 2004;101(7):2082–7. PMCID: 357055.
98. Sato W, Horie Y, Kataoka E, Ohshima S, Dohmen T, Iizuka M, et al. Hepatic gene expression in hepatocyte-specific Pten deficient mice showing steatohepatitis without ethanol challenge. Hepatol Res. 2006;34(4):256–65.
99. Horie Y, Suzuki A, Kataoka E, Sasaki T, Hamada K, Sasaki J, et al. Hepatocyte-specific Pten deficiency results in steatohepatitis and hepatocellular carcinomas. J Clin Invest. 2004;113(12):1774–83. PMCID: 420505.
100. Watanabe S, Horie Y, Kataoka E, Sato W, Dohmen T, Ohshima S, et al. Non-alcoholic steatohepatitis and hepatocellular carcinoma: lessons from hepatocyte-specific phosphatase and tensin homolog (PTEN)-deficient mice. J Gastroenterol Hepatol. 2007;22(Suppl 1):S96–100.
101. Mato JM, Corrales FJ, Lu SC, Avila MA. S-Adenosylmethionine: a control switch that regulates liver function. Faseb J. 2002;16(1):15–26.
102. Lu SC, Alvarez L, Huang ZZ, Chen L, An W, Corrales FJ, et al. Methionine adenosyltransferase 1A knockout mice are predisposed to liver injury and exhibit increased expression of genes involved in proliferation. Proc Natl Acad Sci U S A. 2001;98(10):5560–5. PMCID: 33252.
103. Martinez-Chantar ML, Corrales FJ, Martinez-Cruz LA, Garcia-Trevijano ER, Huang ZZ, Chen L, et al. Spontaneous oxidative stress and liver tumors in mice lacking methionine adenosyltransferase 1A. Faseb J. 2002;16(10):1292–4.
104. Cook WS, Jain S, Jia Y, Cao WQ, Yeldandi AV, Reddy JK, et al. Peroxisome proliferator-activated receptor alpha-responsive genes induced in the newborn but not prenatal liver of peroxisomal fatty acyl-CoA oxidase null mice. Exp Cell Res. 2001;268(1):70–6.
105. Fan CY, Pan J, Usuda N, Yeldandi AV, Rao MS, Reddy JK. Steatohepatitis, spontaneous peroxisome proliferation and liver tumors in mice lacking peroxisomal fatty acyl-CoA oxidase. Implications for peroxisome proliferator-activated receptor alpha natural ligand metabolism. J Biol Chem. 1998;273(25):15639–45.
106. Svegliati-Baroni G, Candelaresi C, Saccomanno S, Ferretti G, Bachetti T, Marzioni M, et al. A model of insulin resistance and nonalcoholic steatohepatitis in rats: role of peroxisome proliferator-activated receptor-alpha and n-3 polyunsaturated fatty acid treatment on liver injury. Am J Pathol. 2006;169(3):846–60. PMCID: 1698833.
107. Kersten S, Seydoux J, Peters JM, Gonzalez FJ, Desvergne B, Wahli W. Peroxisome proliferator-activated receptor alpha mediates the adaptive response to fasting. J Clin Invest. 1999;103(11):1489–98. PMCID: 408372.
108. Yokoyama C, Wang X, Briggs MR, Admon A, Wu J, Hua X, et al. SREBP-1, a basic-helix-loop-helix-leucine zipper protein that controls transcription of the low density lipoprotein receptor gene. Cell. 1993;75(1):187–97.
109. Wang X, Sato R, Brown MS, Hua X, Goldstein JL. SREBP-1, a membrane-bound transcription factor released by sterol-regulated proteolysis. Cell. 1994;77(1):53–62.
110. Shimomura I, Hammer RE, Richardson JA, Ikemoto S, Bashmakov Y, Goldstein JL, et al. Insulin resistance and diabetes mellitus in transgenic mice expressing nuclear SREBP-1c in adipose tissue: model for congenital generalized lipodystrophy. Genes Dev. 1998;12(20):3182–94. PMCID: 317215.
111. Nakayama H, Otabe S, Ueno T, Hirota N, Yuan X, Fukutani T, et al. Transgenic mice expressing nuclear sterol regulatory element-binding protein 1c in adipose tissue exhibit liver histology similar to nonalcoholic steatohepatitis. Metabolism. 2007;56(4):470–5.
112. Chitturi S, Abeygunasekera S, Farrell GC, Holmes-Walker J, Hui JM, Fung C, et al. NASH and insulin resistance: insulin hypersecretion and specific association with the insulin resistance syndrome. Hepatology. 2002;35(2):373–9.

113. Fujiwara M, Ito R, Fukamizu H. Bipedicle flap reconstruction for complex longitudinal dorsal tissue defects of the digits. Hand Surg. 2007;12(3):191–7.
114. Sato R, Uchida K, Kobayashi S, Yayama T, Kokubo Y, Nakajima H, et al. Ossification of the posterior longitudinal ligament of the cervical spine: histopathological findings around the calcification and ossification front. J Neurosurg Spine. 2007;7(2):174–83.
115. Tanaka K, Ito M, Kodama M, Maruyama H, Hoyano M, Mitsuma W, et al. Longitudinal change in renal function in patients with idiopathic dilated cardiomyopathy without renal insufficiency at initial diagnosis. Circ J. 2007;71(12):1927–31.
116. Ito M, Suzuki J, Tsujioka S, Sasaki M, Gomori A, Shirakura T, et al. Longitudinal analysis of murine steatohepatitis model induced by chronic exposure to high-fat diet. Hepatol Res. 2007;37(1):50–7.
117. Toye AA, Lippiat JD, Proks P, Shimomura K, Bentley L, Hugill A, et al. A genetic and physiological study of impaired glucose homeostasis control in C57BL/6J mice. Diabetologia. 2005;48(4):675–86.
118. Biddinger SB, Almind K, Miyazaki M, Kokkotou E, Ntambi JM, Kahn CR. Effects of diet and genetic background on sterol regulatory element-binding protein-1c, stearoyl-CoA desaturase 1, and the development of the metabolic syndrome. Diabetes. 2005;54(5):1314–23.
119. Deng QG, She H, Cheng JH, French SW, Koop DR, Xiong S, et al. Steatohepatitis induced by intragastric overfeeding in mice. Hepatology. 2005;42(4):905–14.
120. Hill-Baskin AE, Markiewski MM, Buchner DA, Shao H, DeSantis D, Hsiao G, et al. Diet-induced hepatocellular carcinoma in genetically predisposed mice. Hum Mol Genet. 2009;18(16):2975–88. PMCID: 2714725.
121. Ogasawara M, Hirose A, Ono M, Aritake K, Nozaki Y, Takahashi M, et al. A novel and comprehensive mouse model of human non-alcoholic steatohepatitis with the full range of dysmetabolic and histological abnormalities induced by gold thioglucose and a high-fat diet. Liver Int. 2011;31(4):542–51.
122. Nomura K, Yamanouchi T. The role of fructose-enriched diets in mechanisms of nonalcoholic fatty liver disease. J Nutr Biochem. 2012;23(3):203–8.
123. Tetri LH, Basaranoglu M, Brunt EM, Yerian LM, Neuschwander-Tetri BA. Severe NAFLD with hepatic necroinflammatory changes in mice fed trans fats and a high-fructose corn syrup equivalent. Am J Physiol Gastrointest Liver Physiol. 2008;295(5):G987–95. PMCID: 4059366.
124. Tappy L, Le KA, Tran C, Paquot N. Fructose and metabolic diseases: new findings, new questions. Nutrition. 2010;26(11–12):1044–9.
125. Spruss A, Kanuri G, Wagnerberger S, Haub S, Bischoff SC, Bergheim I. Toll-like receptor 4 is involved in the development of fructose-induced hepatic steatosis in mice. Hepatology. 2009;50(4):1094–104.
126. Thuy S, Ladurner R, Volynets V, Wagner S, Strahl S, Konigsrainer A, et al. Nonalcoholic fatty liver disease in humans is associated with increased plasma endotoxin and plasminogen activator inhibitor 1 concentrations and with fructose intake. J Nutr. 2008;138(8):1452–5.
127. Bergheim I, Weber S, Vos M, Kramer S, Volynets V, Kaserouni S, et al. Antibiotics protect against fructose-induced hepatic lipid accumulation in mice: role of endotoxin. J Hepatol. 2008;48(6):983–92.
128. Kanuri G, Spruss A, Wagnerberger S, Bischoff SC, Bergheim I. Fructose-induced steatosis in mice: role of plasminogen activator inhibitor-1, microsomal triglyceride transfer protein and NKT cells. Lab Invest. 2011;91(6):885–95.
129. Kanuri G, Spruss A, Wagnerberger S, Bischoff SC, Bergheim I. Role of tumor necrosis factor alpha (TNFα) in the onset of fructose-induced nonalcoholic fatty liver disease in mice. J Nutr Biochem. 2011;22(6):527–34.
130. Tsuchiya H, Ebata Y, Sakabe T, Hama S, Kogure K, Shiota G. High-fat, high-fructose diet induces hepatic iron overload via a hepcidin-independent mechanism prior to the onset of liver steatosis and insulin resistance in mice. Metabolism. 2013;62(1):62–9.

131. Kohli R, Kirby M, Xanthakos SA, Softic S, Feldstein AE, Saxena V, et al. High-fructose, medium chain trans fat diet induces liver fibrosis and elevates plasma coenzyme Q9 in a novel murine model of obesity and nonalcoholic steatohepatitis. Hepatology. 2010;52(3):934–44. PMCID: 2932817.
132. Bieghs V, Verheyen F, van Gorp PJ, Hendrikx T, Wouters K, Lutjohann D, et al. Internalization of modified lipids by CD36 and SR-A leads to hepatic inflammation and lysosomal cholesterol storage in Kupffer cells. PLoS One. 2012;7(3):e34378. PMCID: 3314620.
133. Vlahcevic ZR, Stravitz RT, Heuman DM, Hylemon PB, Pandak WM. Quantitative estimations of the contribution of different bile acid pathways to total bile acid synthesis in the rat. Gastroenterology. 1997;113(6):1949–57.
134. Feng B, Yao PM, Li Y, Devlin CM, Zhang D, Harding HP, et al. The endoplasmic reticulum is the site of cholesterol-induced cytotoxicity in macrophages. Nat Cell Biol. 2003;5(9):781–92.
135. Devries-Seimon T, Li Y, Yao PM, Stone E, Wang Y, Davis RJ, et al. Cholesterol-induced macrophage apoptosis requires ER stress pathways and engagement of the type A scavenger receptor. J Cell Biol. 2005;171(1):61–73. PMCID: 2171235.
136. Matsuzawa N, Takamura T, Kurita S, Misu H, Ota T, Ando H, et al. Lipid-induced oxidative stress causes steatohepatitis in mice fed an atherogenic diet. Hepatology. 2007;46(5):1392–403.
137. Yao ZM, Vance DE. Reduction in VLDL, but not HDL, in plasma of rats deficient in choline. Biochem Cell Biol. 1990;68(2):552–8.
138. Leclercq IA, Farrell GC, Field J, Bell DR, Gonzalez FJ, Robertson GR. CYP2E1 and CYP4A as microsomal catalysts of lipid peroxides in murine nonalcoholic steatohepatitis. J Clin Invest. 2000;105(8):1067–75. PMCID: 300833.
139. Chowdhry S, Nazmy MH, Meakin PJ, Dinkova-Kostova AT, Walsh SV, Tsujita T, et al. Loss of Nrf2 markedly exacerbates nonalcoholic steatohepatitis. Free Radic Biol Med. 2010;48(2):357–71.
140. Larter CZ, Yeh MM, Williams J, Bell-Anderson KS, Farrell GC. MCD-induced steatohepatitis is associated with hepatic adiponectin resistance and adipogenic transformation of hepatocytes. J Hepatol. 2008;49(3):407–16.
141. Weltman MD, Farrell GC, Liddle C. Increased hepatocyte CYP2E1 expression in a rat nutritional model of hepatic steatosis with inflammation. Gastroenterology. 1996;111(6):1645–53.
142. Angulo P. Obesity and nonalcoholic fatty liver disease. Nutr Rev. 2007;65(6 Pt 2):S57–63.
143. Musso G, Cassader M, Rosina F, Gambino R. Impact of current treatments on liver disease, glucose metabolism and cardiovascular risk in non-alcoholic fatty liver disease (NAFLD): a systematic review and meta-analysis of randomised trials. Diabetologia. 2012;55(4):885–904.
144. Kim MS, Kung S, Grewal T, Roufogalis BD. Methodologies for investigating natural medicines for the treatment of nonalcoholic fatty liver disease (NAFLD). Curr Pharm Biotechnol. 2012;13(2):278–91.
145. Larter CZ, Yeh MM. Animal models of NASH: getting both pathology and metabolic context right. J Gastroenterol Hepatol. 2008;23(11):1635–48.
146. Sanyal AJ, Chalasani N, Kowdley KV, McCullough A, Diehl AM, Bass NM, et al. Pioglitazone, vitamin E, or placebo for nonalcoholic steatohepatitis. N Engl J Med. 362(18):1675–85. PMCID: 2928471.
147. Musso G, Gambino R, Cassader M, Pagano G. A meta-analysis of randomized trials for the treatment of nonalcoholic fatty liver disease. Hepatology. 2010;52(1):79–104.
148. Nissen SE, Wolski K. Rosiglitazone revisited: an updated meta-analysis of risk for myocardial infarction and cardiovascular mortality. Arch Intern Med. 2010;170(14):1191–201.
149. Seeff LB, Lindsay KL, Bacon BR, Kresina TF, Hoofnagle JH. Complementary and alternative medicine in chronic liver disease. Hepatology. 2001;34(3):595–603.
150. Dehmlow C, Erhard J, de Groot H. Inhibition of Kupffer cell functions as an explanation for the hepatoprotective properties of silibinin. Hepatology. 1996;23(4):749–54.

151. Federico A, Trappoliere M, Tuccillo C, de Sio I, Di Leva A, Del Vecchio Blanco C, et al. A new silybin-vitamin E-phospholipid complex improves insulin resistance and liver damage in patients with non-alcoholic fatty liver disease: preliminary observations. Gut. 2006;55(6):901–2. PMCID: 1856214.
152. Li Volti G, Salomone S, Sorrenti V, Mangiameli A, Urso V, Siarkos I, et al. Effect of silibinin on endothelial dysfunction and ADMA levels in obese diabetic mice. Cardiovasc Diabetol. 2011;10:62. PMCID: 3152512.
153. Loguercio C, Andreone P, Brisc C, Brisc MC, Bugianesi E, Chiaramonte M, et al. Silybin combined with phosphatidylcholine and vitamin E in patients with nonalcoholic fatty liver disease: a randomized controlled trial. Free Radic Biol Med. 2012;52(9):1658–65.
154. Mira L, Silva M, Manso CF. Scavenging of reactive oxygen species by silibinin dihemisuccinate. Biochem Pharmacol. 1994;48(4):753–9.
155. Salamone F, Galvano F, Cappello F, Mangiameli A, Barbagallo I, Li Volti G. Silibinin modulates lipid homeostasis and inhibits nuclear factor kappa B activation in experimental nonalcoholic steatohepatitis. Transl Res. 2012;159(6):477–86.
156. Schumann J, Prockl J, Kiemer AK, Vollmar AM, Bang R, Tiegs G. Silibinin protects mice from T cell-dependent liver injury. J Hepatol. 2003;39(3):333–40.
157. Ghosh N, Ghosh R, Mandal V, Mandal SC. Recent advances in herbal medicine for treatment of liver diseases. Pharm Biol. 2011;49(9):970–88.
158. Murosaki S, Lee TR, Muroyama K, Shin ES, Cho SY, Yamamoto Y, et al. A combination of caffeine, arginine, soy isoflavones, and L-carnitine enhances both lipolysis and fatty acid oxidation in 3T3-L1 and HepG2 cells in vitro and in KK mice in vivo. J Nutr. 2007;137(10):2252–7.
159. Saller R, Meier R, Brignoli R. The use of silymarin in the treatment of liver diseases. Drugs. 2001;61(14):2035–63.
160. Flora K, Hahn M, Rosen H, Benner K. Milk thistle (*Silybum marianum*) for the therapy of liver disease. Am J Gastroenterol. 1998;93(2):139–43.
161. Maeda H, Hosokawa M, Sashima T, Murakami-Funayama K, Miyashita K. Anti-obesity and anti-diabetic effects of fucoxanthin on diet-induced obesity conditions in a murine model. Mol Med Rep. 2009;2(6):897–902.
162. Woo MN, Jeon SM, Shin YC, Lee MK, Kang MA, Choi MS. Anti-obese property of fucoxanthin is partly mediated by altering lipid-regulating enzymes and uncoupling proteins of visceral adipose tissue in mice. Mol Nutr Food Res. 2009;53(12):1603–11.
163. Woo MN, Jeon SM, Kim HJ, Lee MK, Shin SK, Shin YC, et al. Fucoxanthin supplementation improves plasma and hepatic lipid metabolism and blood glucose concentration in high-fat fed C57BL/6N mice. Chem Biol Interact. 2010;186(3):316–22.
164. Hosokawa M, Miyashita T, Nishikawa S, Emi S, Tsukui T, Beppu F, et al. Fucoxanthin regulates adipocytokine mRNA expression in white adipose tissue of diabetic/obese KK-Ay mice. Arch Biochem Biophys. 2010;504(1):17–25.
165. Beppu F, Hosokawa M, Yim MJ, Shinoda T, Miyashita K. Down-regulation of hepatic stearoyl-CoA desaturase-1 expression by fucoxanthin via leptin signaling in diabetic/obese KK-A(y) mice. Lipids. 2013;48(5):449–55.
166. Ahn J, Lee H, Jung CH, Ha T. Lycopene inhibits hepatic steatosis via microRNA-21-induced downregulation of fatty acid-binding protein 7 in mice fed a high-fat diet. Mol Nutr Food Res. 2012;56(11):1665–74.
167. Lee YS, Cha BY, Saito K, Yamakawa H, Choi SS, Yamaguchi K, et al. Nobiletin improves hyperglycemia and insulin resistance in obese diabetic ob/ob mice. Biochem Pharmacol. 2010;79(11):1674–83.
168. Lee YS, Cha BY, Choi SS, Choi BK, Yonezawa T, Teruya T, et al. Nobiletin improves obesity and insulin resistance in high-fat diet-induced obese mice. J Nutr Biochem. 2013;24(1):156–62.
169. Pu P, Wang XA, Salim M, Zhu LH, Wang L, Chen KJ, et al. Baicalein, a natural product, selectively activating AMPKalpha(2) and ameliorates metabolic disorder in diet-induced mice. Mol Cell Endocrinol. 2012;362(1–2):128–38.

170. Jung CH, Cho I, Ahn J, Jeon TI, Ha TY. Quercetin reduces high-fat diet-induced fat accumulation in the liver by regulating lipid metabolism genes. Phytother Res. 2013;27(1):139–43.
171. Hoek-van den Hil EF, Keijer J, Bunschoten A, Vervoort JJ, Stankova B, Bekkenkamp M, et al. Quercetin induces hepatic lipid omega-oxidation and lowers serum lipid levels in mice. PLoS One. 2013;8(1):e51588. PMCID: 3554747.
172. Marcolin E, San-Miguel B, Vallejo D, Tieppo J, Marroni N, Gonzalez-Gallego J, et al. Quercetin treatment ameliorates inflammation and fibrosis in mice with nonalcoholic steatohepatitis. J Nutr. 2012;142(10):1821–8.
173. Marcolin E, Forgiarini LF, Rodrigues G, Tieppo J, Borghetti GS, Bassani VL, et al. Quercetin decreases liver damage in mice with non-alcoholic steatohepatitis. Basic Clin Pharmacol Toxicol. 2013;112(6):385–91.
174. Jeong SM, Kang MJ, Choi HN, Kim JH, Kim JI. Quercetin ameliorates hyperglycemia and dyslipidemia and improves antioxidant status in type 2 diabetic db/db mice. Nutr Res Pract. 2012;6(3):201–7. PMCID: 3395784.
175. Chen JH, Tipoe GL, Liong EC, So HS, Leung KM, Tom WM, et al. Green tea polyphenols prevent toxin-induced hepatotoxicity in mice by down-regulating inducible nitric oxide-derived prooxidants. Am J Clin Nutr. 2004;80(3):742–51.
176. Chung MY, Park HJ, Manautou JE, Koo SI, Bruno RS. Green tea extract protects against nonalcoholic steatohepatitis in ob/ob mice by decreasing oxidative and nitrative stress responses induced by proinflammatory enzymes. J Nutr Biochem. 2012;23(4):361–7. PMCID: 3704146.
177. Bose M, Lambert JD, Ju J, Reuhl KR, Shapses SA, Yang CS. The major green tea polyphenol, (-)-epigallocatechin-3-gallate, inhibits obesity, metabolic syndrome, and fatty liver disease in high-fat-fed mice. J Nutr. 2008;138(9):1677–83. PMCID: 2586893.
178. Chen YK, Cheung C, Reuhl KR, Liu AB, Lee MJ, Lu YP, et al. Effects of green tea polyphenol (-)-epigallocatechin-3-gallate on newly developed high-fat/Western-style diet-induced obesity and metabolic syndrome in mice. J Agric Food Chem. 2011;59(21):11862–71. PMCID: 3243651.
179. Klaus S, Pultz S, Thone-Reineke C, Wolfram S. Epigallocatechin gallate attenuates diet-induced obesity in mice by decreasing energy absorption and increasing fat oxidation. Int J Obes (Lond). 2005;29(6):615–23.
180. Friedrich M, Petzke KJ, Raederstorff D, Wolfram S, Klaus S. Acute effects of epigallocatechin gallate from green tea on oxidation and tissue incorporation of dietary lipids in mice fed a high-fat diet. Int J Obes (Lond). 2012;36(5):735–43.
181. Luo XY, Takahara T, Hou J, Kawai K, Sugiyama T, Tsukada K, et al. Theaflavin attenuates ischemia-reperfusion injury in a mouse fatty liver model. Biochem Biophys Res Commun. 2012;417(1):287–93.
182. Jeon BT, Jeong EA, Shin HJ, Lee Y, Lee DH, Kim HJ, et al. Resveratrol attenuates obesity-associated peripheral and central inflammation and improves memory deficit in mice fed a high-fat diet. Diabetes. 2012;61(6):1444–54. PMCID: 3357272.
183. Cho SJ, Jung UJ, Choi MS. Differential effects of low-dose resveratrol on adiposity and hepatic steatosis in diet-induced obese mice. Br J Nutr. 2012;108(12):2166–75.
184. Ahn J, Cho I, Kim S, Kwon D, Ha T. Dietary resveratrol alters lipid metabolism-related gene expression of mice on an atherogenic diet. J Hepatol. 2008;49(6):1019–28.
185. Kang W, Hong HJ, Guan J, Kim DG, Yang EJ, Koh G, et al. Resveratrol improves insulin signaling in a tissue-specific manner under insulin-resistant conditions only: in vitro and in vivo experiments in rodents. Metabolism. 2012;61(3):424–33.
186. Chen S, Li J, Zhang Z, Li W, Sun Y, Zhang Q, et al. Effects of resveratrol on the amelioration of insulin resistance in KKAy mice. Can J Physiol Pharmacol. 2012;90(2):237–42.
187. Joe B, Vijaykumar M, Lokesh BR. Biological properties of curcumin-cellular and molecular mechanisms of action. Crit Rev Food Sci Nutr. 2004;44(2):97–111.
188. Shapiro H, Bruck R. Therapeutic potential of curcumin in non-alcoholic steatohepatitis. Nutr Res Rev. 2005;18(2):212–21.

189. Kuo JJ, Chang HH, Tsai TH, Lee TY. Positive effect of curcumin on inflammation and mitochondrial dysfunction in obese mice with liver steatosis. Int J Mol Med. 2012;30(3):673–9.
190. Weisberg SP, Leibel R, Tortoriello DV. Dietary curcumin significantly improves obesity-associated inflammation and diabetes in mouse models of diabesity. Endocrinology. 2008;149(7):3549–58. PMCID: 2453081.
191. Lai YS, Chen WC, Ho CT, Lu KH, Lin SH, Tseng HC, et al. Garlic essential oil protects against obesity-triggered nonalcoholic fatty liver disease through modulation of lipid metabolism and oxidative stress. J Agric Food Chem. 2014;62(25):5897–906.

Chapter 13
Mouse Models to Study Metformin Effects in Carcinogenesis

Abraham Schneider

Abstract Emerging evidence suggests that common drugs that were originally developed for other chronic conditions may also trigger protective actions against cancer development and progression. Early retrospective observational studies in diabetic patients indicated that among the most commonly used oral antihyperglycemic agents, the biguanide compound metformin was associated with a reduced risk of cancer or improved cancer prognosis. Indeed, numerous studies have been conducted in the last decade to investigate whether metformin could be repurposed as an antineoplastic agent. This chapter will discuss promising data gained through the use of mouse models to study the effects of metformin in carcinogenesis. Overall, metformin has consistently demonstrated encouraging antineoplastic activity in preclinical mouse models. While in certain types of cancer these responses are highly influenced by the systemic metabolic status of the host, in others it is more dependent on the cellular and molecular background of the primary tumor. What still remains a critical challenge is extrapolating these findings to the biological and behavioral complexities of neoplastic disease in humans. The positive findings obtained with a drug that is relatively affordable and safe as metformin in experimental mouse models of cancer should continue to guide the design and implementation of clinical trials. These studies should serve to validate and optimize the potential use of metformin in cancer chemoprevention and treatment, and ultimately impact the selection of the most suitable patients who can benefit from metformin in the oncologic setting.

Keywords Metformin · Biguanides · Carcinogenesis · Obesity · Diabetes · AMPK · mTOR · Organic cation transporter

---

A. Schneider (✉)
School of Dentistry, Department of Oncology and Diagnostic Sciences, University of Maryland, 650 West Baltimore Street, Room 7265, Baltimore, MD 21201, USA
e-mail: aschneider@umaryland.edu

N. A. Berger (ed.), *Murine Models, Energy Balance, and Cancer,*
Energy Balance and Cancer, DOI 10.1007/978-3-319-16733-6_13

Metformin: Its Origins and Early Use as Antidiabetic Agent

Metformin is currently the most widely prescribed drug to treat type 2 diabetes mellitus. This compound was originally derived from guanidine, the pharmacologically active component of *Galega officinalis* (*aka*. French Lilac or Goat's Rue). Guanidine was used in medieval times in Europe as herbal remedy to treat an array of medical conditions, including polyuria and polydipsia, now well-known symptoms associated with diabetes. At the beginning of the twentieth century, guanidine and its naturally occurring isoprenyl derivative galegine were used as antidiabetic compounds, but were not easily tolerated due to toxic effects [1]. When metformin was originally synthesized in the 1920s and used in animals as N, N-dimethylbiguanide, it triggered a highly effective blood glucose-lowering action with minimal toxicity. The advent of insulin therapy, though, relegated the study and use of biguanides for many years.

It was not until 1957, when Jean Sterne, a French physician and diabetologist, published his work on the clinical antidiabetic properties of metformin [2]. Since the 1950s, metformin has been in the frontline of therapy for type 2 diabetes (T2D) in Europe. In 1995, the Food and Drug Administration approved its use in the USA. This inexpensive and well-tolerated drug has been for the last 60 years the preferred antidiabetic agent used by more than 100 million people worldwide [3].

Mechanisms of Action of Metformin as Antidiabetic Agent

Metformin exerts its antihyperglycemic effects in type 2 diabetes primarily by reducing glucose output in the liver and restoring tissue responsiveness to insulin [4, 5]. In hepatocytes, intracellular uptake and activity of metformin, a hydrophilic biguanide, is facilitated by the highly expressed organic cation transporter-1 (OCT-1) [6]. Although, long term clinical improvement of the hyperglycemic status is achieved following metformin treatment through reduction in circulating blood glucose and insulin levels, the precise molecular mechanism of action has not been clearly defined, and remains an active area of investigation.

To this end, proposed mechanisms have mainly focused on the role of the AMP-activated protein kinase (AMPK) pathway, a conserved signaling cascade that acts as master sensor of cellular energetics [7]. As a mild and reversible inhibitor of mitochondrial oxidative phosphorylation at complex I of the electron transport chain, metformin decreases the intracellular energy status and increases the AMP/ATP ratio in hepatocytes [8, 9]. Cellular AMP binds to AMPK, making it a better substrate for activation through phosphorylation by its upstream tumor suppressor serine/threonine kinase LKB1 [10]. AMPK is not directly activated by metformin, but as a consequence of the transient and specific decrease in cellular bioenergetics following mitochondrial complex I inhibition in hepatocytes [11].

Metformin-induced AMPK signaling also appears to control de novo hepatic fatty acid synthesis. In the context of obesity-induced insulin resistance, metformin treatment has been recently shown to restore insulin sensitivity by reducing hepatic lipid content via a direct AMPK-dependent inhibitory phosphorylation of acetyl-CoA carboxylase 1 (Acc1) and Acc2, key rate-controlling enzymes in the biosynthesis of fatty acids. These studies show that reduction in hepatocyte lipid storage inhibits Protein Kinase C-ε (PKC-ε)-mediated suppression of insulin receptor signaling. In turn, this inhibition leads to increase in AKT levels and a more effective blockade of FOXO1-mediated gluconeogenesis [12, 13].

Some groups have challenged AMPK's indispensable role in mediating the inhibitory effects of metformin on hepatic gluconeogenesis [14]. In this regard, Miller et al. have recently proposed that metformin reduces hepatic glucose production via the inhibition of glucagon-induced cyclic AMP (cAMP)/protein kinase A (PKA) signaling, an important gluconeogenic mechanism used to primarily maintain blood glucose levels in response to fasting or starvation [15]. This work also demonstrated that the inhibitory effect of metformin on glucagon-induced cAMP/PKA signaling remained functional when tested in AMPK-negative hepatocytes, thus supporting the dispensability of AMPK signaling.

Moreover, evidence arising from studies conducted by Madiraju et al. also disputes the central role of metformin-induced AMPK in reducing plasma glucose concentration. This report demonstrates that metformin inhibits the redox shuttle mitochondrial enzyme glycerophosphate dehydrogenase underscoring the direct mitochondrial effects of metformin. This inhibition resulted in a reduction in gluconeogenesis following an altered mitochondrial redox state in hepatocytes, and decreased conversion of lactate and glycerol to glucose. Indeed, these findings position mitochondrial glycerophosphate dehydrogenase as a potential new therapeutic target of metformin-induced inhibition of hepatic gluconeogenesis in T2D [16].

Recent major breakthroughs continue to shed light on the complex mechanisms underlying the blood glucose lowering effects of metformin. Seminal in vitro and in vivo studies argue in favor or against the role of AMPK as an essential intracellular factor mediating the inhibition of hepatic gluconeogenesis. Indeed, it is plausible that these novel mechanisms are acting in parallel or as specific mechanisms responding to metformin in either a fed or fasting state. Inconsistency in the results may also reflect variability in the experimental design and methodology relative to in vitro cellular models utilizing a wide range of metformin concentrations together with in vivo murine models, where different doses of metformin, routes of administration, and treatment duration were used.

Metformin in Cancer Chemoprevention and Treatment

The American Association of Clinical Endocrinologists (AACE) and the American College of Endocrinology (ACE) in a recent consensus statement concluded that based on epidemiological data, obesity is associated with a higher risk of malignancies. Increased endogenous levels of insulin, insulin-like growth factors, proinflammatory cytokines, and other factors act either directly or indirectly to promote tumor growth and progression [17]. Although the role of persistently high plasma glucose levels in diabetic patients is more elusive, current observational data tend to support hyperglycemia as a tumor growth-promoting factor as well. Diabetes increases the risk of total cancer, and of site-specific cancers of the liver, pancreas, colorectum, breast and others [18]. According to the AACE/ACE task force, the lack of data from large-scale randomized studies does not support an association between antihyperglycemic drugs and increased cancer risk. Likewise, not enough strong clinical evidence is available to conclude that these medications positively affect cancer development and progression. Additional research endeavors with improved study designs are still needed to address unresolved concerns surrounding the relationship between diabetes and cancer [17].

The current interest on determining the impact of antidiabetic drugs on cancer was in part triggered by retrospective epidemiological studies reporting the potential protective effects of metformin on cancer outcomes in diabetics when compared to other medications [19]. Recent systematic reviews and meta-analyses on overall cancer risk and mortality in diabetic patients treated with metformin revealed significant positive results in liver, pancreatic, colorectal, and breast cancers. Since these findings have been primarily based on retrospective population studies, the need for long-term randomized clinical trials to more strongly validate the potential benefits of metformin in neoplastic disease has been encouraged [20–22].

In fact, as of June 2014, more than 200 clinical trials are actively recruiting patients, still ongoing or have been completed to evaluate the effects of metformin alone or in combination with other drugs on different types of cancer which were listed in the clinicaltrials.gov database. Many of these clinical trials were based on promising data obtained through experimental studies carried out in a variety of mouse models of cancer, including spontaneous tumors in transgenic mice, chemically-induced carcinogenesis and ectopic or orthotopic tumors. These animal models serve not only as preclinical tools to evaluate potential antineoplastic effects of metformin, but also to gain insight into key underlying whole organismal, cellular, and molecular factors that could play a decision-making role in predicting how patients would ultimately respond to metformin in the clinical oncology setting.

Metformin in Mouse Models of Liver Cancer

The liver is the primary site of action of metformin to trigger its antihyperglycemic effects. This therapeutic action is facilitated by the intracellular uptake of metformin through highly OCT-1 expressing hepatocytes, as evidenced by the lack of a glucose-lowering effect of metformin in OCT-1-deficient mice [6]. Because of its proven direct uptake of metformin, the liver appears as an ideal organ to examine the emerging antitumoral properties of metformin. Primary liver cancer, which is largely diagnosed as hepatocellular carcinoma (HCC), is the fifth most common cancer and the third most common cause of cancer mortality worldwide [23]. In the USA, it is estimated that there are more than 33,000 new cases of primary liver cancer, and 23,000 deaths associated with the disease occurred in 2014 [24]. Based on the observational association of increased incidence of liver cancer in diabetes, and the positive effect of metformin intake in this malignancy, targeting HCC with metformin appears promising [20, 22].

A recent study in a chemically-induced HCC mouse model presented compelling evidence for a chemopreventive action of metformin in hepatic carcinogenesis [25]. Bhalla et al. induced HCC by injecting intraperitoneally (i.p.) 2-week old non-diabetic male C57BL/6J immunocompetent mice with 25 mg/kg body weight diethylnitrosamine (DEN), a liver-specific carcinogen. After weaning at 4 weeks of age, the mice were fed a control or Metformin-containing chow at a dose of 250 mg/kg body weight for 24 or 36 weeks. Although no visible surface liver tumors were present in either control- or metformin-treated mice after 24 weeks, histological examination revealed that metformin-treated mice developed 57 % fewer tumors than controls. At 24 weeks, tumor size from metformin-treated mice was also reduced to nearly 37 % as compared to control mice. In the 36-week group, metformin treatment also induced a very significant reduction in the number and size of tumors. Most significantly, visible tumors on the surface of the liver were reduced approximately by 80 %.

The study by Bhalla et al. suggested that metformin inhibits liver tumorigenesis in an AMPK-independent manner, since no AMPK activation occurred in the livers of mice treated with metformin. Despite the absence of hepatic AMPK activation, key intermediary factors within the hepatic de novo lipogenic pathway were significantly reduced at the mRNA and protein levels in the metformin-treated mice. When the effect of metformin on the expression of lipogenic genes was examined in HCC cell lines, it was found that overexpressing SREBP1c, a key transcription factor driving hepatic lipogenic gene expression, was sufficient to rescue the cell growth inhibiting effects of metformin. These results led the authors to conclude that the chemopreventive effect of metformin in HCC is in part mediated through the downregulation of lipogenic gene expression. In spite of these promising findings, these studies were only conducted in non-obese, non-diabetic mice.

Based on preclinical and clinical studies, nonalcoholic fatty liver disease (NAFLD), in particular, its most aggressive nonalcoholic steatohepatitis (NASH) form has been closely associated with obesity and insulin resistance. Moreover, NASH may potentially progress to cirrhosis and later to HCC [26, 27]. When investigating the long-term effects of metformin on high-fat diet induced hepatic tumorigenesis in a C57BL/6J mouse model of NASH, Tajima et al. found that metformin, administered daily at a dose of 250 mg/kg in the drinking water for 60 weeks concurrently with a high-fat diet, prevented liver tumorigenesis by suppressing hepatic fat accumulation only in the early stages of the disease before the onset of NAFLD [28]. Noteworthy, the scores for adipose tissue inflammation and fibrosis were significantly improved in the metformin-treated high-fat diet mice compared to those in the high-fat diet only group. Surprisingly, metformin did not significantly affect expression levels of lipogenic enzymes as reported in the chemically-induced

HCC model [25]. Based on the results of their study, Tajima et al. proposed that the antitumor effect of metformin against high-fat diet induced liver tumorigenesis is most likely associated with a delayed inflammatory response in the hepatic adipose tissue [28].

Certain conflicting findings highlight the pleiotropic nature of metformin in two different mouse models of liver tumorigenesis. Most likely, distinct triggering oncogenic factors underlying the process of cellular transformation induced by a chemical carcinogen such as DEN or dietary intervention through high-fat diet intake make it puzzling to determine the specific mechanisms responsible for the antitumoral activity of metformin in liver carcinogenesis [29].

Metformin in Mouse Models of Pancreatic Cancer

In 2014, more than 46,000 new cases of cancer of the pancreas and about 40,000 deaths were predicted to occur in the USA, placing pancreatic cancer as the fourth leading cause of cancer-related death in both men and women [24]. Both environmental and genetic factors make pancreatic cancer, largely pancreatic ductal adenocarcinoma (PDAC), a highly lethal malignancy with an overall 5-year survival rate of only 3–5 % [30]. Although cigarette smoking accounts for almost 25 % of cases, obesity and diabetes are increasingly becoming key negative factors promoting the disease and affecting clinical outcomes [31].

A recent meta-analysis of observational epidemiological studies determined that the use of metformin is associated with a significant lower risk of pancreatic cancer in patients with T2D [32]. Disease stage appears to play a role in the response to metformin since in diabetic patients diagnosed with pancreatic cancer, metformin treatment was significantly associated with better outcomes and longer survival only in patients with non-metastatic disease [33]. However, metformin was not found to be associated with improved survival in a retrospective cohort study that included subjects with advanced PAC and a preexisting diagnosis of T2D [34]. Although better designed clinical studies are required [35], early preclinical work in hamsters fed a high-fat diet had already pointed to a promising chemopreventive role of metformin in a chemically-induced model of pancreatic carcinogenesis using the pancreatic carcinogen N-nitrosobis-(2-oxopropyl)amine. Whereas 50 % of the high-fat diet-fed hamsters developed pancreatic malignant lesions, none were found in high-fat diet-fed hamsters administered metformin in the drinking water [36].

More recently, studies performed in athymic nude mice bearing subcutaneous tumor xenografts developed from human-derived PDAC cell lines (i.e., PANC-1 and MiaPaCa-2) showed that metformin administration significantly inhibited tumor growth [37]. The authors demonstrated that metformin given i.p. once daily in various doses (50–250 mg/kg) decreased tumor growth in a dose-dependent manner. These findings were also replicated when metformin was administered orally in the drinking water (2.5 mg/mL) either prior or following tumor cell implantation.

Orally administered metformin also inhibited the growth of orthotopic tumors developed in the pancreas with MiaPaCa-2 cells. According to Kisfalvi et al., signaling pathways inhibited by metformin treatment in the tumors included the mammalian target of rapamycin (mTOR) and extracellular signal-regulated kinase (ERK) mitogenic pathways. Previously, the same research group had also proposed that in PDAC cell lines and tumor xenografts, metformin treatment disrupts the crosstalk between insulin/IGF-1 receptor and G-protein coupled receptor (GPCR) signaling in an AMPK-dependent manner [38, 39].

In orthotopic tumor xenografts established with the highly aggressive L3.6pL human-derived PDAC cells, Nair et al. demonstrated that a daily dose of metformin (250 mg/kg) significantly decreased tumor growth when compared to control mice. According to this report, the inhibitory action of metformin resulted in part by downregulating specificity protein (Sp) transcription factors Sp1, Sp2, and Sp4. Sp factors regulate the transcriptional activation of several cancer-related genes including *Bcl-2*, *survivin*, *cyclin D1*, fatty acid synthase, and vascular endothelial growth factor [40].

Emerging evidence also indicates that metformin might be affecting pancreatic cancer cell stemness. To this end, recent studies showed that metformin treatment decreased cancer stem cell markers including *CD44*, *CD133*, *aldehyde dehydrogenase isoform 1* (*ALDH1*), *EZH2*, *Notch-1*, *Nanog*, *Oct4*, and epithelial cell adhesion molecule (EPCAM) [41–44]. Gou et al. examined the in vitro effect of low concentrations of metformin (100–200 μM) on different subpopulations of pancreatic cancer cells and found that these selectively inhibited the proliferation of $CD133^{(+)}$, but not $CD24^{(+)}CD44^{(+)}ESA^{(+)}$ cells. Indeed, $CD133^{(+)}$ pancreatic cancer cells are regarded as cancer stem cells that contribute to the progression, recurrence, and treatment resistance in pancreatic cancer.

To investigate the effect of metformin on pancreatic cancer in vivo, tumor xenograft experiments in immunodeficient mice were conducted. This report highlighted that the amount of drug diluted in the drinking water (800 mg/L) of metformin-treated mice was equivalent to a human dose of 20 mg/kg. Moreover, plasma concentration of metformin was approximately 20 μM, simulating those measured in the plasma of individuals administered a recommended therapeutic dose of metformin. Metformin significantly inhibited the growth of pancreatic cancer xenografts; however, these tumors were developed from whole cell populations of pancreatic cancer cells, and not only from the $CD133^{(+)}$ cancer stem cell subpopulation. Overall, this work implied that metformin might be triggering antitumoral effects in vivo by selectively inhibiting $CD133^{(+)}$ cells [42].

The chemopreventive effects of metformin have also been tested on pancreatic intraepithelial neoplasia (PanIN) and its progression to PDAC in a previously described conditional $p48^{Cre/+}$ LSL-KrasG$12^{D/+}$ transgenic mice [45]. This well-established mouse model closely recapitulates the stepwise progression from PanIN to PDAC as observed in humans. Transgenic mice were fed ad libitum, a control diet or a diet containing metformin at 1000 ppm (~200 mg/kg body weight) or 2000 ppm, for 38 weeks and sacrificed after 44 weeks. While mice which were fed the control diet showed 80 and 62% incidence of PDAC in males and females, respectively,

PDAC incidence with metformin treatments ranged from 0–26 % overall. Metformin treatment also caused suppression of carcinoma in situ lesions by 28–39 %, and significantly prevented carcinoma spread in the pancreas.

Most notably, a marked decrease in cancer stem cell markers (i.e., *CD24*, *CD44*, *EPCAM*, *CD133* and *ALDH1*) was observed in PanIN lesions from transgenic mice which were fed the metformin-supplemented diet when compared to controls. When analyzing pancreatic tissue or serum, there was a significant inhibition of mTOR, activated ERK and IGF-1 in the metformin-treated mice as well as activation of the AMPK pathway. The authors concluded that long-term dietary metformin delayed the progression of PanIN lesions without toxic effects, and ultimately decreased the incidence of invasive PDAC, suggesting that the chemopreventive action is in part mediated through a direct effect on mTOR signaling and CSCs [43].

Metformin in Mouse Models of Colorectal Cancer

Colorectal cancer remains as the third most commonly diagnosed cancer and the third leading cause of cancer death in both men and women in the USA. Approximately 137,000 new cases and 50,000 deaths were expected in 2014 [24]. Observational, population-based studies investigating the association between metformin and colorectal cancer incidence or survival have reported contradictory findings ranging from decreased risk, no difference or a dose-dependent increased risk. Indeed, the use of different methodologies to assess outcomes may have been responsible for these inconclusive results [46].

In spite of these significant yet paradoxical findings, experimental preclinical studies have consistently pointed to metformin as a safe and promising antineoplastic agent in mouse models of colorectal cancer. Although, the basic mechanisms underlying the association of obesity and diabetes with colorectal cancer still remain unclear, several studies have proposed that excess caloric intake, insulin, insulin-like growth factors, and/or adipose-derived factors are linked with increased risk of colorectal cancer [47–50].

In this regard, Algire et al. provided early preclinical evidence demonstrating a positive effect of metformin in the context of a high-energy diet as promoter of colon cancer cell growth [51]. In their study, the volume of syngeneic tumor grafts derived from MC38 colon carcinoma cells and implanted in C57Bl/6 mice was found to be significantly larger in animals fed with a high-energy diet compared to mice in a control diet. Together with hyperinsulinemia, tumors from mice in the high-energy diet had elevated expression of phosphorylated AKT (pAKT) and fatty acid synthase (FASN), a key enzyme in long-chain fatty acid biosynthesis. After 12 weeks in the experimental diets, metformin was given daily for 5 weeks at a dose of 50 mg/kg body weight in the drinking water to a subgroup of mice from each dietary group. At the end of the experimental period, metformin significantly diminished tumor growth in mice fed with the high-energy diet, but no significant effect was found in tumors from mice in the control diet.

These findings suggest that the host's metabolic status markedly impacts the antineoplastic response to metformin. Metformin treatment also reduced insulin levels, induced AMPK activation, and downregulated pAKT and FASN expression levels in the tumors from high-energy diet-fed mice. Since AMPK activation was observed in neoplastic tissue, but the tumor inhibitory action of metformin was only significant in mice with high-energy diet intake, this study and others indicate that the antitumoral response to metformin appears to be positively driven by the molecular background (i.e., intact LKB1/AMPK signaling; p53 mutation status) of a metabolically defined subset of colon cancers [51–55].

Intriguingly, studies performed in a chemically-induced colorectal carcinogenesis mouse model are more supportive of the direct action. Hosono et al. used i.p. azoxymethane to induce colorectal carcinogenesis in BALB/c mice. Then, mice were treated with or without metformin (250 mg/kg daily i.p.) for either 6 weeks to assess aberrant crypt foci formation or 32 weeks to investigate polyp formation [56]. Their findings showed a significant reduction in aberrant crypt foci formation, indicative of colorectal cancer chemoprevention, as evidenced by metformin-induced AMPK activation and mTOR inhibition in colonic tissue independent of circulating insulin and lipid levels [56]. The effects of metformin on colon polyp inhibition were less evident.

Previously, the same research group had demonstrated that in the $Apc^{Min/+}$ mice, which have a germ-line mutation in the Apc tumor suppressor gene and are predisposed to the development of multiple adenomas in the small intestine and colon, metformin restricted intestinal polyp growth by directly activating AMPK and inhibiting mTOR, pointing to metformin as an attractive candidate drug in colorectal carcinogenesis [57]. Overall, current evidence suggests that the antineoplastic activity of metformin in colorectal cancer is mediated through a direct action at the primary colonic tumor and/or via an indirect modification of the host metabolism. Nevertheless, these promising responses appear to be dependent on the type of experimental carcinogenesis mouse model utilized in the investigation.

Metformin in Mouse Models of Breast Cancer

Compelling evidence obtained from observational studies and meta-analyses underscore the link between obesity and diabetes with an increased risk of breast cancer [17, 18, 58]. This malignancy remains as the second cause of cancer-related deaths in the USA with predictions of more than 40,000 women dying as a consequence of this disease in 2014 [24]. The use of an affordable and relatively safe drug like metformin would be an invaluable benefit to a large number of patients at risk of or diagnosed with breast cancer. In fact, many exploratory clinical studies designed as "window-of-opportunity" biomarker trials have provided encouraging results. Many of the positive effects of metformin in breast cancer treatment, though, appear to be associated with a subset of patients having specific tumor or host metabolic characteristics (reviewed in [59]).

Indeed, the justification for these clinical trials was sparked by promising results obtained through the use of metformin in mouse models of mammary carcinogenesis. Early findings reporting beneficial effects of metformin in breast cancer came from studies performed in transgenic HER-2/neu mice, where the incidence and size of mammary adenocarcinoma significantly decreased in response to metformin administered for a maximum of 9 months in the drinking water at a dose of 100 mg/kg body weight [60]. The *HER-2/neu* oncogene belongs to the epidermal growth factor receptor family, and it is overexpressed in approximately 25–40 % of human breast cancers. This occurrence is strongly associated with overall poor prognosis in breast cancer patients [61]. It has also been shown that metformin treatment significantly inhibited the subcutaneous growth of HER-2/neu-derived mammary carcinoma cells transplanted into FVB/N HER-2/neu male mice [62].

More recent studies have also provided interesting information regarding the potential selective targeting of breast cancer stem cells in response to metformin treatment. Hirsch et al. demonstrated that the combination of metformin (100 μg/ml i.p.) with the well-known chemotherapeutic drug doxorubicin (4 mg/kg i.p.) given for 15 days significantly reduced the mass and delayed recurrence of tumor xenografts more efficiently than either drug administered alone. Tumor xenografts were derived from the human mammary non-transformed epithelial cell line MCF10A containing an integrated fusion of the v-Src oncoprotein with the ligand-binding domain of the estrogen receptor (ER). These cells were transformed upon tamoxifen treatment, and approximately 10 % of the cell population was characterized as cancer stem cells due to their expression of $CD44^{high}/CD24^{low}$ markers and the ability to develop mammospheres in vitro. Surprisingly, cancer stem cells were nearly absent from tumor xenografts in mice treated with the combination therapy; however, they were noticeable in tumors from doxorubicin only-treated mice. These results suggest that metformin likely possesses a therapeutic advantage in targeting and killing cancer stem cells when compared to standard chemotherapy [63].

Since overexpression of the *HER-2/neu* oncogene has been recently associated with carcinogenesis in part by regulating the expansion of the mammary stem cell population, Zhu et al. tested the chemopreventive effects of metformin (250 mg/kg body weight i.p. daily for 10 weeks) on the cancer stem cells/tumor-initiating cell population in HER-2/neu transgenic mice. They found that metformin treatment elicited a significant and selective inhibition of a subpopulation of $CD61^{high}/CD49^{high}$ tumor-initiating cells in preneoplastic mammary glands. Likewise, metformin also downregulated HER-2 and EGFR expression and decreased the phosphorylation of EGFR family members, IGF-1 receptor, AKT, mTOR, and STAT3 in premalignant tissue [64]. These results and work from others highlight the role of cancer stem cells in mammary carcinogenesis, but in particular shed light into the promising chemopreventive properties of metformin in premalignant mammary disease [65].

Metformin in Mouse Models of Prostate Cancer

Prostate cancer continues to be a leading cause of cancer-related illness and the second most common cause of cancer-associated mortality, resulting in 233,000 new cases and approximately 30,000 deaths per year in the USA [24]. Retrospective, case control population studies in diabetic men remain inconclusive about the positive effects of metformin on prostate cancer risk [66–70]. However, recent experimental studies exploring the chemopreventive effects of metformin in a mouse model of prostate carcinogenesis reported promising outcomes.

Because of the major role of the *c-MYC* oncogene on early prostate epithelia cell transformation and growth, Akinyeke et al. investigated the effect of metformin in Hi-Myc transgenic mice overexpressing c-myc relative to the progression of prostate intraepithelial neoplasia (PIN) to invasive prostate adenocarcinoma. After 4 weeks of treatment of Hi-Myc mice, metformin (200 mg/kg body weight in drinking water) affected the development of early prostatic premalignant lesions as evidenced by a marked reduction in PIN formation together with decreased Ki67 proliferation marker and androgen receptor expression levels. According to the authors, inhibition of PIN onset in the Hi-Myc mouse model is most likely linked to inhibition of c-myc signaling, underscoring that the chemopreventive action of metformin in prostate carcinogenesis is in part a consequence of an inhibitory activity on c-myc function [71].

Previously, Ben Sahra et al. had reported that i.p. or orally-administered metformin resulted in a marked inhibition of tumor growth by 35 and 55 %, respectively, in subcutaneous tumor xenografts derived from human prostate cancer cell lines. Interestingly, they found a significant reduction in tumor cyclin D1 protein levels implying that metformin decreases tumor growth by downregulating cyclin D1 protein levels [72].

Metformin in Mouse Models of Lung Cancer

Lung cancer is the most common cause of cancer death worldwide. As the top leading cancer type in the USA for both sexes, estimates for the year 2014 predicted more than 224,000 new cases and approximately 160,000 deaths [24]. A tremendous impact on public health would occur if the risk of lung cancer is reduced, largely by eliminating exposure to tobacco carcinogens, or if progression is prevented in high-risk individuals [73]. Meta-analyses studies in conjunction with large retrospective cohort and case-control investigations have given inconclusive results relative to the effects of metformin and the risk of lung cancer in type 2 diabetics. Some of these studies indicate a benefit of metformin intake in risk reduction [74–76], while others report no association between metformin use and decreased risk of lung cancer [70, 77, 78]. Different methodological analyses and shortcomings are possibly responsible for the inconsistency of the results.

However, promising findings on the antitumoral effects of metformin have been shown in a tobacco-induced mouse model of lung carcinogenesis through the exposure of non-diabetic, non-obese A/J mice to the tobacco carcinogen 4-(methylnitrosamino)-1-(3-pyridil)-1-butanone (NNK) [79]. NNK is known for inducing K-ras-mediated tumorigenesis [80]. Lung tumors that develop in these mice also show hyperactivation of the mTOR pathway. Following inducing tumors with NNK, Memmott et al. found that metformin significantly decreased tumor burden and mTOR activity in lung tissue via AMPK-independent effects when administered daily at a dose of 250 mg/kg i.p. for 13 weeks. In lung tumor tissue metformin did not activate AMPK but inhibited phosphorylation of IGF-1 receptor/insulin receptor, AKT, ERK, and mTOR. In addition, circulating levels of insulin and IGF-1 were reduced in metformin-treated mice, suggesting that the effects on lung tumor growth may have not been directly induced at the primary tumor site, but secondary to reduced hormonal control derived from the hepatic uptake of metformin [79].

More recently, Quinn et al. reported that metformin was also effective in reducing NNK-induced lung tumorigenesis in liver IGF-1 deficient (LID) mice, which in this case suggests an antineoplastic effect of metformin via IGF-1 independent mechanisms. In lung tumors, metformin reduced phosphorylation of various receptor tyrosine kinases as well as GTP-bound Ras via an AMPK independent manner [81].

As with previous reports in other types of cancer, metformin also appears to attenuate the stimulatory effect of a high-energy diet on the growth of tumor syngeneic grafts. When comparing the effect of orally-administered metformin (50 mg/kg body weight per day) on the subcutaneous growth of Lewis lung LLC1 cells in C57BL/6J mice fed a high-energy diet versus a control standard rodent diet, Algire et al. found that metformin treatment significantly decreased tumor growth rate and tumor size in mice fed the high-energy diet [82]. However, no significant effect of metformin was observed in tumors developed in mice receiving the control diet. Moreover, the authors found increased tyrosine phosphorylation of the insulin receptor (IR) in tumor tissue from the high-energy diet mice not receiving metformin. In contrast, a marked reduction in IR phosphorylation and insulin secretion was found in response to metformin in the high-energy dietary group. Intriguingly, metformin-induced AMPK activation occurred in neoplastic tissue in mice on both dietary groups suggesting that tumor growth inhibition was independent of AMPK signaling.

Altogether, these studies demonstrate that metformin was capable of inducing antitumoral effects in different murine models of lung tumorigenesis. However, in some circumstances the tumor inhibitory effect was restricted, as previously mentioned, to a metabolically defined subset of tumors. These findings also raise the question on whether the antineoplastic activity of metformin is triggered directly at the primary site of tumor development, indirectly at extratumoral sites, or both. This important concept will be covered in more detail when discussing the role of metformin in head and neck cancer.

Metformin in Mouse Models of Head and Neck Cancer

More than 500,000 cases of head and neck cancer, mainly affecting mucosa of the oral cavity and pharynx, are diagnosed yearly worldwide. In the USA, more than 42,000 new cases of oral cancer, largely oral squamous cell carcinomas (OSCC), were estimated to be diagnosed in 2014, and 8390 deaths were predicted to occur in the same year [24]. Small improvement has occurred in OSCC patient survival rates in the last four decades, particularly when associated with main risk factors such as heavy tobacco use and alcohol consumption. Also, the increasing incidence of a distinct form of oropharyngeal SCC associated with human papillomavirus (HPV) infection in the USA and Europe may potentially become a critical public health burden in future years [83, 84].

The recent identification of the mTOR signaling pathway as a highly prevalent molecular signature underlying HPV− and HPV+ OSCC pathogenesis has provided opportunities to investigate novel strategies to deter disease progression [85–89]. The potential clinical benefit of mTOR inhibitors, such as rapamycin and its analogs in the treatment of head and neck/oral cancer has already been explored [89–94]. Rapamycin and its analogs represent an appropriate therapeutic option in the management of patients with fully established or recurrent OSCC tumors; however, their long-term use may trigger undesirable side effects due to immunosuppressing activity. Long-term interventions with well-tolerated, low-cost drugs like metformin may offer a better alternative to prevent the progression of a cancer type where locoregional invasion, nodal metastasis, and chemoresistance are hallmarks of advanced disease [95].

Recent studies show that metformin selectively affects tumor cell proliferation by perturbing mTOR signaling translational control of a specific subset of mRNAs encoding cell cycle regulators [96]. As described previously, these inhibitory effects are mostly controlled by the AMPK signaling pathway [97, 98]. Activated AMPK phosphorylates and activates the tumor suppressor protein TSC2, which then negatively regulates mTOR activity [99]. Through an oral-specific, chemical carcinogenesis mouse model, metformin has been shown to markedly prevent the conversion of premalignant oral dysplastic lesions into OSCC tumors possible via mTOR inhibition [100].

When 4-nitroquinoline 1-oxide (4NQO), a DNA adduct-forming agent mimicking tobacco exposure, was administered orally for 14 weeks to C57BL/6J mice (50 μg/mL in drinking water) it induced oral tumors closely resembling human multistage oral carcinogenesis. Interestingly, 4NQO exposure caused primarily low-grade and high-grade oral epithelial dysplasias affecting mostly the tongue, an intraoral site affected in 20–40 % of all human oral cancers [95]. It is evident from this work and studies from other groups that this mouse model provides a predictable preclinical strategy to simulate human OSCC carcinogenesis [101].

In the study by Vitale-Cross et al., mice were randomly assigned to two groups following 14 weeks of 4NQO administration in drinking water. To this end, one group received daily i.p. injections of metformin (50 mg/kg/day) and the other an equal volume of sterile saline (control group) for 8 weeks. By week 22, mice exposed to 4NQO predictably developed sizable oral tumors. However, the number of visible oral lesions, tumor multiplicity and OSCC multiplicity was significantly diminished in mice treated with metformin. Relative to mTOR signaling, metformin inhibited this oncogenic pathway primarily in the basal/suprabasal proliferating dysplastic epithelium. Interestingly, no impact on metabolic markers and serum components, including glucose, insulin, and IGF-1, was found. These results suggest that metformin prevented malignant transformation non-obese, non-diabetic mice directly at the site of oral cancer development [100].

A critical translationally relevant question that still remains elusive in the field is whether metformin exerts its antineoplastic effects by directly targeting primary tumor cells or indirectly acting at extratumoral sites. To this end, elucidating largely unexplored uptake and pharmacokinetic mechanisms will provide key information to ultimately maximize the chemopreventive and therapeutic action of metformin. This question holds significant clinical implications since a range of OCTs polymorphic genetic variants with reduced uptake function have been reported [102, 103]. As polyspecific cell membrane transporters, OCT-1, OCT-2, and OCT-3 mediate the influx and efflux of metformin and other biguanides together with a number of other structurally diverse, small hydrophilic endogenous and exogenous organic cationic compounds [6, 104–107]. OCT-1 and OCT-2 expression is highly restricted to liver and kidney, respectively. In this regard, patients harboring different OCT-1 or OCT-2 genotypes and/or expression levels may respond differently to the therapeutic effects of metformin. The distribution of OCT-3 in tissues is much broader. In fact, OCT-3 has been considered a potential determinant of the peripheral effects of metformin on skeletal muscle [107].

In certain types of human cancer, such as colorectal, gastric, and renal, and their derived cell lines, OCT-3 expression is significantly elevated [108, 109]. On the other hand, OCT-1 downregulation in hepatocellular carcinoma, most likely as a consequence of DNA methylation, has been associated with tumor progression and worse patient survival [110, 111]. A recent study also demonstrated that in human oral epithelial dysplasias and well-differentiated OSCC, only OCT-3 was highly expressed [112]. Undetectable or weak OCT-3 expression was found in less differentiated OSCC tumors, suggesting a previously unidentified link between the degree of tumor cell differentiation and OCT-3 expression [112].

Detection of OCT-3 protein expression patterns in 4NQO-induced mouse oral epithelial dysplasia is comparable to human lesions (Fig. 13.1). In addition, similar to human oral premalignant lesions and OSCC, OCT-3 expression levels are highly elevated in oral epithelial dysplasia, but reduced in less differentiated OSCC tumor foci (Fig. 13.2). Most importantly, AMPK activation as evidenced by increased pACC levels is found in OCT-3 positive, 4NQO-induced dysplastic epithelium as well as residual OSCC cells derived from mice treated with metformin, but not in

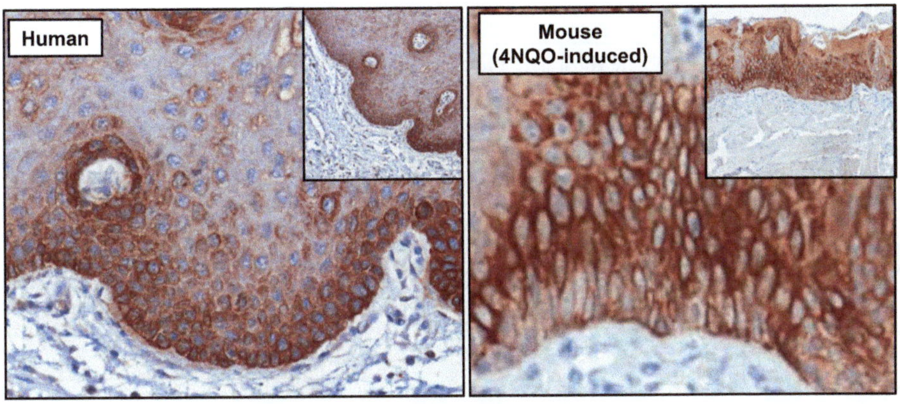

Fig. 13.1 OCT-3 immunostaining in oral epithelial dysplasia

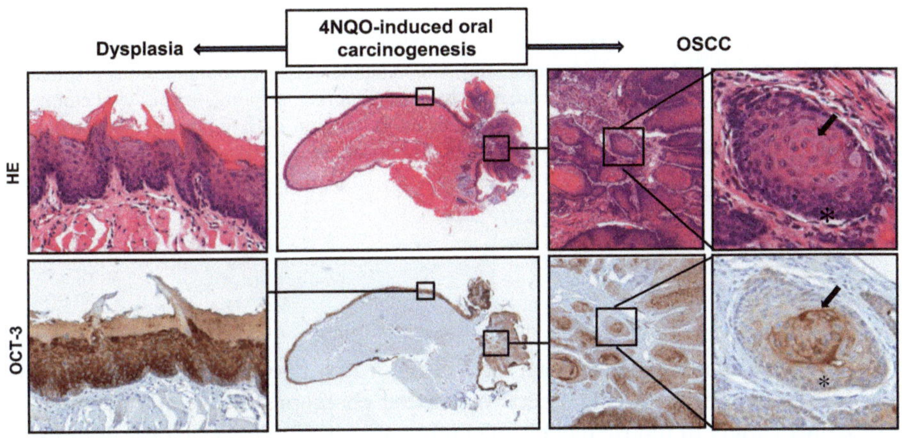

Fig. 13.2 Differential OCT-3 expression in 4NQO-induced oral carcinogenesis. Strong OCT-3 expression is observed in murine dysplastic tissue (*bottom left* panel). In comparison to highly differentiated oral squamous cell carcinoma cells (*arrows*), OCT-3 expression is progressively lost in moderately to poorly differentiated cancer cells (*asterix*)

lesions from the control mice (Fig. 13.3). These findings imply that in oral carcinogenesis, metformin may be preventing tumor progression via AMPK activation and mTOR inhibition. Moreover, these effects appear to be dependent on OCT-3 uptake and activity at the site of oral cancer development. This emerging information underscores the potential use of metformin as a novel targeted chemopreventive agent in oral oncology, and positions OCT-3 as a potential tumor biomarker to predict the chemopreventive activity of metformin in oral precancerous lesions.

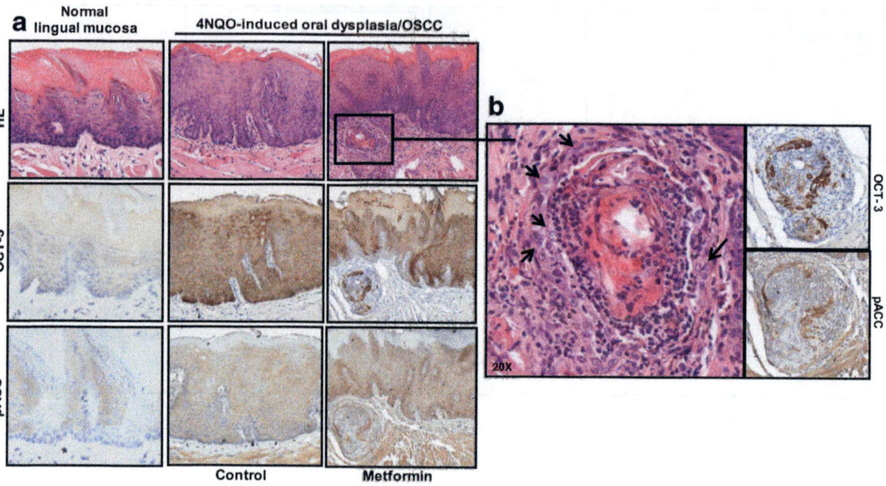

Fig. 13.3 Metformin triggers AMPK activation in 4NQO-induced oral dysplastic and cancer tissues. (A) 4NQO-induced lingual dysplastic and oral squamous cell carcinoma lesions responded to metformin by activating AMPK, as evidenced by pACC immunostaining, in epithelial compartments harboring OCT-3 positive cells. (B) Higher magnification of a tumor site shown in (A) displaying positive OCT-3 and pACC immunostaining in cancer cells (*arrows*)

Conclusions

Compelling evidence indicates that metformin, a first line oral biguanide medication for the treatment of T2D, exerts promising chemopreventive and therapeutic effects in a variety of mouse models of carcinogenesis (Table 13.1). Less optimism, however, is conveyed by a number of retrospective population studies conducted in different settings in diabetic patients, which makes it more challenging to extrapolate the preclinical experimental findings to the human condition. Clearly, some of the positive antitumoral effects of metformin depend on cancer type-specific cellular, molecular, and/or metabolic vulnerabilities as well as host metabolic characteristics (Fig. 13.4). Future research endeavors must be supported to fill gaps that still remain elusive. Elucidating the direct or indirect mechanism of metformin action, identifying tumoral and extratumoral predictive markers for improved outcomes, validating the role of OCTs in metformin pharmacokinetics and defining optimal dosages of metformin in the oncology setting are all important areas of research needed to be explored in order to strongly validate the role of metformin in human neoplastic diseases.

Table 13.1 Antineoplastic effects of metformin in various mouse models of carcinogenesis

Type of tumor	Mouse model	Dose of metformin	Route	Reference
Liver	Chemically-induced	250 mg/kg body weight	Oral	[25]
	Dietary intervention	250 mg/kg body weight	Oral	[28]
Pancreas	Xenograft	50–250 mg/kg body weight	i.p.	[37]
		2.5 mg/l	Oral	[37]
	Xenograft	250 mg/kg body weight	Oral	[40]
	Xenograft	800 mg/l	Oral	[42]
	$p48^{Cre/+}$ LSL-KrasG12$^{D/+}$transgenic	1000–2000 ppm	Oral	[43]
Colorectal	Syngeneic grafts/dietary intervention	50 mg/kg body weight	Oral	[51]
	Chemically-induced	250 mg/kg body weight	i.p.	[56]
	$Apc^{Min/+}$ transgenic	250 mg/kg body weight	Oral	[57]
Breast	Her-2/neu transgenic	100 mg/kg body weight	Oral	[60]
	Her-2/neu transgenic	250 mg/kg body weight	i.p.	[65]
	Xenograft	100 μg/ml	i.p.	[63]
Prostate	Hi-Myc transgenic	200 mg/kg body weight	Oral	[71]
	Xenograft	200 μg/ml	Oral	[72]
		1 mg/day	i.p.	[72]
Lung	Chemically-induced	250 mg/kg body weight	i.p.	[79]
	Chemically-induced/LID transgenic	5 mg/ml	Oral	[81]
	Syngeneic grafts/dietary intervention	50 mg/kg body weight	Oral	[82]
Head and neck	Chemically-induced	50 mg/kg body weight	i.p.	[100]

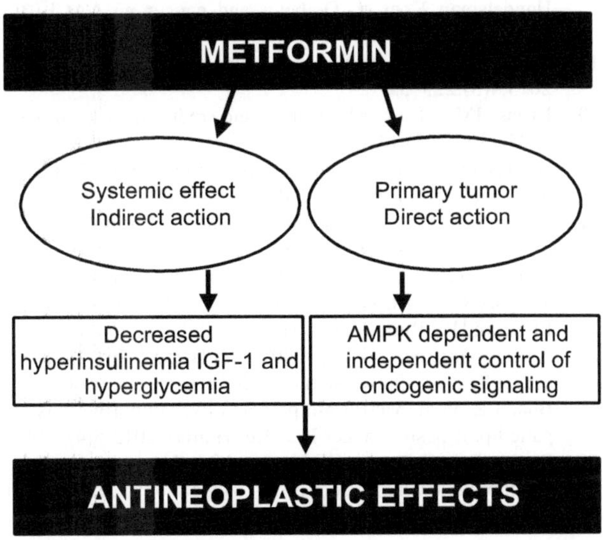

Fig. 13.4 Potential mechanisms of action underlying the effects of metformin in carcinogenesis. The antineoplastic activity of metformin is associated with indirect systemic responses mostly affecting a subset of insulin-dependent tumor types, or through direct effects via AMPK dependent and independent signaling pathways in the primary tumor

References

1. Bailey CJ, Day C. Traditional plant medicines as treatments for diabetes. Diabetes Care. 1989;12(8):553–64.
2. Hardie DG. Metformin-acting through cyclic AMP as well as AMP? Cell Metab. 2013;17(3):313–4.
3. Bailey CJ. Metformin: a multitasking medication. Diab Vasc Dis Res. 2008;5(3):156.
4. Hundal RS, et al. Mechanism by which metformin reduces glucose production in type 2 diabetes. Diabetes. 2000;49(12):2063–9.
5. Viollet B, et al. Cellular and molecular mechanisms of metformin: an overview. Clin Sci (Lond). 2012;122(6):253–70.
6. Shu Y, et al. Effect of genetic variation in the organic cation transporter 1 (OCT1) on metformin action. J Clin Invest. 2007;117(5):1422–31.
7. Gowans GJ, Hardie DG. AMPK: a cellular energy sensor primarily regulated by AMP. Biochem Soc Trans. 2014;42(1):71–5.
8. El-Mir MY, et al. Dimethylbiguanide inhibits cell respiration via an indirect effect targeted on the respiratory chain complex I. J Biol Chem. 2000;275(1):223–8.
9. Zhou G, et al. Role of AMP-activated protein kinase in mechanism of metformin action. J Clin Invest. 2001;108(8):1167–74.
10. Shaw RJ, et al. The kinase LKB1 mediates glucose homeostasis in liver and therapeutic effects of metformin. Science. 2005;310(5754):1642–6.
11. Stephenne X, et al. Metformin activates AMP-activated protein kinase in primary human hepatocytes by decreasing cellular energy status. Diabetologia. 2011;54(12):3101–10.
12. Fullerton MD, et al. Single phosphorylation sites in Acc1 and Acc2 regulate lipid homeostasis and the insulin-sensitizing effects of metformin. Nat Med. 2013;19(12):1649–54.
13. Shaw RJ. Metformin trims fats to restore insulin sensitivity. Nat Med. 2013;19(12):1570–2.
14. Foretz M, et al. Metformin inhibits hepatic gluconeogenesis in mice independently of the LKB1/AMPK pathway via a decrease in hepatic energy state. J Clin Invest. 2010;120(7):2355–69.
15. Miller RA, et al. Biguanides suppress hepatic glucagon signalling by decreasing production of cyclic AMP. Nature. 2013;494(7436):256–60.
16. Madiraju AK, et al. Metformin suppresses gluconeogenesis by inhibiting mitochondrial glycerophosphate dehydrogenase. Nature. 2014;510(7506):542–6.
17. Handelsman Y, et al. Diabetes and cancer–an AACE/ACE consensus statement. Endocr Pract. 2013;19(4):675–93.
18. Noto H, et al. Latest insights into the risk of cancer in diabetes. J Diabetes Investig. 2013;4(3):225–32.
19. Evans JM, et al. Metformin and reduced risk of cancer in diabetic patients. BMJ. 2005;330(7503):1304–5.
20. Noto H, et al. Cancer risk in diabetic patients treated with metformin: a systematic review and meta-analysis. PLoS One. 2012;7(3):e33411.
21. Soranna D, et al. Cancer risk associated with use of metformin and sulfonylurea in type 2 diabetes: a meta-analysis. Oncologist. 2012;17(6):813–22.
22. Zhang P, et al. Association of metformin use with cancer incidence and mortality: a meta-analysis. Cancer Epidemiol. 2013;37(3):207–18.
23. El-Serag HB, Rudolph KL. Hepatocellular carcinoma: epidemiology and molecular carcinogenesis. Gastroenterology. 2007;132(7):2557–76.
24. Siegel R, et al. Cancer statistics, 2014. CA Cancer J Clin. 2014;64(1):9–29.
25. Bhalla K, et al. Metformin prevents liver tumorigenesis by inhibiting pathways driving hepatic lipogenesis. Cancer Prev Res (Phila). 2012;5(4):544–52.
26. Starley BQ, Calcagno CJ, Harrison SA. Nonalcoholic fatty liver disease and hepatocellular carcinoma: a weighty connection. Hepatology. 2010;51(5):1820–32.

27. Nakamura A, et al. Protection from non-alcoholic steatohepatitis and liver tumourigenesis in high fat-fed insulin receptor substrate-1-knockout mice despite insulin resistance. Diabetologia. 2012;55(12):3382–91.
28. Tajima K, et al. Metformin prevents liver tumorigenesis induced by high-fat diet in C57Bl/6 mice. Am J Physiol Endocrinol Metab. 2013;305(8):E987–98.
29. Pollak M, Gonzalez-Angulo AM Metformin and hepatic carcinogenesis. Cancer Prev Res (Phila). 2012;5(4):500–2.
30. Yue W, et al. Repurposing of metformin and aspirin by targeting AMPK-mTOR and inflammation for pancreatic cancer prevention and treatment. Cancer Prev Res (Phila). 2014;7(4):388–97.
31. Li D, Abbruzzese JL. New strategies in pancreatic cancer: emerging epidemiologic and therapeutic concepts. Clin Cancer Res. 2010;16(17):4313–8.
32. Wang Z, et al. Metformin is associated with reduced risk of pancreatic cancer in patients with type 2 diabetes mellitus: A systematic review and meta-analysis. Diabetes Res Clin Pract. 2014;106(1)19–26.
33. Sadeghi N, et al. Metformin use is associated with better survival of diabetic patients with pancreatic cancer. Clin Cancer Res. 2012;18(10):2905–12.
34. Hwang AL, et al. Metformin and survival in pancreatic cancer: a retrospective cohort study. Pancreas. 2013;42(7):1054–9.
35. Pollak M. Metformin and pancreatic cancer: a clue requiring investigation. Clin Cancer Res. 2012;18(10):2723–5.
36. Schneider MB, et al. Prevention of pancreatic cancer induction in hamsters by metformin. Gastroenterology. 2001;120(5):1263–70.
37. Kisfalvi K, et al. Metformin inhibits the growth of human pancreatic cancer xenografts. Pancreas. 2013;42(5):781–5.
38. Kisfalvi K, et al. Metformin disrupts crosstalk between G protein-coupled receptor and insulin receptor signaling systems and inhibits pancreatic cancer growth. Cancer Res. 2009;69(16):6539–45.
39. Rozengurt E, Sinnett-Smith J, Kisfalvi K. Crosstalk between insulin/insulin-like growth factor-1 receptors and G protein-coupled receptor signaling systems: a novel target for the antidiabetic drug metformin in pancreatic cancer. Clin Cancer Res. 2010;16(9):2505–11.
40. Nair V, et al. Metformin inhibits pancreatic cancer cell and tumor growth and downregulates Sp transcription factors. Carcinogenesis. 2013;34(12):2870–9.
41. Bao B, et al. Metformin inhibits cell proliferation, migration and invasion by attenuating CSC function mediated by deregulating miRNAs in pancreatic cancer cells. Cancer Prev Res (Phila). 2012;5(3):355–64.
42. Gou S, et al. Low concentrations of metformin selectively inhibit CD133(+) cell proliferation in pancreatic cancer and have anticancer action. PLoS One. 2013;8(5):e63969.
43. Mohammed A, et al. Antidiabetic Drug Metformin Prevents Progression of Pancreatic Cancer by Targeting in Part Cancer Stem Cells and mTOR Signaling. Transl Oncol. 2013;6(6):649–59.
44. Lonardo E, et al. Metformin targets the metabolic achilles heel of human pancreatic cancer stem cells. PLoS One. 2013;8(10):e76518.
45. Mohammed A, et al. The epidermal growth factor receptor inhibitor gefitinib prevents the progression of pancreatic lesions to carcinoma in a conditional LSL-KrasG12D/ + transgenic mouse model. Cancer Prev Res (Phila). 2010;3(11):1417–26.
46. Smiechowski B, et al. The use of metformin and colorectal cancer incidence in patients with type II diabetes mellitus. Cancer Epidemiol Biomarkers Prev. 2013;22(10):1877–83.
47. Brown KA, Samarajeewa NU, Simpson ER. Endocrine-related cancers and the role of AMPK. Mol Cell Endocrinol. 2013;366(2):170–9.
48. Ollberding NJ, et al. Genetic variants, prediagnostic circulating levels of insulin-like growth factors, insulin, and glucose and the risk of colorectal cancer: the Multiethnic Cohort study. Cancer Epidemiol Biomarkers Prev. 2012;21(5):810–20.

49. Gao Y, et al. Serum IGF1; IGF2 and IGFBP3 and risk of advanced colorectal adenoma. Int J Cancer. 2012;131(2):E105–13.
50. Cannata D, et al. Type 2 diabetes and cancer: what is the connection? Mt Sinai J Med. 2010;77(2):197–213.
51. Algire C, et al. Metformin blocks the stimulative effect of a high-energy diet on colon carcinoma growth in vivo and is associated with reduced expression of fatty acid synthase. Endocr Relat Cancer. 2010;17(2):351–60.
52. Mashhedi H, et al. Metformin abolishes increased tumor (18)F-2-fluoro-2-deoxy-D-glucose uptake associated with a high energy diet. Cell Cycle. 2011;10(16):2770–8.
53. Algire C, et al. Diet and tumor LKB1 expression interact to determine sensitivity to antineoplastic effects of metformin in vivo. Oncogene. 2011;30(10):1174–82.
54. Buzzai M, et al. Systemic treatment with the antidiabetic drug metformin selectively impairs p53-deficient tumor cell growth. Cancer Res. 2007;67(14):6745–52.
55. Takahashi H, et al. Colon epithelial proliferation and carcinogenesis in diet-induced obesity. J Gastroenterol Hepatol. 2013;28 (Suppl 4):41–7.
56. Hosono K, et al. Metformin suppresses azoxymethane-induced colorectal aberrant crypt foci by activating AMP-activated protein kinase. Mol Carcinog. 2010;49(7):662–71.
57. Tomimoto A, et al. Metformin suppresses intestinal polyp growth in Apc(Min/+) mice. Cancer Sci. 2008;99(11):2136–41.
58. Larsson SC, Mantzoros CS, Wolk A. Diabetes mellitus and risk of breast cancer: a meta-analysis. Int J Cancer. 2007;121(4):856–62.
59. Pollak MN. Investigating metformin for cancer prevention and treatment: the end of the beginning. Cancer Discov. 2012;2(9):778–90.
60. Anisimov VN, et al. Effect of metformin on life span and on the development of spontaneous mammary tumors in HER-2/neu transgenic mice. Exp Gerontol. 2005;40(8–9):685–93.
61. Moasser MM. The oncogene HER2: its signaling and transforming functions and its role in human cancer pathogenesis. Oncogene. 2007;26(45):6469–87.
62. Anisimov VN, et al. Metformin extends life span of HER-2/neu transgenic mice and in combination with melatonin inhibits growth of transplantable tumors in vivo. Cell Cycle. 2010;9(1):188–97.
63. Hirsch HA, et al. Metformin selectively targets cancer stem cells, and acts together with chemotherapy to block tumor growth and prolong remission. Cancer Res. 2009;69(19):7507–11.
64. Zhu P, et al. Metformin selectively targets tumor-initiating cells in ErbB2-overexpressing breast cancer models. Cancer Prev Res (Phila). 2014;7(2):199–210.
65. Economopoulou P, Kaklamani VG, Siziopikou K. The role of cancer stem cells in breast cancer initiation and progression: potential cancer stem cell-directed therapies. Oncologist. 2012;17(11):1394–401.
66. Hitron A, et al. The influence of antidiabetic medications on the development and progression of prostate cancer. Cancer Epidemiol. 2012;36(4):e243–50.
67. Margel D, et al. Association between metformin use and risk of prostate cancer and its grade. J Natl Cancer Inst. 2013;105(15):1123–31.
68. Margel D, et al. Metformin use and all-cause and prostate cancer-specific mortality among men with diabetes. J Clin Oncol. 2013;31(25):3069–75.
69. Preston MA, et al. Metformin use and prostate cancer risk. Eur Urol. 2014;66(6):1022–3.
70. Tsilidis KK, et al. Metformin does not affect cancer risk: A cohort study in the U.K. clinical practice research datalink analyzed like an intention-to-treat trial. Diabetes Care. 2014;37(9):2522–32.
71. Akinyeke T, et al. Metformin targets c-MYC oncogene to prevent prostate cancer. Carcinogenesis. 2013;34(12):2823–32.
72. Ben Sahra I, et al. The antidiabetic drug metformin exerts an antitumoral effect in vitro and in vivo through a decrease of cyclin D1 level. Oncogene. 2008;27(25):3576–86.
73. Greenberg AK, et al. Chemoprevention of lung cancer: prospects and disappointments in human clinical trials. Cancers (Basel). 2013;5(1):131–48.

74. Lai SW, et al. Antidiabetes drugs correlate with decreased risk of lung cancer: a population-based observation in Taiwan. Clin Lung Cancer. 2012;13(2):143–8.
75. Franciosi M, et al. Metformin therapy and risk of cancer in patients with type 2 diabetes: systematic review. PLoS One. 2013;8(8):e71583.
76. Wu Y, et al. Conventional hypoglycaemic agents and the risk of lung cancer in patients with diabetes: a meta-analysis. PLoS One. 2014;9(6):e99577.
77. Bodmer M, et al. Metformin does not alter the risk of lung cancer: a case-control analysis. Lung Cancer. 2012;78(2):133–7.
78. Smiechowski BB, et al. The use of metformin and the incidence of lung cancer in patients with type 2 diabetes. Diabetes Care. 2013;36(1):124–9.
79. Memmott RM, et al. Metformin prevents tobacco carcinogen–induced lung tumorigenesis. Cancer Prev Res (Phila Pa). 2010;3(9):1066–76.
80. Granville CA, et al. Identification of a highly effective rapamycin schedule that markedly reduces the size, multiplicity, and phenotypic progression of tobacco carcinogen-induced murine lung tumors. Clin Cancer Res. 2007;13(7):2281–9.
81. Quinn BJ, et al. Inhibition of lung tumorigenesis by metformin is associated with decreased plasma IGF-I and diminished receptor tyrosine kinase signaling. Cancer Prev Res (Phila). 2013;6(8):801–10.
82. Algire C, et al. Metformin attenuates the stimulatory effect of a high-energy diet on in vivo LLC1 carcinoma growth. Endocr Relat Cancer. 2008;15(3):833–9.
83. Gillison ML, et al. Prevalence of oral HPV infection in the United States, 2009–2010. JAMA. 2012;307(7):693–703.
84. Lui VW, Grandis JR. Primary chemotherapy and radiation as a treatment strategy for HPV-positive oropharyngeal cancer. Head Neck Pathol. 2012;6(Suppl 1):S91–7.
85. Amornphimoltham P, et al. Mammalian target of rapamycin, a molecular target in squamous cell carcinomas of the head and neck. Cancer Res. 2005;65(21):9953–61.
86. Molinolo AA, et al. Dissecting the Akt/mammalian target of rapamycin signaling network: emerging results from the head and neck cancer tissue array initiative. Clin Cancer Res. 2007;13(17):4964–73.
87. Nathan CO, et al. Mammalian target of rapamycin inhibitors as possible adjuvant therapy for microscopic residual disease in head and neck squamous cell cancer. Cancer Res. 2007;67(5):2160–8.
88. Patel V, et al. Cellular systems for studying human oral squamous cell carcinomas. Adv Exp Med Biol. 2011;720:27–38.
89. Molinolo AA, et al. mTOR as a molecular target in HPV-associated oral and cervical squamous carcinomas. Clin Cancer Res. 2012;18(9):2558–68.
90. Czerninski R, et al. Targeting mammalian target of rapamycin by rapamycin prevents tumor progression in an oral-specific chemical carcinogenesis model. Cancer Prev Res (Phila Pa). 2009;2(1):27–36.
91. Hu M, et al. Efficacy and comparative effectiveness of sirolimus as an anticancer drug. Laryngoscope. 2011;121(5):978–82.
92. Patel V, et al. Decreased lymphangiogenesis and lymph node metastasis by mTOR inhibition in head and neck cancer. Cancer Res. 2011;71(22):7103–12.
93. Price KA, Cohen EE. Current treatment options for metastatic head and neck cancer. Curr Treat Options Oncol. 2012;13(1):35–46.
94. Nguyen SA, et al. mTOR Inhibitors and its Role in the Treatment of Head and Neck Squamous Cell Carcinoma. Curr Treat Options Oncol. 2012;13(1):71–81.
95. Regezi JA, Sciubba JJ, Jordan RCK. Oral pathology: Clinical pathologic correlations. 5th ed. St. Louis: Saunders Elsevier; 2008.
96. Larsson O, et al. Distinct perturbation of the translatome by the antidiabetic drug metformin. Proc Natl Acad Sci U S A. 2012;109(23):8977–82.
97. Dowling RJ, et al. Metformin inhibits mammalian target of rapamycin-dependent translation initiation in breast cancer cells. Cancer Res. 2007;67(22):10804–12.
98. Zakikhani M, et al. Metformin is an AMP kinase-dependent growth inhibitor for breast cancer cells. Cancer Res. 2006;66(21):10269–73.

99. Inoki K, Zhu T, Guan KL. TSC2 mediates cellular energy response to control cell growth and survival. Cell. 2003;115(5):577–90.
100. Vitale-Cross L, et al. Metformin Prevents the Development of Oral Squamous Cell Carcinomas from Carcinogen-Induced Premalignant Lesions. Cancer Prev Res (Phila). 2012;5(4):562–573.
101. Vitale-Cross L, et al. Chemical carcinogenesis models for evaluating molecular-targeted prevention and treatment of oral cancer. Cancer Prev Res (Phila). 2009;2(5):419–22.
102. Jonker JW, Schinkel AH. Pharmacological and physiological functions of the polyspecific organic cation transporters: OCT1, 2, and 3 (SLC22A1–3). J Pharmacol Exp Ther. 2004;308(1):2–9.
103. Gupta S, Burckhardt G, Hagos Y. SLC22 transporter family proteins as targets for cytostatic uptake into tumor cells. Biol Chem. 2011;392(1–2):117–24.
104. Shu Y, et al. Evolutionary conservation predicts function of variants of the human organic cation transporter, OCT1. Proc Natl Acad Sci U S A. 2003;100(10):5902–7.
105. Shu Y, et al. Effect of genetic variation in the organic cation transporter 1, OCT1, on metformin pharmacokinetics. Clin Pharmacol Ther. 2008;83(2):273–80.
106. Sakata T, et al. Functional analysis of human organic cation transporter OCT3 (SLC22A3) polymorphisms. J Pharmacol Sci. 2010;113(3):263–6.
107. Chen L, et al. Role of organic cation transporter 3 (SLC22A3) and its missense variants in the pharmacologic action of metformin. Pharmacogenet Genomics. 2010;20(11):687–99.
108. Yokoo S, et al. Significance of organic cation transporter 3 (SLC22A3) expression for the cytotoxic effect of oxaliplatin in colorectal cancer. Drug Metab Dispos. 2008;36(11):2299–306.
109. Shnitsar V, et al. Expression of human organic cation transporter 3 in kidney carcinoma cell lines increases chemosensitivity to melphalan, irinotecan, and vincristine. Cancer Res. 2009;69(4):1494–501.
110. Schaeffeler E, et al. DNA methylation is associated with downregulation of the organic cation transporter OCT1 (SLC22A1) in human hepatocellular carcinoma. Genome Med. 2011;3(12):82.
111. Heise M, et al. Downregulation of organic cation transporters OCT1 (SLC22A1) and OCT3 (SLC22A3) in human hepatocellular carcinoma and their prognostic significance. BMC Cancer. 2012;12:109.
112. Patel H, et al. Differential expression of organic cation transporter OCT-3 in oral premalignant and malignant lesions: potential implications in the antineoplastic effects of metformin. J Oral Pathol Med. 2013;42(3):250–6.

Index

N. A. Berger (ed.), *Murine Models, Energy Balance, and Cancer*,
Energy Balance and Cancer, DOI 10.1007/978-3-319-16733-6

MIX
Papier aus verantwortungsvollen Quellen
Paper from responsible sources
FSC® C105338

If you have any concerns about our products,
you can contact us on
ProductSafety@springernature.com

In case Publisher is established outside the EU,
the EU authorized representative is:
Springer Nature Customer Service Center GmbH
Europaplatz 3, 69115 Heidelberg, Germany

Printed by Libri Plureos GmbH
in Hamburg, Germany